Biocatalysis and Biodegradation

MICROBIAL TRANSFORMATION OF ORGANIC COMPOUNDS

Biocatalysis and Biodegradation

MICROBIAL TRANSFORMATION OF ORGANIC COMPOUNDS

BY

Lawrence P. Wackett
Department of Biochemistry, Molecular Biology & Biophysics
and Biological Process Technology Institute
University of Minnesota, St. Paul, Minnesota 55108-1030

AND

C. Douglas Hershberger
Biological Process Technology Institute
University of Minnesota, St. Paul, Minnesota 55108-1030

ASM PRESS WASHINGTON, D.C.

American Society for Microbiology
1752 N Street, N.W.
Washington, DC 20036–2904

Library of Congress Cataloging-in-Publication Data

Wackett, Lawrence Philip.
Biocatalysis and biodegradation : microbial transformation of organic compounds / by
Lawrence P. Wackett and C. Douglas Hershberger.
p. cm.
Includes bibliographical references and index.
ISBN 1-55581-179-5
1. Biodegradation. 2. Biotransformation (Metabolism) 3. Organic compounds—Biodegradation. I. Hershberger, C. Douglas. II. Title.

QP517.B5 W33 2000
572′.429—dc21

00-061836

Address editorial correspondence to: ASM Press, 1752 N St., N.W., Washington, DC 20036-2904, U.S.A.

Send orders to: ASM Press, P.O. Box 605, Herndon, VA 20172, U.S.A.
Phone: 800-546-2416; 703-661-1593
Fax: 703-661-1501
Email: books@asmusa.org
Online: www.asmpress.org

Contents

Foreword

This book was written during one of the most exciting periods in the history of the biological sciences. It is the first of its genre and represents the collective thoughts of its authors on the birth and growth of biodegradation and biocatalysis as related areas in the disciplines of microbiology, biochemistry, and chemistry. Emphasis is placed on the transition period between traditional microbial chemistry and the emerging fields of genomics (structural and functional), proteomics, and bioinformatics. The elements of discovery and prediction pervade each chapter, with emphasis placed on the function and future developments of novel individual and groups of biocatalysts rather than, as in more conventional texts, on descriptions of the pathways used by microorganisms to degrade specific classes of organic compounds.

In this Foreword, I have chosen Sir Frederick Gowland Hopkins as the catalyst and Cambridge University as the reaction vessel to develop the themes of biodegradation and biocatalysis. Hopkins was the most distinguished biochemist at Cambridge University from the early 1900s through the end of World War II. It was Hopkins who encouraged Marjory Stephenson to write her classic book *Bacterial Metabolism*, which was first published in 1930 and followed by second and third editions in 1940 and 1949. Stephenson traced her interest in things microbial to Pasteur's 1857 paper, "Mémoire de la Fermentation dite lactique." This initial work was criticized by Berzelius and Liebig, both leading chemists at the time. Nevertheless, Pasteur held firm to his interpretation of lactic acid fermentation as a process resulting from, and necessary for, the growth and multiplication of living cells. These observations marked the birth of bacterial metabolism, which was defined in 1949 by Stephenson as "the interpretation of the physiological life of the bacterial cell in terms of biochemistry and biophysics both in respect of its own growth and reproduction and also of its action on its chemical and biological environment."

Hopkins also encouraged Ernest Baldwin, a faculty member in the School of Biochemistry, to write *Dynamic Aspects of Biochemistry*, first published in 1947. There is surprisingly little overlap between the two books in spite of Stephenson and Baldwin being in the School of Biochemistry at the same time. Baldwin's textbook was used to introduce the palpably cold equations of equilibrium thermodynamics to beginning biochemistry students. *Dynamic Aspects of Biochemistry* achieved this objective admirably and, together with Stephenson's *Bacterial Metabolism*, provided substantial evidence to support Hopkins' celebrated aphorism, "Life is a dynamic equilibrium in a polyphasic system."

Baldwin was a dedicated teacher and found that students comprehended and remembered reactions if they were presented pictorially. His "whirligigs" were the forerunners of the elegant multicolored cartoons that pervade the pages of microbiology and biochemistry textbooks today. In this regard it is of interest to note that in England soon after the second world war, there were three weekly comic papers, *The Beano*, *The Wizard*, and *The Dandy*. Each contained three or four episodes of ongoing stories designed to attract the attention of children in the 7- to 10-year age group and each episode was presented in full text without a single picture or cartoon. This was at a time when British children yearned for the so-called "American comics" that in the postwar era were finding their way to the shelves of British newsagents. It was the widespread fear of parents that the glorious cartoons in *Superman* would somehow bring down the whole British educational system by stifling the imagination of students in their formative years. Retrospective analysis shows that these apprehensions had no basis in fact. Although Baldwin's whirligigs could not compare with the pictorial presentation of Superman in flight, they have survived in the representation of the structure of ATP as A-P~P~P. Baldwin's squiggle (~) was used to identify a "high-energy" phosphate bond and, as the late Stanley Dagley noted, "in one fell swoop raised ATP to the level of a high explosive." Of course Baldwin did state that the importance of ATP "lies in the ease with which its terminal phosphate radical can be transferred with a part or all of the 11,000 calories of free energy with which it is associated, to other molecules," and he went on to use the A-P~P~P structure to show how ATP plays a central role in energy transactions in living cells. It is remarkable that the squiggle representing a high-energy phosphate bond can be found in almost every biochemistry textbook from 1947 to the present day and that, almost without exception, it is accompanied by an explanation that denies the existence of high-energy phosphate bonds.

The two decades following the second world war saw unprecedented growth in biochemistry and all areas of the biological sciences. The determination of the structure and function of DNA, RNA, and messenger RNA provided the basis for the development of the field of molecular biology. Albert Lehninger at Johns Hopkins University recognized, like Baldwin before him, the need for a textbook that correlated thermodynamic principles with metabolic activities. His book, *Bioenergetics: the Molecular Basis of Biological Energy Transformations*, was written for undergraduates beginning the study of molecular biology. This text, in elegant and simple presentations, showed that thermodynamic principles "comprise a central and unifying theme in biology." In this context, metabolism can be thought of as the total chemical activities of the cell, with the terms "catabolism" and "anabolism" representing the degradation and synthesis of cellular constituents, respectively.

The relationship between catabolism and anabolism can be seen in the carbon cycle, which, at the global level, represents an equilibrium between the processes of photosynthesis and respiration. Light energy from the sun is used by plants, algae, and certain bacteria to drive the synthesis of cellular constituents from carbon dioxide and an appropriate electron donor. The carbon sequestered in these molecules is ultimately released to the atmosphere as carbon dioxide, mainly by the catabolic activities of heterotrophic microorganisms. For example, glucose is catabolized via the Embden-Meyerhoff pathway in a stepwise sequence of reactions to pyruvate. In the process, ATP for biosynthesis is produced by substrate-level phosphorylation, and oxidation reactions are mediated by the removal of hydride ions as NADH. Continuation of glucose degradation is dependent on the regeneration of oxidized NAD^+. In fermentative microorganisms, this is achieved by the transfer of electrons to organic compounds. For example, yeast regenerates NAD^+ by transferring reducing equivalents to acetaldehyde. The end product of this reaction is ethanol. The same strategy is used by mammals when oxygen concentrations are low. In this case, however, hydrogen and electrons are used to reduce pyruvate to lactate. In contrast, aerobic organisms utilize the tricarboxylic acid cycle to oxidize pyruvate to carbon dioxide and water. Hydrogen and electrons are transferred to oxygen by a sequence of redox proteins, and ATP is generated by oxidative phosphorylation. In anaerobic environments, the acids and alcohols generated by fermentation can serve as growth substrates for other groups of bacteria, which convert them to acetate, hydrogen, and carbon dioxide. These products are converted to methane by methanogenic bacteria. Methane is oxidized to CO_2 by methylotrophic bacteria and thus completes the carbon cycle. Although oxygen is the terminal electron acceptor in aerobic organisms from bacteria to humans, an alternate lifestyle is embraced by iron-, sulfate-, and nitrate-reducing bacteria, and the search for a phosphate-reducing organism continues (R. S. Wolfe, personal communication, 2000).

The catabolic and anabolic processes used by microorganisms for the degradation and synthesis of carbohydrates, lipids, proteins, and other cellular constituents are reasonably well known. Together, they represent the field of central or intermediary metabolism. Thus it seems appropriate to ask how the pathways of central metabolism relate to biodegradation and biocatalysis. In this context, it is important to recognize that catabolism and anabolism are powerful terms that directly relate to the fundamental activities of living cells, whereas biodegradation and biocatalysis are used to identify individual specific objectives within the area encompassed by metabolism. In its broadest sense, biodegradation refers to the microbial catabolism of a compound into molecules that can enter the central metabolic pathways used by aerobic and anaerobic microorganisms. For example, lignocellulose is the main structural component of woody plants, and lignocellulose biodegradation is the most important process in the recycling of terrestrial biosynthetic compounds. This is accomplished by a cascade of thermodynamically favorable reactions mediated by microorganisms. Certain fungi use extracellular hydrogen peroxide-dependent enzymes to depolymerize lignocellulose. The aromatic alcohols and ethers liberated by ligninases, together with free cellulose, serve as carbon and energy sources for a variety of microorganisms, and ultimately the carbon sequestered in lignocellulose is returned to the atmosphere as carbon dioxide. Biocatalysis, in contrast to biodegradation, is a term that is usually used to identify catabolic or anabolic reactions that lead to the accumulation of specific metabolic products. The reduction of acetaldehyde to ethanol by alcohol dehydrogenase and the generation of high-fructose corn syrup by glucose-6-phosphate isomerase are classic examples of biocatalysis.

Although examples of biodegradation and biocatalysis abound in nature, it is only in the 38 years since the publication of Rachel Carson's *The Silent Spring* that biodegradation has achieved serious recognition as a research area that focuses on the pathways and reactions used by microorganisms to degrade environmental pollutants. In this context, it is desirable to have an understanding of the nature of environmental pollutants. There are, however, no unifying concepts to satisfy all situations. In simple terms, a compound can be considered a pollutant if its concentration reaches a level that has deleterious effects on life forms at any stage from the bottom to the top of the food chain. At the turn of the century, London hatters used mercury to provide a sheen for top hats, and the frequency of dementia in this group of individuals was very high. The Mad Hatter in Lewis Carroll's *Alice in Wonderland* provides anecdotal evidence for this condition, which is due to the formation of neurogenic dimethyl mercury by intestinal bacteria. A more recent example is seen in the high incidence of dimethyl mercury poisoning of the inhabitants of Minamata, a small fishing village in Japan. Mercury from a chlorine plant was released into the estuary near the village. Methanogens in anaerobic sediments produced dimethyl mercury, which moved through the food chain to fish and finally to the fishermen and their families.

The cycle of carbon in nature operates on the assumption that all biosynthetic organic compounds are biodegradable. There is, however, a second major source of organic compounds that has to be considered in terms of carbon recycling and biodegradation. Every year since the Precambrian era, a small amount of the plankton in ancient oceans has been buried before it can be completely converted to carbon dioxide by microorganisms. This material undergoes thermodynamic stabilization by diagenesis, and the end product is graphite. The early stages of diagenesis occur in shallow sediments where organic material is subjected to extremes in pH and redox conditions. During this time, carbon-carbon bonds are broken and oxygen and other elements are displaced, leading to the temporary formation of new and often more complex compounds (humic acids) even though overall stabilization of the system takes place. Eventually, most of this material is converted to kerogen, a complex polymer of unknown structure. As sediments sink to lower levels, kerogen is subjected to increases in temperature and pressure. These conditions, in conjunction with catalytic changes due to the composition of the rock, can eventually lead to the formation of crude oil. It is important to realize that crude oil represents an intermediate stage on the geological time scale between kerogen and graphite. Thus no two oils have the same chemical composition.

Today, petroleum derived from crude oil drives the economies of industrial nations, and not surprisingly, crude oil and products formed from it find their way into environments where their presence is not anticipated or appreciated. At the beginning of the new millennium, the public is still enamored of stories relating to the use of bacteria to clean up oil spills. However, work supported by the American Petroleum Institute reveals the chemical complexity of oil and the realization that the timeframes required for complete biodegradation make it unlikely that significant bioremediation technology for catastrophic events such as the *Exxon Valdez* spill will ever be developed. Most crude oils contain significant quantities of linear alkanes. Bacteria that degrade these hydrocarbons are ubiquitous in the environment. The introduction of nitrogen and phosphorus sources to environments polluted by oil to accelerate biodegradation by indigenous microorganisms (biostimulation) has met with some success. This is probably due to stimulation of bacterial growth on linear alkanes and some of the smaller alicyclic and aromatic components.

The use of microorganisms to remedy other anthropogenic insults to the environment has met with more success. In each case the solution was found in the results of basic research. For example, in the 1960s, Liverpudlians in Lancashire, England, were treated to the sight of large plumes of foam drifting on the surface of the River Mersey. The culprit was found to be detergents that contained hydrophobic branched alkyl side chains. An environmental solution was found in the substitution of linear alkyl side chains for their branched-chain counterparts. Detergent activity was maintained and the visible pollution problem was eliminated. It is of interest to note that quaternary methyl-substituted alkanes are products of the petroleum industry and are not found in nature, whereas, as noted previously, linear alkanes are ubiquitous. Thus one possible explanation for the observed results is the fact that microorganisms have had millions of years to evolve enzyme systems for the degradation of linear alkanes whereas they have only been in contact with quaternary-substituted alkanes for less than a century. More recent interpretations of the evolution of biodegradation pathways are discussed in later chapters of this book.

Unfortunately, not all cases of environmental pollution are as visible as the foam from non-biodegradable detergents. DDT was used as an insecticide for many years before it was identified as the agent responsible for the almost total demise of the bald eagle and other birds of prey. In addition, analyses of environmental samples for the presence of DDT led to the identification of polychlorinated biphenyls (PCBs) as a major group of ubiquitous environmental pollutants. The identification of DDT and PCBs in soil and sediment samples was made possible by the combined techniques of gas chromatography and high-resolution mass spectrometry. Today these techniques are used routinely to monitor the concentration and fate of specific pollutants in different environments. It is worth noting, however, that a little more than 50 years ago Kern isolated chrysene from soil. This was the first polycyclic hydrocarbon to be detected in the environment. Two decades later it was possible to say that Kern's 1947 soil sample almost certainly contained individual polycyclic compounds numbering in the 10^5 range. This was due to advances in analytical resolution by several orders of magnitude. It is also an example of what the late Max Blumer called "pure ignorance," namely, "the ignorance of which we are not even aware." For instance, with each advance in analytical technique, the boundary of the unknown is not brought closer as expected, but in fact recedes rapidly, and there is no reason for us to feel complacent that the boundary will be reached in the near future. The fact that our present knowledge of the organic compounds in nature is incomplete makes it extremely difficult to predict the biological impact of chemicals in the environment. This situation is compounded by our "pure ignorance" of the composition of bacterial communities in different ecological niches. Until recently, most of our current knowledge in the fields of biodegradation and biocatalysis resulted from studies with pure cultures and single substrates. It is now widely acknowledged that more than 99% of the bacteria in a single soil or sediment sample have yet to be isolated and characterized. The presence of 4,000 different bacteria and 100,000 organic compounds of unknown structure and physiological activity in a single soil or sediment sample is a daunting example of bacterial and chemical diversity which exposes our "pure ignorance" and challenges our imagination and perseverance.

There is no doubt that advances in analytical techniques for the isolation, separation, and identification of organic compounds have played a major role in the developing study of bacterial metabolism in general and biodegradation and biocatalysis in particular. By 1970, significant advances had been made in identifying the pathways used by bacteria to degrade compounds regarded as environmental pollutants. This was achieved by isolating and identifying intermediate compounds, followed by demonstrating the presence of enzymes responsible for their formation and further metabolism. Most of this work was done with pure cultures and single substrates. For example, aromatic hydrocarbons and related compounds were recognized as environmental pollutants emanating from the refinement and use of crude oil and the procedures used to manufacture paper products from lignocellulose. Initial studies showed that certain bacteria channel many simple aromatic compounds through arene *cis*-diols to catechol or protocatechuate. The initial reaction involves the addition of dioxygen to the aromatic nucleus by multicomponent enzyme systems belonging to a family of enzymes known as Rieske non-heme iron oxygenases. The resulting arene *cis*-diols are rearomatized by pyridine nucleotide-dependent dehydrogenases, which replace the NAD(P)H used in the initial dihydroxylation reaction. The products formed, catechols or protocatechuate, are substrates for ring-fission dioxygenases which cleave the aromatic nucleus between the hydroxyl groups (*ortho*-cleavage) or at a site adjacent to one of the hydroxyl groups (*meta*-cleavage). Ring-fission products formed by *ortho*-cleavage are channeled to a common intermediate, β-ketoadipate enol lactone, whereas in most cases *meta*-ring-fission products are converted to 2-oxo-4-hydroxyvalerate. The end products from both pathways are then metabolized to intermediates that can enter the tricarboxylic acid cycle.

These and other catabolic pathways were elucidated, as mentioned above, by studying individual reactions and their attendant enzymes. The results obtained then served as building blocks for the assembly of each pathway. Stanley Dagley was an avid proponent of this reductionist approach. This can be seen in his statement that "the most sensible questions to ask of nature are the simplest; but oversimplified answers must then be expected and these will provoke further questions." A classic example of this prediction can be seen in the convergent pathways used by bacteria to degrade aromatic compounds. These observations stimulated Gunsalus and his colleagues to explore the organization and expression of the genes encoding the enzymes responsible for aromatic metabolism, since they regarded metabolic convergence as a mechanism for reducing the total genetic load of the cell. Their studies in this area led to the discovery of the NAH, SAL, and TOL transmissible catabolic plasmids and paved the way for new molecular approaches in the fields of biodegradation and biocatalysis.

For many years pseudomonads were regarded as the most omnivorous microorganisms in the environment. Support for this conclusion can be traced back to the 1926 thesis of der Dooren de Jong in Delft, The Netherlands, which lists 80 compounds that will support the growth of *Pseudomonas putida*. Further support was provided by the detailed classification of the pseudomonads by Stanier, Palleroni, and Doudoroff at the University of California at Berkeley. Today, however, we can trace the dominance of the pseudomonads in biodegradation studies to their preferential isolation by classical enrichment culture techniques and also to pseudomicrobiologists who identified all gram-negative motile rods as pseudomonads. Comparative sequencing of 16S ribosomal RNA has led to a recognition of the catabolic versatility exhibited by species of *Rhodococcus, Sphingomonas, Comamonas, Burkholderia, Ralstonia,* and other genera. New organisms lead to the identification of new metabolic capabilities. For example, the unwritten but general assumption that aromatic hydrocarbons cannot be degraded under anaerobic conditions has fallen under the onslaught of nitrate-, sulfate-, and iron-reducing bacteria. Another example resulted from the discovery that recalcitrant highly chlorinated biphenyls undergo reductive dehalogenation in the anaerobic sediments of the upper Hudson River and other PCB-contaminated environments. One common feature of "natural" biodegradation is the slow rate of contaminant removal. This serves to remind us that Nature marches to the beat of her own drum and is often ill prepared to withstand surges of excessive amounts of chemicals resulting from human activities.

Fortunately, the molecular techniques generated and driven by recombinant DNA technology have had a significant impact on all aspects of biodegradation and biocatalysis. Today it is possible to optimize the conditions necessary for the microbial degradation of many environmental pollutants. Examples listed by Timmis and Pieper include the isolation of new strains with desired phenotypes and the construction of novel pathways that eliminate nonproductive side reactions, reduce the formation of toxic intermediates, and, where necessary, improve enzyme performance. In addition, the advent of the World Wide Web has facilitated the development of databases that can be used to predict pathways for the biodegradation of new chemicals. Discussion of these and other aspects of biodegradation permeates the later chapters of this book and does not need to be repeated here.

Recent developments in biocatalysis stem from industry's realization of the need for environmentally benign procedures for the synthesis of commercial chemical products. The ideal synthetic system would incorporate Trost's suggestion of "atom economy," in which all atoms in the reactants are present in the product. This would favor addition reactions over elimination reactions, a situation that is not always feasible from a synthetic chemist's point of view. In contrast, there are many biocatalytic routes to commercial products that are addition reactions. These include asymmetric dihydroxylation reactions catalyzed by Rieske non-heme iron dioxygenases. It is here, and also in many chapters throughout this book, that biodegradation and biocatalysis meet. Bacteria that degrade toluene are readily detected in contaminated groundwater, and at least five distinct pathways of toluene degradation have been elucidated. The "dihydrodiol" pathway used by *Pseudomonas putida* F1 is initiated by toluene dioxygenase. This enzyme adds both atoms of dioxygen to the aromatic nucleus to form homochiral (+)-*cis*-(1*S*,2*R*)-dihydroxy-3-methylcyclohexa-3,5-diene (toluene *cis*-dihydrodiol). The dihydrodiol has found use as a chiral synthon in the synthesis of biologically active compounds such as prostaglandin $E_{2\alpha}$. The most remarkable feature of toluene dioxygenase and the related enzyme naphthalene dioxygenase is their ability to oxidize more than 200 substrates to single enantiomer arene *cis*-diols. The synthetic potential of these and other chiral products from biocatalysis is virtually untapped and should serve synthetic chemists well in the future search for environmentally benign syntheses.

In 1949 Marjory Stephenson predicted that in 25 to 50 years a biochemical description of cell growth would be almost complete, leading her to suggest that "biochemistry and microbiology as we know them will become cold stars (black holes)," to be replaced with "fresh fields whose character we can only dimly guess at." One can only wonder at the words of this prescient scientist in light of the fact that June 26, 2000, saw the announcement of the DNA sequence of the human genome. This came just 5 years after the publication of the first complete genome, that of *Haemophilus influenzae*. The entire genomes of more than 30 organisms have been determined to date. These include the genome of *Enterococcus faecium* (2.98 million base pairs), which was sequenced in a single day. Robotic nucleotide sequencing has clearly revolutionized biology and given birth to genomics, structural genomics, proteomics, toxicogenomics, and no doubt many more "-omics" still to come. These qualify as the fresh fields Stephenson envisaged. They are also powerful tools that will enable scientists to determine a clear picture of bacterial growth in terms of the activity of regulatory proteins that control the timing and levels of gene expression throughout the cell growth cycle. The ability to conduct these and related experiments is extolled weekly in the scientific press—accompanied by articles warning of the difficulties to be encountered in the emerging fields of postgenomic biology.

One of the many attractive features of this book is the integration of the birth and growth of biodegradation and biocatalysis into the pre- and potential postgenomic eras. It thus seems appropriate to conclude this introduction with a few words on Arthur Kornberg's "Ten Commandments: Lessons from the Enzymology of DNA Replication," a Guest Commentary published in the July 2000 issue of the *Journal of Bacteriology*. Kornberg presents a powerful statement for the value of studies on pure enzymes in biology and cites Frederick Gowland Hopkins (1931) as an early spokesman for this approach (see above). I have always been hesitant to invoke the thoughts and commandments of a higher being in my experiments. Nevertheless, Kornberg's statement that poly P is "an inorganic polymer of hundreds of phosphate residues linked by 'high-energy' anhydride bonds" suggests an 11th commandment: *Thou shalt not ignore thermodynamics when considering the birth and future development of any field of biology.*

David T. Gibson
Department of Microbiology
The University of Iowa
Iowa City, Iowa

Suggested Reading

Baldwin, E. 1952. *Dynamic Aspects of Biochemistry*. Cambridge University Press, London, United Kingdom.

Blumer, M. 1975. Organic compounds in nature: limits of our knowledge. *Angew. Chem. Int. Ed. Eng.* **14:**507–514.

Dagley, S. 1984. Introduction, p. 1–10. *In* D. T. Gibson (ed.), *Microbial Degradation of Organic Compounds*. Marcel Dekker, New York, N.Y.

Dagley, S. 1987. Lessons from biodegradation. *Annu. Rev. Microbiol.* **41:**1–23.

Dooley, J. E., C. J. Thompson, D. E. Hirsch, and C. C. Ward. 1974. Analyzing heavy ends of crude. *Hydrocarb. Process* **53:**93–100.

Gibson, D. T. 1975. Microbial degradation of hydrocarbons, p. 667–696. *In* E. D. Goldberg (ed.), *The Nature of Seawater. Physical and Chemical Sciences Research Report*. Dahlem Konferenzen, Berlin, Germany.

Gibson, D. T. 1988. Microbial metabolism of aromatic hydrocarbons and the carbon cycle, p. 33–58. *In* S. R. Hagedorn, R. S. Hanson, and D. A. Kunz (ed.), *Microbial Metabolism and the Carbon Cycle*. Harwood Academic Publishers, New York, N.Y.

Gibson, D. T., and R. E. Parales. 2000. Aromatic hydrocarbon dioxygenases in environmental biotechnology. *Curr. Opin. Biotechnol.* **11:**236–243.

Gibson, D. T., and V. Subramanian. 1984. Microbial degradation of aromatic hydrocarbons, p. 181–252. *In* D. T. Gibson (ed.), *Microbial Degradation of Organic Compounds*. Marcel Dekker, Inc., New York, N.Y.

Gunsalus, I. C. 1984. Learning. *Annu. Rev. Microbiol.* **38:**xiii-xliv.

Hudlicky, T., D. Gonzales, and D. T. Gibson. 1999. Enzymatic dihydroxylation of aromatics in enantioselective synthesis: expanding asymmetric methodology. *Aldrichim. Acta* **32:**35–62.

Kornberg, A. 2000. Ten commandments: lessons from the enzymology of DNA replication. *J. Bacteriol.* **182:**3613–3618.

Lehninger, A. L. 1965. *Bioenergetics*. W.A. Benjamin, Inc., New York, N.Y.

Ourisson, G., P. Albrecht, and M. Rohmer. 1984. The microbial origin of fossil fuels. *Sci. Am.* **251:**44–51.

Reineke, W. 1998. Development of hybrid strains for the mineralization of chloroaromatics by patchwork assembly. *Annu. Rev. Microbiol.* **52:**287–331.

Stephenson, M. 1949. *Bacterial Metabolism*. Longmans, Green and Co., London, United Kingdom.

Timmis, K. N., and D. H. Pieper. 1999. Bacteria designed for bioremediation. *Trends Biotechnol.* **17:**201–204.

Preface

This book takes the view that it is more important to define the emerging questions in biodegradation and biocatalysis than to contribute to the pretense that there are answers to many of the most important questions in the field. James Thurber expressed it more simply, "It is better to know some of the questions than all of the answers." This book frames questions such as, "What is the extent of microbial metabolism on Earth?"; "What microorganisms participate in certain types of metabolic transformations?"; "How does one define microbial metabolism in light of widespread genome sequencing?"; "How will microbial biocatalysis and biodegradation be applied commercially in the future?" Some of the answers will be found in the short term; some may remain elusive for a long time. In fact, it is important to question whether microbial metabolism can ever be fully appreciated; might not microbes evolve new catabolic activities faster than they can be studied by scientists?

To begin to answer the global questions posed in this book, a reductionist approach is taken. This goes against the perception of some people who feel that global questions require non-reductionist, or global, methods of experimentation. Throughout this book, the underlying molecular basis of microbial biocatalysis is detailed. One must reduce the complexity of a system, understand its parts, and then see how the parts act together to make a functional whole. To understand microbial biocatalysis globally, it is important both to delve deeply *and* to step back periodically for the broad view.

A reductionist approach also flows naturally from the duality of biocatalysis and biodegradation: the chemistry of the substances being transformed and the microbiology of how those transformations occur in the world. This necessitates going back and forth between microbiology and chemistry. The first part of the book, chapters 2 through 4, focuses on microorganisms and laboratory methods for their isolation and study. Then, chapter 5 focuses on chemical compounds, their origins, and their distribution on Earth. In chapters 6 through 8, microbiology and chemistry blend with coverage of enzymes, evolution, and metabolic logic. Last, in chapters 9 to 13, issues currently attracting strong interest are covered: predicting biodegradative metabolism, genomics, and industrial applications of biocatalysis and biodegradation.

The book also seeks to deal with the complexity of biocatalysis and biodegradation, viewing the global microbial ecosystem and its collective metabolic activities as part of the beautiful tapestry of nature. Edward O. Wilson described complexity in human terms as follows, "The love of complexity without reductionism makes art; the love of complexity with reductionism makes science." This book stresses the complexity of organic molecules, over ten million of which are currently known, and the complementary complexity of microbial metabolism that has evolved to transform those organic compounds.

The book's focus on events occurring at the molecular level differentiates it from a number of other recent books on biodegradation which stress bioremediation engineering or other aggregate processes such as microbiological aspects of soils and waters that influence biodegradation. Other treatments of biodegradation stress different chemical compounds and how each is metabolized by microorganisms. The present book differs in that it seeks to describe the *logic* of microbial biocatalysis. By logic, we mean the conceptual framework at the molecular level: how enzymes work, how pathways interact, and how physiological systems support biodegrading organisms. Overall, we envision this book to be useful for graduate students, for whom it might be suitable as a textbook, and for specialists in academia and industry interested in microbial biocatalysis and biodegradation. With microbial genome sequencing projects providing a new approach for studying individual prokaryotes, we anticipate that microbial catalysis will enter a new golden age. It is more important than ever, for better annotation of gene sequences and for applying biocatalysis to practical problems, that the underlying logic of microbial metabolism be better appreciated.

This book would not have reached its current form without the help of many people, and we wish to acknowledge them here. Much of the manuscript was written at the Yoss Ranch in Zumbro Falls, Minn. (and we thank Kathe and Robert Yoss for their hospitality), and at the Dow Chemical Company in San Diego, Calif. At the latter site, Mani Subramanian kindly provided one of the authors a quiet place to write and Davetta Adams provided assistance in getting materials.

We also thank the following people for review of and helpful commentary on the manuscript: Alasdair Cook, Mervyn deSouza, David Gibson, Jack Richman, Jennifer Seffernick, Alfred Spormann, Lisa Strong, Mani Subramanian, and Gregg Whited. An important source of inspiration for this book was the University of Minnesota Biocatalysis/Biodegradation Database (UM-BBD). The UM-BBD has been broadly used; it has received over 2,000,000 accesses from users in 75 countries. The UM-BBD has achieved this success because of the work of many people, most notably Lynda Ellis, the co-Director, but many volunteers and coworkers have contributed and we wish to thank them all for their efforts. Finally, we wish to express our appreciation to our families for their patient understanding of the time required of us to bring this project to completion. Special thanks are due to Deborah Allan, Adrian Wackett, Katherine Wackett, Yifat Bar-Dagan, Charlie Hershberger, and Joy Hershberger.

Lawrence P. Wackett
Department of Biochemistry, Molecular Biology & Biophysics
and Biological Process Technology Institute

AND

C. Douglas Hershberger
Biological Process Technology Institute
240 Gortner Laboratory
1479 Gortner Avenue
University of Minnesota
St. Paul, MN 55108-1030

1

Definition of Terms

Further Resources in Biocatalysis and Biodegradation

General Concepts in Biodegradation and Biocatalysis

Every living thing is a sort of imperialist, seeking to transform as much as possible of its environment into itself.

— *Bertrand Russell*

Bacteria take relentlessly from their environments, but they do so selectively, because the chemical composition of a bacterium differs markedly from that of its environment. This is illustrated in Table 1.1. The elemental composition of the Earth does not reflect that of the universe. The elemental composition of an *Escherichia coli* cell does not reflect that of the Earth's crust. We use *E. coli*, referred to by Stanley Dagley as the scientist's "honorary life form," because it is comparatively well studied. It is generally representative, in elemental composition, of many prokaryotic cells. In a prokaryotic cell, most of the mass of hydrogen and oxygen is in the form of water. The other major element, comparatively rare in the universe and the Earth's crust, is carbon. Carbon, however, is what we often think of as the building block of life. The genetic tape, DNA or RNA, is carbon based, as are the cell membrane; the catalysts, or enzymes;

Table 1.1 Average elemental compositions[a]

Element	Percent elemental composition[b]		
	The Universe	Earth's crust	*E. coli*
Hydrogen	60	1	16
Helium	37	Negligible	Negligible
Oxygen	1	50	62
Neon	0.7	Negligible	Negligible
Carbon	0.3	<1	15
Silicon	Negligible	26	Negligible
Aluminum	Negligible	7	Negligible

[a]Derived from data in references 1 and 2.
[b]Percentage of mass.

and the energy storage molecules, such as ATP. When we study the chemistry of life, carbon is at the center of the action. Living things transform carbon-based compounds voraciously, and microorganisms, as Earth's most prolific and earliest-evolved life forms, do so most avidly. The carbon cycle on Earth is largely dependent on microbiological processes, and biodegradation constitutes one-half of the carbon cycle. So biodegradation and microbial-biocatalysis research, and this book, naturally focus on the microbial transformation of organic compounds.

Definition of Terms

> The language of science is incredibly interesting; it's a natural language under stress. The language is put under stress to explain things that are difficult to explain. . . . In this language simple words . . . acquire a host of alternative meanings.
>
> — *Roald Hoffmann*

Most often, the term microorganism is used without definition. To avoid confusion, we will define what we mean in this book when we describe microbial biodegradation. Our definition is not necessarily the most accepted, and we understand the inherent impreciseness in the term. Most simply, we consider a microorganism to be (i) an organism which it would require a microscope to observe over its entire life cycle or (ii) a macroscopic fungus. The latter is included because numerous fungi have unicellular yeast phases, they produce microscopic spores, and they are important in biodegradation. The former must be defined more precisely. For example, some prokaryotes, such as *Myxobacteria* spp. and *Streptomyces* spp., form macroscopic assemblages, but we include all prokaryotes among microorganisms.

In this book, we use the terms *biotransformation, biodegradation,* and *biocatalysis.* While they have similar meanings, they are used in different contexts. First, *biotransformation* is technically the same as metabolism. It is most often used for biological transformations of exotic compounds, such as drugs or carcinogens in mammals and environmental pollutants in prokaryotes. The term *biocatalysis* is used similarly to biotransformation, but it has the additional connotation of metabolism for the purpose of making a useful compound. Thus, the synthesis of antibiotics and acrylamide by microorganisms is often referred to as biocatalysis. The term *biodegradation,* by contrast, is most typically used when the focus is on taking a compound away, ideally, a process by which a potentially toxic compound is transformed into a nontoxic one. *Bioremediation* is a more recently coined term which refers to the application of biodegradation reactions to the practical cleanup of a compound or compounds.

The terms mentioned above derive from the interests of humans in understanding and using microbial metabolism for learning about nature, manufacturing specialty chemicals, or eliminating pollutants from the environment. The term *metabolism* is used to describe the enzyme-catalyzed reactions occurring in living cells for the purpose of making new chemical structures or deriving chemical energy to maintain the cell. The term *anabolism,* or biosynthesis, is used to denote the reactions which construct new molecules, typically more complex structurally, from some precursor compounds. This includes making important building blocks that assemble into structural and catalytic cellular macromolecules. For example, amino acids, nucleotides, and lipids are made by anabolic reactions when a microbe is grown on a single carbon source, such as glucose. In that process, the microbes break down, or catabolize, the glucose in order to capture chemical energy as ATP and biosynthesize the amino acids, nucleotides, and lipids necessary to make a functional cell. Many compounds are catabolized by microorganisms. Catabolism is a central theme of this book, which focuses on the diverse microbes, chemicals, enzymes, and genes which are involved in this process.

At the fundamental level, biodegradation and biocatalysis are inextricably linked by the biotransformation reactions and the enzymes catalyzing the reactions. This is best illustrated by specific examples. As shown in Fig. 1.1, naphthalene dioxygenase is prototypical of an important class of aromatic hydrocarbon dioxygenases which initiate the biotransformation of thousands of naturally occurring and industrial hydrocarbons. The reaction with naphthalene and dioxygen yields *cis*-1,2-dihydroxydihydronaphthalene. This is thought of as a biodegradation reaction, as it typically initiates the metabolism of naphthalene and related compounds to produce carbon dioxide, cell carbon, and chemical energy captured as ATP. However, the same enzyme, and indeed others of the same class, act on the heterobicyclic

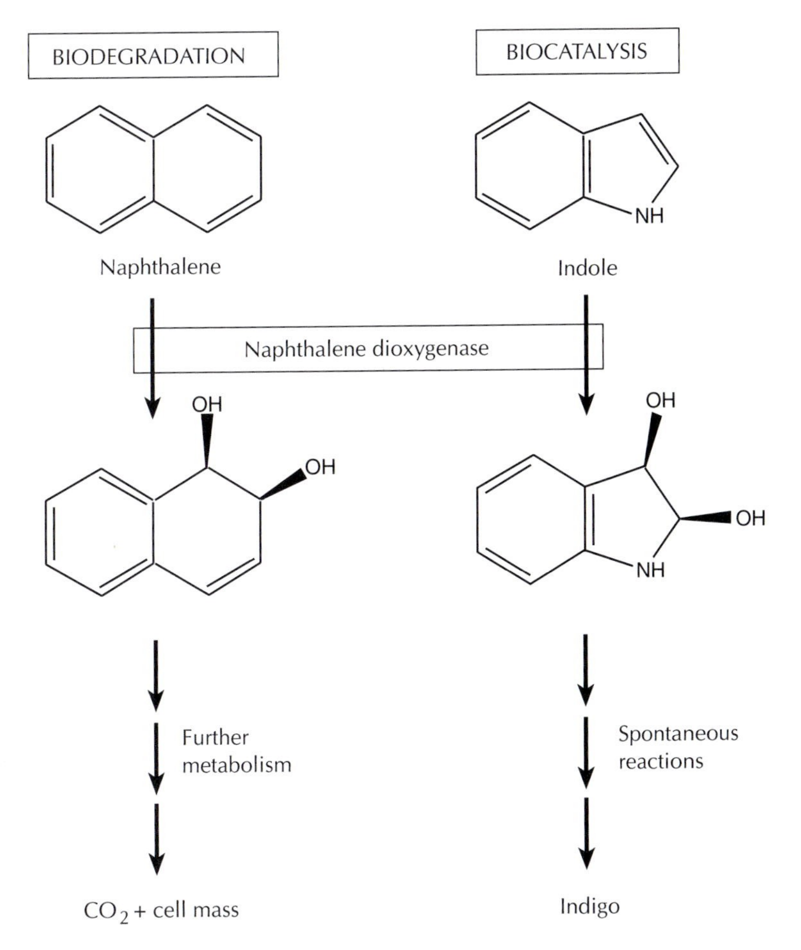

Figure 1.1 Biodegradative and biocatalytic reactions catalyzed by the same enzyme. Naphthalene dioxygenase transforms naphthalene and other polycyclic aromatic hydrocarbons for biodegradation, and indole is oxidized to the commercial blue-jean dye indigo.

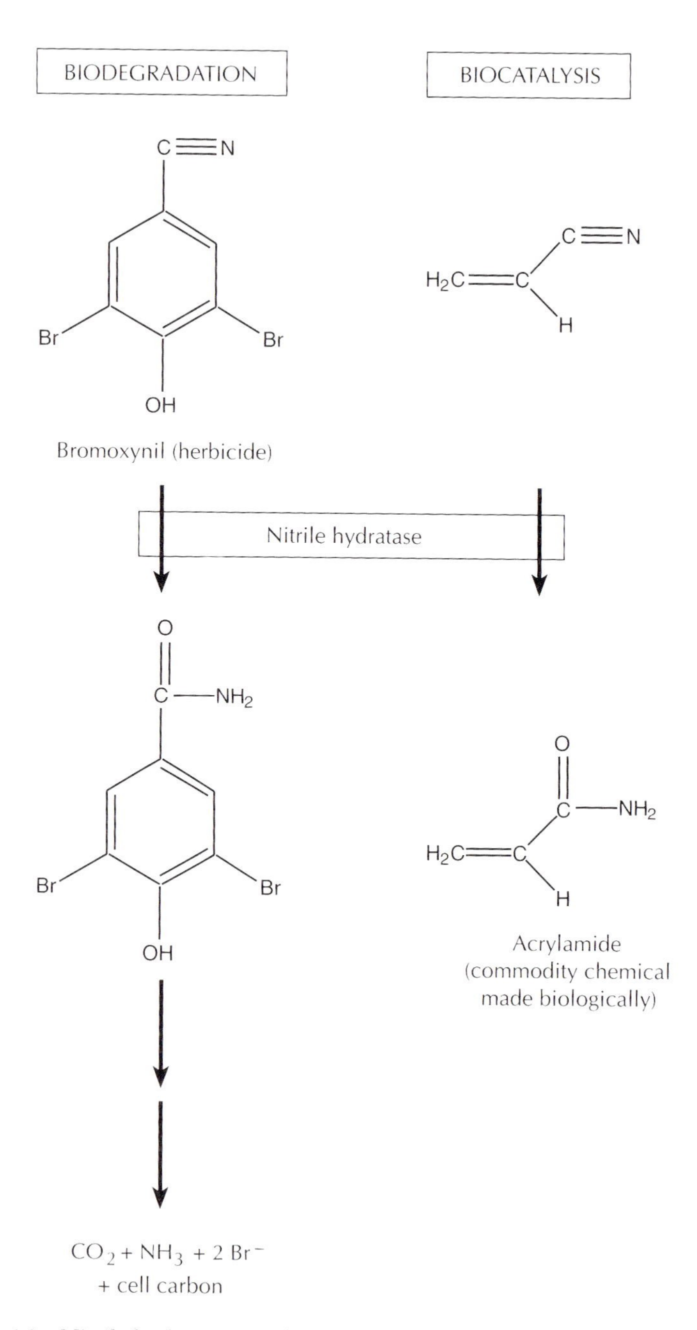

Figure 1.2 Nitrile hydratase catalyzes biodegradation of the herbicide bromoxynil and biotransformation of the feedstock acrylonitrile to yield the commodity chemical acrylamide.

aromatic compound indole to yield a corresponding dihydrodiol. In that case, the resultant intermediate is unstable and undergoes spontaneous dehydration to yield 3-hydroxyindole, commonly known as indoxyl. Indoxyl is known to undergo spontaneous dimerization to yield the blue-jean dye indigo. This observation has been developed industrially to generate a process for making indigo commercially. To date, 50,000-liter fermentation mixtures have been used for biological indigo production, although the low cost of indigo on the world market has precluded the wide-scale adoption of this biotechnology.

Another example of the dichotomy between biocatalysis and biodegradation is found with the enzyme nitrile hydratase (Fig. 1.2). This enzyme has been studied for its role in biodegradation of the nitrile-containing herbicide bromoxynil (3). A more prominent use of the enzyme is in the biotechnological production of the commodity chemical acrylamide. This application is discussed in some detail in chapter 4. The biodegradation and biocatalysis reactions are the same—the hydration of a nitrile to yield an amide. The major difference, as shown in Fig. 1.2, is that people desire the further metabolism of the intermediate for biodegradation and the accumulation of the intermediate for biocatalysis.

Further Resources in Biocatalysis and Biodegradation

It is the goal of this book to provide a comprehensive resource for microbial biocatalysis and biodegradation while acknowledging that we cannot cover all the topics which deserve coverage. Thus, each chapter contains a list of references cited in the chapter. Each list has a few recommended books or review articles which are of broad interest to those wishing to learn more about the main topics of the chapter. These are marked with asterisks. Additionally, appendixes are included at the end of this book that list Internet resources, general books, and journals relevant to biocatalysis and biodegradation.

References

1. **Hoffmann, R., and V. Torrence.** 1993. *Chemistry Imagined: Reflections on Science.* Smithsonian Institution Press, Washington, D.C.

2. **Ingraham, J. L., O. Maaloe, and F. C. Neidhardt.** 1983. *Growth of the Bacterial Cell.* Sinhauer Publishers, Sunderland, Mass.

3. **Vokounova, M., O. Vacek, and F. Kunc.** 1992. Degradation of the herbicide bromoxynil in *Pseudomonas putida. Folia Microbiol.* **37:**122–127.

2

A History of Concepts in Biodegradation and Microbial Catalysis

. . . progress in a scientific discipline can be measured by how quickly its founders are forgotten.

— *Edward O. Wilson*

It is true that scientific disciplines move inexorably forward and researchers pay the most attention to recent findings. However, a person who has a knowledge of the history of a discipline is much better able to see the context of recent work. Moreover, with the expansion of scientific knowledge, we increasingly see examples of rediscovery—cases where decades-old reports that have been overlooked are found to offer lucid explanations of new observations. This is especially true now that recent journals are available electronically but old versions are not. In this context, a historical view of microbial biodegradation and biocatalysis is presented in this chapter. One should study the past to anticipate the future.

The Beginnings of Biodegradation on Earth

Biodegradation and biocatalysis are as old as life itself. Virtually all theories of early evolution posit a prebiotic soup of organic molecules that served as precursors for the molecules that constituted the first life (26). However, these molecules must also have served as the energy sources to drive the first life processes. Self-replication requires energy, and that energy most likely came from preformed molecules in the prebiotic soup. By definition, a thermodynamically favorable chemical reaction involves the transformation of one or more molecules into a thermodynamically simpler molecule(s). The previous sentence is a simple but elegant description of biodegradation, both at the inception of life and now. Viewed in this way, biodegradation is one of the oldest life functions and was likely initiated 3.6 billion years ago.

It is thought that an explosion of life may have rapidly consumed most of the organic molecules that formed in the primordial soup during the first 1 billion years of Earth's history (26). At that point, the richest source of

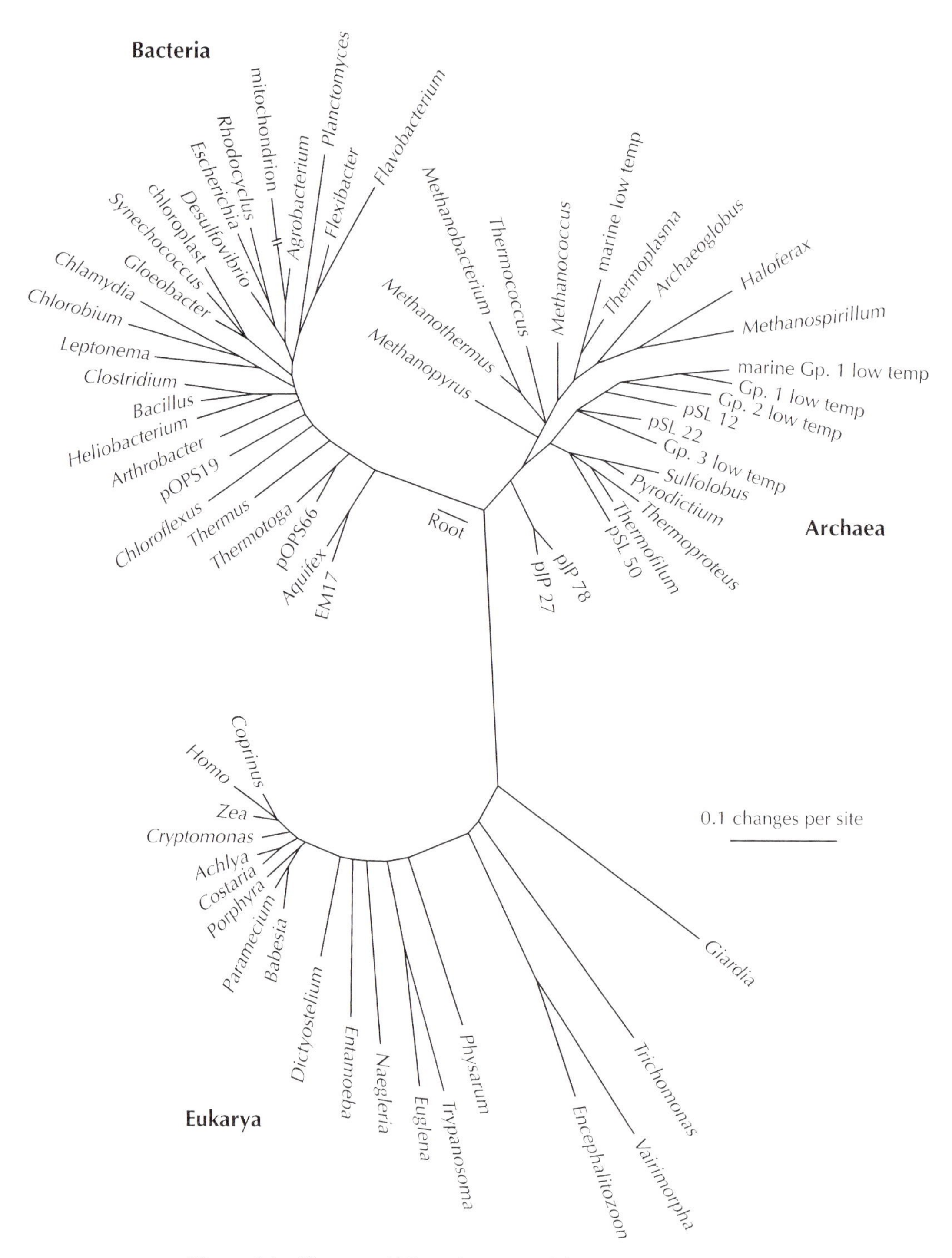

Figure 2.1 The tree of life as determined from comparison of 16S rRNA sequences. (Reproduced from reference 33 with permission of the American Association for the Advancement of Science.)

food for life was other forms of life. This continues today. Microorganisms produce lipases, proteases, cellulases and ligninases that decompose living organisms or their remains after death. It is thought that photosynthesis was an important development on Earth's surface that allowed much greater biomass production and hence generated more molecules to be biodegraded.

The taxonomic division of living things depicts three branches, or kingdoms (Fig. 2.1). Two of them are constituted of single-celled prokaryotes, and the third, *Eukaryota,* also contains many single-celled forms. The modern prokaryote kingdoms are as extensively evolved as the eukaryota, but the latter have expended considerable evolutionary capital to develop multicellular structures and elaborate behaviors. The prokaryotes have invested much of their evolutionary development in elaborate metabolism, and metabolically, many bacteria have very different catabolic pathways in addition to those involved in the breakdown of amino acids and nucleotide bases. In part, a modern bacterium's niche in a crowded bacterial world is defined by its catabolism. For example, there are delignifying bacteria, microbes that decompose the aromatic acids the delignifiers release, cellulolytic bacteria, and a complex web of anaerobes that handle the end products. So while an individual bacterium may be limited to several thousand genes, the collective catabolic gene pool on Earth is likely to be enormous.

Early Human Observations of Biodegradation and Biocatalysis

Humans have always "known" about biodegradation at some level. People surely observed rotting trees (Fig. 2.2) and suffered from food spoilage long before written accounts documented those observations. Food spoilage was most critical for survival and thus spurred technological innovation. The preservation of foods by drying, smoking, or pickling proved effective, even in the absence of a knowledge of the underlying mechanisms.

Similarly, biocatalysis has been part of human experience for thousands of years in the context of food fermentation. The experience was analogous to food rotting and was initially derived from observations that led to desirable changes in food preparation and preservation. Probably the most notable case was the fermentation of juices and other liquids to yield alcoholic beverages due to yeasts which were on the fruit or adventitiously fell into the liquid. Over time, humans learned to control the process, and the production of beer, wine, and other alcoholic beverages formally became an industry.

The wine industry proved to play a pivotal role in the development of microbiology and microbial biocatalysis when science was needed to save a business plagued by undesirable wine fermentations. This led to a protracted and vehement scientific debate about the underlying basis of the fermentation process, the potential for life to be generated spontaneously, and the ubiquity of microorganisms and their biochemical transformations. This is discussed in greater detail below.

Figure 2.2 Fungus growing on wood in the forest. (Reproduced from the Microbial World website [http://helios.bto.ed.ac.uk/bto/microbes/armill.htm] by permission of J. Deacon.)

Early Scientific Studies of Biodegradation and the Spontaneous-Generation Debate

> Nihil e nihilo [Nothing ever comes from nothing.]
>
> — *Lucretius*

Antonie van Leeuwenhoek, who is generally credited as the discoverer of microbes, observed microbes in infusions and spoiled organic matter in the late 1600s (Fig. 2.3) (15). Van Leeuwenhoek is thought to have believed that microbes were widely present in the environment, but serious questions about the origins of the microbes found in spoiled material persisted for the next 2 centuries (Fig. 2.3).

A clearer understanding of microbial degradation of foods and other organic matter likely came from macroscopic fungal systems. In fact, early studies with macroscopic fungi conducted by Pier Antonio Micheli, published in 1729, were important forerunners of later cultivation of microscopic microorganisms, yeast, and prokaryotes (7). These studies involved cultivation of the fungi *Mucor, Botrytis,* and *Aspergillus* on freshly cut slices of melon, quince, and pear. Micheli serially transferred the fungi and observed that they showed preferences for certain fruits used as the cultivating medium. The quality of the fruit for human consumption was observed to decline, and ultimately the fruit would be visibly consumed by the microbes.

The studies by Micheli were continued by a 19th century German botanist, Oskar Brefeld (7). Brefeld focused on pure-culture methodologies, which he conducted successfully using clear gelatin medium and inoculation of nutrient media with single spores. His results were probably known to Robert Koch, who pursued similar goals with prokaryotes.

In many cases, macroscopic fungi were not associated with the transformation of food and other organic substances. In those cases, the spoilage was typically attributed to chemical rather than biological processes. The cell theory of Mathias Schlier and Theodor Schwann, explaining the conversion of sucrose to ethanol as a biological process, was ridiculed in an anonymous paper, which most people attributed to the chemist Justus von Liebig. Liebig concluded that this must be a purely chemical process (Fig. 2.3) (8, 21). He understood that microorganisms existed and that they were found in spoiled foods and in wines. But Liebig explained that yeasts and other microorganisms were generated from fermentation products. That idea seems unimaginable given our current understanding that microorganisms transform sugars to ethanol and other fermentation products. But Liebig's claim must be viewed against the backdrop of the heated debate about spontaneous generation that raged in his time.

Many respected scientists in the 18th and early 19th centuries believed that putrefying meat and spoiled food gave rise to life de novo. The formation of life from clearly nonliving materials that were rotting was denoted by the term *spontaneous generation.* Even the appearance of macroscopic life was explained in this way; the spores of mycelial fungi and small eggs of insects were typically not observed. The Italian scientist Lazzaro Spallanzani disputed this view. He heated samples and sealed them immediately afterward, producing organic material that was preserved indefinitely. However, the results were not unequivocal, as some heated samples that likely contained endospores or other heat-resistant forms did spoil. Moreover, some scientists explained Spallanzani's results by saying that the sealed vessels lacked air and thus would not give rise to spontaneously generated life.

These and other objections were effectively countered in a series of elegant experiments conducted by Louis Pasteur over the period 1860 to 1862 (34, 35). Pasteur, a chemist by training, had originally been hired by the French wine industry to investigate the basis of spoiled products. Pasteur examined "good" and "bad" ferments under the microscope and found one dominant microbial population in the former and different or-

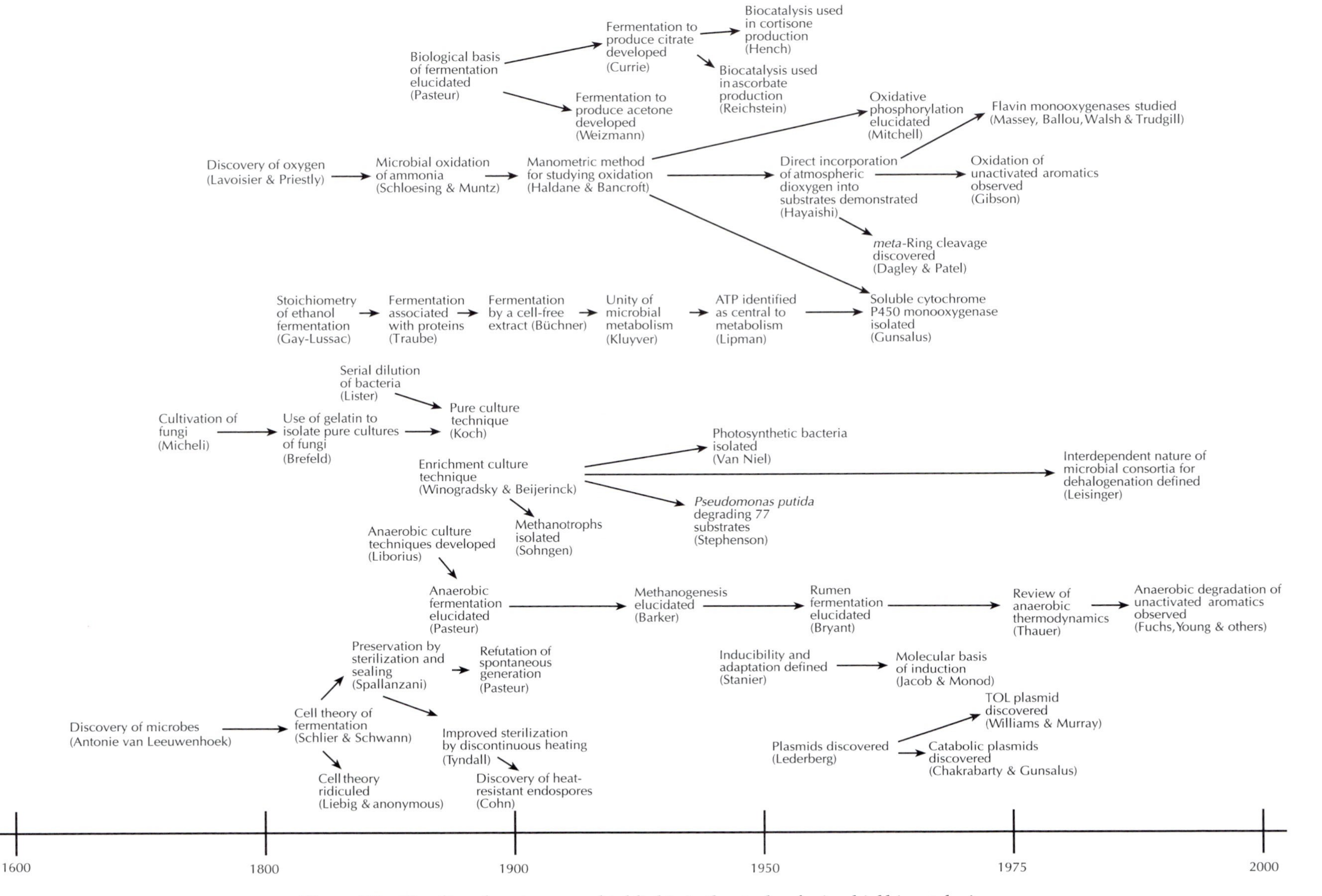

Figure 2.3 Timeline showing some highlights in the study of microbial biocatalysis over the last 400 years.

ganisms associated with the latter. The solution for obtaining good wines was to use only the single dominant organism leading to good ferments. This established the idea in Pasteur's mind that fermentation products, and by analogy putrefaction, resulted from the growth of the microorganisms. Further experiments established the principle more broadly and argued strongly against spontaneous generation. First, Pasteur observed that bacteria were ubiquitous in air, in water samples, and on solid surfaces. He further showed that microorganisms from the air could cause spoilage of food in sterilized media, whereas uninoculated controls showed no spoilage. To go one step further, it was necessary to rule out the possibility that the failure to observe growth in uninoculated media was explainable by the absence of air in a sealed vessel. To do that, Pasteur modified flasks to have horizontally positioned long necks, curved like that of a swan, in which dust particles and free microbes from the air would settle out in the neck and not reach the broth. Such uninoculated swan-neck flasks remained sterile for years, although air clearly had access to the medium.

Pasteur's results were buttressed by the observations of the Englishman John Tyndall (Fig. 2.3). Tyndall found that he could reproducibly sterilize media by discontinuous heat treatments (44). The currently accepted explanation for the effectiveness of such a heating regimen is that endospore-forming organisms will germinate after a heat treatment and be killed in subsequent heatings. Further support of Tyndall's findings came from the laboratory of the German scientist Ferdinand Cohn, who described endospores of *Bacillus subtilis* and their surprising heat resistance (12). Another important observation of Cohn's was that bacteria have characteristic morphologies, a property that was not well accepted at the time. Most investigators at that time dealt with mixed cultures, and the succession of organisms observed in flasks was often thought to be the changing forms of a single organism.

These experiments by different investigators working in different countries clearly established the fact that a medium, scrupulously freed of microorganisms, would not spontaneously generate life. Natural fermentations or rotting material arose via inoculation with microorganisms; these microorganisms were ubiquitously distributed in the environment, and thus inoculation occurred unless stringent conditions were taken to prevent it.

Microbial Pure Cultures from Nature

The groundwork laid in the third quarter of the 19th century touched off, in the last quarter of the century, the quest for convenient cultivation of pure cultures of microorganisms (11). In 1878, Joseph Lister had an early success when he made serial dilutions of his microbial solutions to extinction such that he would have only one cell in some of his dilutions used for inoculation. The shortcoming of this method was that one could not obtain numerically minor bacteria from a mixed population. By 1880, Robert Koch was convinced of the need for better methods of obtaining and transferring bacteria in pure culture and turned to using solid media. He first used natural materials, such as potato slices, for cultivating bacteria, but this was clearly selective for only certain bacteria. He then tried meat extracts hardened with gelatin. Moreover, he initiated the use of a platinum wire that had been sterilized by flame to spread bacteria on the gelatin. This proved effective, but a significant number of bacteria were able to liquefy the gelatin, and the cultures would be lost. A major advance was the use of the algal polysaccharide agar-agar, a material still very much used in microbiology today.

The development of reliable pure-culture techniques ushered in a golden age of discovery in microbiology. By the late 1800s, bacteria could be obtained from diverse sources and their properties could be studied. The greatest excitement was in the arena of pathogenic microbiology. Within a short time, many of the bacterial agents causing infectious diseases were identified. This allowed the development of vaccines and greatly improved methods of diagnosis. The path was also open for discovering the bacteria that fill the soils, waters, and airways of the Earth. The vast majority of bacteria do not cause human disease but instead have a niche(s) somewhere in the soil or water of the planet.

The advancement of environmental microbiology required its own set of techniques to cull one type of bacterium from its association with myriad other microbial types. Thus, environmental microbiology presented a challenge different from that of obtaining a pathogenic bacterium from the site of an infection. For example, Robert Koch had cultivated and studied *Bacillus anthracis* before he developed pure-culture methods because the organism had completely outcompeted other bacteria in the bodies of experimental animals. This type of situation is rare in soils and waters. As a result, many early studies of environmental microorganisms were conducted with mixed cultures. This presented an impediment to understanding the roles of different classes of bacteria in natural environments, since the properties studied were composites of those of different organisms that could not be readily duplicated in other laboratories or even in the same laboratory on a different occasion.

Figure 2.4 Sergei Winogradsky, a pioneer in environmental microbiology and one of the developers of the enrichment culture technique. (Courtesy American Society for Microbiology [ASM] Archives.)

This problem was alleviated by the technique of enrichment culture, often attributed to Sergei Winogradsky (Fig. 2.4) (46) and Martinus Beijerinck (Fig. 2.5) (1), working independently of each other. Enrichment culture is so important to the study of biodegradation and biocatalysis that chapter 3 is entirely devoted to its principles and practice. It is mentioned here briefly merely to explain the contributions of the environmental microbiologists of the 20th century. In general, enrichment culture entails the use of selective medium conditions that favor one or a small group of organisms that one desires to obtain from an environmental sample. By inoculating the selective laboratory medium and continuing to transfer the culture in that medium, one enriches, or increases in number, the population of the favored bacterium or bacteria. When this is coupled with plating techniques, pure cultures can sometimes be obtained and confirmed as pure.

Using selective culturing, Winogradsky, Beijerinck, Kaserer, and N. L. Sohngen isolated and characterized bacteria which oxidized hydrogen sulfide, sulfur, methane, ferrous ion, and hydrogen and other groups of bacteria which reduced nitrate and diatomic nitrogen, more often known as nitrogen fixation (11, 21). These independent studies, taken together, forwarded the idea that bacteria were instrumental in the cycling of organic and inorganic compounds in the environment. This is a critical concept which fundamentally captures the role of prokaryotes in the world as we currently understand it.

Figure 2.5 Martinus Beijerinck helped reveal the richness of the microbial world and develop the methods for isolating bacteria via enrichment culture. (Courtesy ASM Archives.)

Early History of the Study of Diverse Metabolic Activities of Microbes

In the early 1800s, the chemist Joseph-Louis Gay-Lussac established the reaction stoichiometry for the fermentation of glucose to ethanol,

$C_6H_{12}O_6 \rightarrow 2CO_2 + 2C_2H_5OH$, even before it was established as a biological process (Fig. 2.3) (8). During the 1800s, there was no clear acceptance of the biological nature of alcohol fermentation reactions and enzymes were just beginning to be appreciated. A prescient opinion was offered by Isidor Traube, a chemist who worked in the wine industry, who suggested that fermentation was associated with proteins (11). Pierre Berthelot isolated the enzyme invertase from yeast in 1860 and postulated that it was involved in fermentation. But there was no concept of enzymes acting in concert to form metabolic pathways.

A major event which paved the way toward understanding the basis of metabolism occurred in 1897 when Eduard Büchner (6) prepared a cell extract from *Saccharomyces cerevisiae* that transformed glucose to ethanol (Fig. 2.3). This was not only important methodologically, but it conceptually showed that metabolism could occur in the absence of a living cell. Several years later, cell-free enzyme activities were measured by in vitro assays, and the general principles of enzyme kinetics were rapidly established. But studies of enzymes in the early 20th century were largely confined to a small set of highly stable enzymes catalyzing hydrolytic reactions. A knowledge of cosubstrates, like nicotinamide adenine dinucleotide, and of coenzymes to assist catalysis had not yet emerged.

However, the ability of whole cells of pure cultures of bacteria to catalyze novel reactions did become known in the early 20th century. For example, even in the 19th century it was known that ammonia was readily oxidized to nitrate in the soil (16). Pasteur even suspected that this was a microbial process, but proof of this awaited the isolation of pure-culture nitrifiers by Winogradsky (47) via enrichment culture. Today, the ammonia monooxygenase (31) and hydroxylamine oxidoreductase (23) of *Nitrosomonas europaea* have been studied by gene sequencing and X-ray structure determination, respectively. Another facet of the nitrogen cycle, the nitrogen fixation reaction, now known to be catalyzed by nitrogenase, was also first studied by Winogradsky (47) using pure microbial cultures isolated via enrichment culture.

Winogradsky is perhaps best known for his insight that bacteria can grow via the oxidation of inorganic compounds and can obtain carbon exclusively from carbon dioxide (46). The ^{14}C labeling methods of the present day could not be employed; in fact, physicists did not suspect that ^{14}C existed until the 1930s. Despite this, Winogradsky was able to be convincing by excluding all sources of carbon except carbon dioxide and carbonate salts from the growth medium (4).

Unity of Metabolism in Living Things

As the 20th century proceeded, biochemical investigations of microbes revealed two important and seemingly contradictory views of the biochemistry of life. First, the idea of the unity of biochemistry was established. That is, the intermediary metabolic pathways of microbes, plants, and animals were observed to be similar. At the same time, the diversity of microbial metabolism was continuously recorded as new microbes were identified and studied biochemically. The two seemingly contradictory ideas are compatible if one realizes that much of metabolism is common while a fraction of a microbe's metabolism is unique to it and a few related strains. Moreover, microbial genetic diversity is enormous, and this translates into global biochemical diversity for that fraction of the genome devoted to unique reactions.

A. J. Kluyver (Fig. 2.6), a microbiologist from Delft, The Netherlands, stressed the underlying unity of microbial metabolism as early as 1924 (21). This proved prescient as the outlines of intermediary metabolism were revealed. The glycolytic pathway, the tricarboxylic acid cycle, fatty acid metabolism, and biosynthetic pathways for amino acids are examples of metabolic pathways which are similar in many taxonomically diverse bacteria. As coenzymes were discovered, these too proved to be common to many systems, although some bacteria can biosynthesize all of them and others require them to be provided, either in laboratory growth media or by cross feeding in the wild. By the time Fritz Albert Lipmann (29) described the structure and function of ATP, the major energy currency of *Escherichia coli* and humans, scientists broadly accepted the idea of the unity of biochemistry in all living things on Earth. As Jacques Monod put it, "What is true for *E. coli* is true for the elephant."

Figure 2.6 The Delft microbiologist A. J. Kluyver recognized both the diversity of microbes and the unity of their fundamental metabolic processes later known as intermediary metabolism. (Courtesy ASM Archives.)

It was thus something of a surprise to some people when the methanogenic bacteria were subjected to intense in vitro study and a whole series of novel coenzymes were discovered (14). Despite the different structures, the fundamental chemistry of some of them, such as methanopterin and coenzyme F420, resembled biochemistry already described in eubacteria. Most recently, with microbial genomics uncovering a greater percentage of previously unrecognized sequences than anticipated, there is a resurgence of interest in microbial genetic and catalytic diversity. This will be discussed further in chapter 11.

Oxygen and Oxygenases

The element oxygen was discovered in experiments conducted by Antoine-Laurent Lavoisier and Joseph Priestly in the late 1700s (21). The term *oxygen* is also used to denote the atmospheric gas which has the molecular formula O_2. To avoid confusion, we will refer to the oxygen molecule as dioxygen and to relevant reactive species of oxygen as monoatomic or diatomic oxygen.

In 1775, Priestly implicated oxygen as an element necessary to sustain the life of a mouse (21). The notion of dioxygen in air as a life-sustaining element led to misinterpretation during the spontaneous-generation debate, as Spallanzani attributed the failure of heated sealed vessels to support life to the exclusion of air. As mentioned above, Pasteur disproved this idea when he heated media in swan-necked flasks which were open to the atmosphere yet remained sterile indefinitely (35).

The role of oxygen in microbially mediated transformation of chemicals was perhaps first suggested by the work of Jean Jacques Schloesing and A. Muntz (39). That work indicated that amines and ammonia were transformed to nitrate in sewage water, coinciding with the growth of microorganisms. The oxidation of ammonia was inhibited by the addition of compounds like chloroform that inhibited microbial growth. As indicated above, the process was firmly attributed to specific bacteria when Winogradsky isolated ammonia-oxidizing bacteria following enrichment culturing experiments.

In the first 3 decades of the 20th century, the overall metabolism of bacteria was studied extensively using the manometric method developed by J. B. S. Haldane and Joseph Bancroft and elaborated on by Otto Warburg

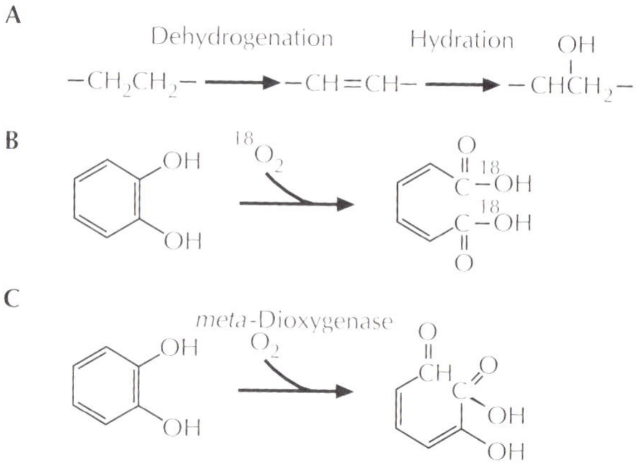

A

Dehydrogenation — Hydration

$-CH_2CH_2- \longrightarrow -CH{=}CH- \longrightarrow -CH(OH)CH_2-$

B

OH, OH — $^{18}O_2$ — $C(=O)-^{18}OH$, $C(=O)-^{18}OH$

C

meta-Dioxygenase

OH, OH — O_2 — CH=O, C(=O)OH, OH

Figure 2.7 The various processes whereby microbial metabolism introduces oxygen atoms into organic substrates. (A) Dehydrogenation of a carbon-to-carbon single bond followed by addition of water; (B) *ortho*-dioxygenative ring cleavage of a catechol to yield a dicarboxylic acid product; (C) *meta*-dioxygenative ring cleavage of catechol to yield a carboxyaldehyde.

(21). The manometric technique was based on dioxygen consumption, and it offered a quantitative measure of metabolism. It was used to monitor the catabolism of sugars, amino acids, nucleotides, and fatty acids. The process of respiration, in its outlines, was thus firmly established well in advance of the present understanding of oxidative phosphorylation coupled to a proton gradient, as elaborated by Peter Mitchell in 1961 (32).

Our understanding of oxygen metabolism was perhaps retarded by a rigorous doctrine, established in the 1920s, which asserted that the oxygen atoms present in biological organic molecules are invariably derived from water. While oxidation reactions clearly occurred, the overall process of oxidation and hydration could neatly account for the incorporation of oxygen atoms (Fig. 2.7A). This in fact occurs, for example, in the oxidation of fatty acids, where a methylene carbon is formally oxidized to an alcohol by intermediate oxidation to an alkene followed by a hydration reaction. The perception that this chemistry alone explained oxygen incorporation was held for over 3 decades because its originator, Heinrich Weiland, was a Nobel prize-winning chemist whose views held sway over his colleagues.

Figure 2.8 Osamuri Hayaishi, discoverer of oxygenase enzymes. (Courtesy ASM Archives.)

In 1955, O. Hayaishi (19a) (Fig. 2.8) and H. S. Mason (30) published papers which changed this perception of oxygen incorporation forever. Taking advantage of the availability of mass spectrometry, Hayaishi had decided to test the hypothesis of oxygen incorporation directly, using a ring cleavage enzyme as the test case. The reaction was the conversion of catechol to *cis,cis*-muconic acid, in which a carbon-to-carbon bond is broken and two atoms of oxygen are incorporated to generate two carboxylic acid groups (Fig. 2.7B). With the availability of mass spectrometry and oxygen isotopes, Hayaishi designed a simple and decisive experiment. The reaction could be run in the presence of heavy dioxygen, $^{18}O_2$, or with heavy monoatomic oxygen in water, $H_2{}^{18}O$. The muconic acid product could then be isolated and analyzed to determine which oxygen source led to an increase in mass. The result, which Hayaishi did not find surprising, was that the atoms of atmospheric dioxygen exclusively were incorporated into the substrate. This was the first demonstration of an oxygenase enzyme activity.

Many years later, when Hayaishi was asked why he had the insight as a young graduate student to defy the common wisdom about oxygen incorporation, he responded humbly that he acted out of ignorance of the entrenched ideas and just did the obvious experiment.

In the same year as Hayaishi's publication, Mason and coworkers reported on an enzyme they called phenolase (30). Phenolase catalyzed the transformation of 3,4-dimethylphenol to yield 4,5-dimethylcatechol. The enzymatic reaction was run in the presence of $^{18}O_2$, the product was pyrolyzed, the carbon dioxide liberated was analyzed by mass spectrometry, and the pyrolysate was found to contain one incorporated ^{18}O atom per product molecule. This was consistent with an enzymatic incorporation of ^{18}O into the substrate during the reaction.

Mason's experiments were carried out with a monooxygenase enzyme, and Hayaishi's used a dioxygenase. With dioxygenases, both atoms from a single molecule of dioxygen are incorporated into the organic substrate. Moreover, both aromatic ring cleavage reactions yielded products in which the ring was cleaved between the carbons bearing the hydroxyl groups, now denoted *ortho* cleavage. This was thought to occur exclusively until S. Dagley and colleagues reported evidence for another type of reaction (Fig. 2.7C), in which the ring was cleaved between a phenolic carbon and an adjacent carbon containing a hydrogen atom (13). This type of reaction, now known as *meta* ring cleavage, is known to occur widely in aerobic soil bacteria.

Figure 2.9 Irwin Gunsalus and his coworkers first studied bacterial cytochrome P450 monooxygenases and deduced key features of heme monooxygenase reaction mechanisms. (With permission from *Annu. Rev. Microbiol.* vol. 38, © 1984, by Annual Reviews www.AnnualReviews.org.)

Around the same time, a cytochrome P450 monooxygenase activity was described in both mammalian and bacterial systems. In monooxygenase reactions, one atom of dioxygen is incorporated into the organic substrate and the other atom is reduced to water. Moreover, cytochrome P450 monooxygenases contain a heme prosthetic group. Other cytochromes had long been known for their roles in electron transport, but cytochrome P450 had novel absorption properties which were intriguing. Mammalian cytochrome P450 monooxygenases were membrane-bound enzymes, but a *Pseudomonas putida* strain which had a soluble P450 enzyme system was isolated in the laboratory of Irwin Gunsalus (Fig. 2.9) (20). The bacterial cytochrome P450cam was shown to initiate catabolism of the plant terpenoid compound camphor via a hydroxylation reaction. Gunsalus immediately saw the value of studying a soluble *Pseudomonas* P450 system, and this rapidly became the paradigm for understanding this broad class of enzymes. Thousands of cytochrome P450 enzymes are now known to carry out a wide range of catalytic reactions in bacteria, plants, and animals.

Figure 2.10 David T. Gibson and his coworkers revealed the mechanisms of aromatic hydrocarbon metabolism and discovered the aromatic hydrocarbon dioxygenases exemplified by naphthalene dioxygenase. (Courtesy ASM Archives.)

Subsequently, the richness of oxygenase catalysis continued to be revealed. In addition to alkanes and hydroxylated aromatic ring compounds, unactivated aromatic hydrocarbons were shown by David Gibson (Fig. 2.10) and his coworkers (19) to undergo dihydroxylation by dioxygenases to yield a *cis*-dihydrodiol intermediate. Subsequent dehydrogenation typically yields a catechol, which undergoes further dioxygenation. The prototypical member of this class is naphthalene dioxygenase, for which an X-ray structure is now available. Naphthalene dioxygenase is known to oxidize over 70 different compounds; for a compilation available on the World Wide Web, see http://umbbd.ahc.umn.edu/naph/ndo.html. Interestingly, naphthalene dioxygenase catalyzes not only dioxygenation of aromatic hydrocarbons but also monooxygenation of carbon centers and sulfur atoms, as well as desaturation reactions. The last is surprising until one considers that fatty acid desaturases generate reactive iron-bound oxygen species and thus bear a mechanistic similarity to oxygenases (18).

Another class of oxygenase discovered during the active midcentury period was those containing flavin adenine dinucleotide to activate dioxygen. Studied by Vincent Massey, David Ballou, Chris Walsh, Peter Trudgill, and others, the flavin monooxygenases serve, in one set of enzymes, to hydroxylate activated aromatic rings. Another important class of the flavin monooxygenases is those catalyzing Baeyer-Villiger monooxygenation reactions with ketone substrates, yielding ester products which are amenable to hydrolysis. The utilization of these reactions in metabolizing specific classes of compounds will be discussed in greater depth in chapter 8.

One indicator of the importance of oxygenases in microbial catabolism, and in the reactions principally studied for biodegradation and biocatalysis, is their high level of occurrence in the Biocatalysis/Biodegradation Database (http://umbbd.ahc.umn.edu). Over 60% of the enzymes covered in the database are oxidoreductases, and the greatest number of those are oxygenases.

History of Anaerobic Biocatalysis

Most of the early uses of microbiology involved anaerobic processes, for example, the production of ethanol, vinegar, or acetic acid. But it was not until the second half of the 19th century that Louis Pasteur popularized the idea that fermentation was microbial metabolism in the absence of air. In fact, the observation that the yeast *Saccharomyces cerevisiae* produces much less ethanol when aerated is commonly known as the Pasteur effect.

While anaerobic metabolism was important in commercial fermentation processes, as discussed below, the details of anaerobic reactions were not well known because of difficulties in culturing anaerobic bacteria in pure culture. This derived partly from the predilection of microbiologists for agar plate culturing and the difficulty of maintaining plates under sufficiently anoxic conditions. Paul Liborius was one of the first to devise methods dedicated to anaerobic culture which allowed him to differentiate between obligate anaerobes, obligate aerobes, and facultative organisms (28). He used tubes with stoppers and displaced air with hydrogen gas. Other developments were to use chemical substances which scavenge oxygen, such as pyrogallol (22). Much later, the development of anaerobic glove boxes was a major advance for isolating, maintaining, and studying biocatalytic reactions found in strictly anaerobic bacteria.

In addition to the intrinsic difficulties of maintaining anaerobic conditions, it is also difficult to separate anaerobic bacteria and study their individual metabolisms. The classic example of this is the description of the organism *Methanobacillus omelianskii*, which catabolizes ethanol to produce methane (5). This culture was transferred and studied for years before it was realized that it was a tight assemblage of an ethanol-oxidizing organism and a methanogen capable of generating methane from carbon dioxide and dihydrogen gas. This began to illustrate the very close metabolic interdependence in anaerobic microbial ecosystems. Much more recently, numerous investigators have cultured defined anaerobic consortia and determined the biochemical role of each member in metabolizing a given compound.

In one example, the industrial solvent dichloromethane is metabolized both aerobically and anaerobically. The aerobic metabolism of dichloromethane is carried out by some methylotrophic bacteria, and the metabolic logic of this type of C1 metabolism will be discussed in chapter 8. The anaerobic biodegradation of dichloromethane was first demonstrated with a consortium of organisms (3). Initial efforts by Thomas Leisinger's group to isolate a pure culture that grew with dichloromethane as the sole carbon and energy source were unsuccessful. Subsequently, the group isolated from their consortium a gram-positive endospore-forming rod and a gram-negative endospore-forming homoacetogen which could be cocultured on solid medium with dichloromethane as the sole carbon and energy source. Thermodynamic considerations suggested an acetogenic fermentation of dichloromethane involving interspecies formate transfer. A knowledge of the biochemical interplay between species ultimately allowed Leisinger's group to begin to characterize the biochemical mechanism of dichloromethane metabolism by the strictly anaerobic bacterium *Dehalobacterium formicoaceticum*.

The idea of using a thermodynamic approach to anaerobic biotransformations has been a powerful one. For determining overall free energy changes, it is not necessary to know the biochemical pathway operating, merely the energy level of the starting material and the end products. These are conditions which can often be met, even when considering complex microbial consortia. One needs to know the starting compound, the organic end products, and the final electron acceptor. The latter can be crucial. When oxygen is not available, common alternative electron acceptors, such as nitrate, sulfate, or carbon dioxide, differ dramatically in their oxidation-reduction potentials and hence in how much energy can be generated in the metabolic oxidation reactions which are coupled to their reduction. A classic treatment of the thermodynamics of anaerobic metabolism was published by Rolf Thauer and colleagues in 1977 (42). This review also elegantly described the thermodynamic logic behind interspecies hydrogen transfer. The *M. omelianskii* example given above was an instance in which the metabolism of one organism is overall thermodynamically unfavorable unless a closely associated organism containing a hydrogenase is able to maintain hydrogen partial pressure at a very low level. This is a clear example of symbiosis, generated by coevolution and guided by a thermodynamic imperative.

Most recently, the study of anaerobic microorganisms has continued to reveal startling biochemical transformations. For example, during a period in the 1960s and 1970s, it was known that anaerobes could transform benzoic acid (17), but many thought that unactivated aromatic hydrocarbons were metabolically inaccessible to anaerobes. The work of George Fuchs (43), Lily Young (2), and others led the way to dispelling this idea. Anaerobes have come up with a defining metabolic solution for benzenoid rings: the carboxylation of aromatic hydrocarbons to yield benzoic acid compounds. Benzoic acid may be a central funneling point in anaerobic benzene ring metabolism, similar to the common intermediacy of catechols during the aerobic metabolism of these compounds. These issues are discussed in more detail in chapter 8, dealing with metabolic logic.

Molecular Genetics and Regulation

Experiments examining the expression of metabolic activities in microbial cells led to the idea of catabolite control, or induction, of enzyme activity. This is a logical consequence of the intense competition of microbes for

Figure 2.11 Roger Stanier was a pioneer in studying *Pseudomonas* and related soil bacteria and deducing features of their broad catabolic metabolism. (Courtesy ASM Archives.)

Figure 2.12 Ananda Chakrabarty pioneered studies of transduction and catabolic plasmid transfer among *Pseudomonas* strains. (Courtesy ASM Archives.)

nutrients and the insight of A. H. Stouthamer (41) that more than half the ATP used for the synthesis of cell material by *E. coli* is used to polymerize amino acids into proteins. Henning Karstrom (21) made the distinction between enzymes which were expressed at all times, or constitutive, and those which were made only when required, called adaptive or inducible. Many of the catabolic pathways of interest in the context of biocatalysis and biodegradation, particularly in the genus *Pseudomonas,* proved to be in the inducible class. This principle was used to advantage by Roger Stanier (Fig. 2.11) and his colleagues. The method they used was denoted simultaneous adaptation, and it tested whether a compound occurs as an intermediate in a given pathway (40). Compounds which are intermediates will be metabolized immediately by cells grown with the starting compound, which acts as an inducer for all the enzymes in its pathway of metabolism. Compounds oxidized by other pathways or that occur earlier in a pathway than the inducing growth compound are metabolized only after a lag period, presumably the time required for it to act as an inducer. While this did not universally hold, it did serve to provide a rough sketch of metabolites for a number of metabolic pathways.

The molecular basis of induction was revealed somewhat later with studies of the *lac* operon described by Francois Jacob and Jacques Monod (24). This was an example of negative regulation and established the idea of derepression. Later data showed that other operons are regulated in a positive manner, with the inducer binding to a regulatory protein to facilitate transcription of the gene.

Our perception of the organization of catabolic genes was greatly influenced by the discovery of plasmids by Joshua Lederberg (27). The term *plasmid* was used by Lederberg to denote any extrachromosomal genetic particle. We often think of plasmids as autonomously replicating genetic elements which are smaller than the chromosome(s) of the organism. Starting around 1970, plasmids began to be heavily used in molecular genetic research and biotechnology. It also became recognized that plasmids carried genes which conferred bacterial resistance to antibiotics and thus that they were very important in medical microbiology.

In the late 1960s, catabolic plasmids were uncovered, deriving from investigations by Ananda Chakrabarty (Fig. 2.12) and Irwin Gunsalus into developing genetic systems for *Pseudomonas* sp. (10). Subsequently, a transmissible plasmid encoding camphor oxidation, including the cytochrome P450 monooxygenase, was discovered in *P. putida* (37). Shortly thereafter, an octane catabolic plasmid was demonstrated in *Pseudomonas* (9). One year later, the TOL plasmid was reported by P. A. Williams and K. Murray (45).

In general, catabolic plasmids are self-transmissible and have broad host ranges. They are often dispensable if the organism is supplied with glucose or low-molecular-weight organic acids, such as tricarboxylic acid intermediates, as carbon sources. In fact, the loss of ability to catabolize "exotic" compounds after transfer in rich medium is often diagnostic for the presence of a large catabolic plasmid. Many such plasmids are large, containing 50 to 500 kb of DNA, and thus represent a considerable energy demand on the cell for their replication.

The spread of catabolic plasmids is thought to be analogous to the spread of R plasmids, which confer antibiotic resistance. In both cases, there is a selective pressure operating under certain conditions, either making a bacterium more competitive with respect to metabolizing many carbon or nitrogen sources or allowing survival when antibiotics are applied. It is not

Industrial Applications: Two Early Examples

Louis Pasteur's work on wine fermentation helped an established industry that was in trouble, but in 1915, Chaim Weizmann saw the need to develop a completely new fermentation industry. It was based on a bacterium, *Clostridium acetobutylicum* (Fig. 2.13), which produces acetone and *n*-butanol as fermentation products. Those two end products are industrial chemicals, and in the early 1900s they were produced by the chemical industry. But the most well-developed chemical industries were in Germany, and in 1915 England was engaged in war with Germany. As a result, acetone was scarce at the very time that it was needed in large volume for manufacturing the military explosive cordite.

Chaim Weizmann was both an accomplished scientist and politically astute; he deduced his nation's need and the bacterium which could potentially meet that need. He obtained the right organism, but he also needed to find the right starting material, which turned out to be corn starch. There was also the hurdle of scaling up a microbial process to make not grams of material, but tons. Weizmann was up to the task, and his acetone fermentation met the war needs of England. The process moved beyond England and achieved an impressive scale. In Canada in 1918, there were 22 fermentors operating to produce acetone, each with a capacity of 30,000 gallons (25).

There were important political ramifications of Weizmann's scientific accomplishments. The British secretary of munitions during the war, David Lloyd George, was very impressed with Weizmann and the value of the fermentation process he had developed. Lloyd George subsequently went on to become prime minister. Weizmann had declined receiving any personal honors based on his work, but he did advance the idea that the Jewish people around the world needed a homeland in the Middle East. The British controlled Palestine at the time, and Lloyd George spoke to the foreign secretary, Earl Balfour, about the problem. The result was the well-known Balfour Declaration of 1917, which paved the way for the creation of the State of Israel. Weizmann served as the first president of Israel. He later acted as a director of the Weizmann Institute, which has become a leading international center of scientific research.

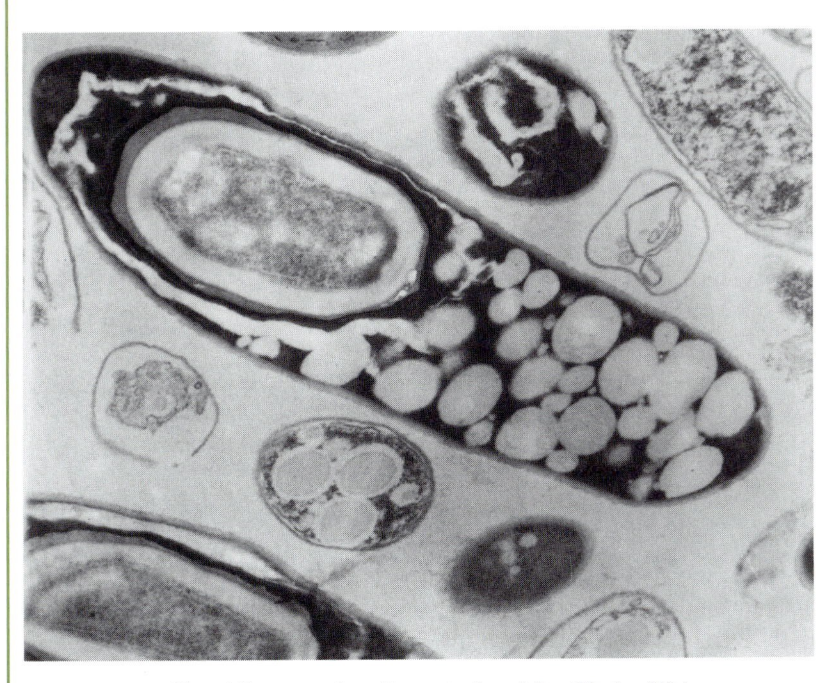

Figure 2.13 *Clostridium acetobutylicum,* isolated by Chaim Weizmann, was used to produce acetone for the British war effort. (From *Power Unseen: How Microbes Rule the World,* by B. Dixon, © 1994, with permission.)

World War I prompted the development of another microbial transformation, the production of citric acid. In the early 1900s, Italy met much of the world's citric acid demand using an extraction process with lemons and limes as the starting material. Difficulties in cultivating the citrus trees during the war years led to an excessively high price for this commodity chemical. The demand was still high because of the compound's important applications in the food industry. This set the stage for innovative technology. A key observation was made by J. N. Currie, who had observed that the filamentous fungus *Aspergillus niger* could be made to accumulate citric acid in the growth medium. Tweaking growth conditions, specifically by using a carefully controlled level of iron, led to the accumulation of reasonably high levels of citric acid by the fungus. In 1923, the Pfizer Company developed a commercial process for manufacturing citric acid from sucrose using *A. niger.* European companies soon joined in the competition, also using the fungus but starting with sugar beet molasses. The use of fungi in citric acid production continues today.

entirely clear how catabolic plasmids are organized, although they are now being mapped and sequenced at an accelerating pace. With the advent of wide-scale genome sequencing, plasmids will be completely sequenced to reveal their fine structure and evolutionary origins (38).

Catabolic plasmids are assisted in gene dissemination by catabolic transposons, often denoted by the letters Tn. Transposons are mobile genetic elements which can insert at random into plasmids or the bacterial chromosome, independent of the host cell recombination system. The smallest bacterial transposons are called insertion sequence elements. Insertion sequence elements are now thought to be important in the assembly of plasmids, and catabolic plasmids are increasingly found to contain such elements. Some examples of catabolic transposons are reviewed by R. Campbell Wyndham and his colleagues (48); for example, they carry genes to metabolize toluene (Tn*4651* and Tn*4653*), naphthalene (Tn*4655*), chlorobenzoate (Tn*5271*), chlorobenzene (Tn*5280*), and chlorobiphenyl (Tn*4371*).

Microbes in Organic Synthesis

The previous examples of chemical manufacture by fermentation use the metabolic pathways of microorganisms to accumulate products. These are fermentative metabolism end products, such as ethanol or acetone, or compounds made for other purposes under unbalanced growth conditions, such as citric acid or antibiotics. Some of these processes, such as ethanol fermentation, have been used for centuries.

A much more modern development is the use of microorganisms to accomplish a single discrete step in a chemical synthesis, usually sought because chemical methods were lacking with respect to yield, cost, or optical purity. One such application of biocatalysis arose in the 1930s. The issue of optical purity was important in developing a commercial synthesis of vitamin C, or ascorbic acid, which involved incorporation of a biotransformation reaction into a synthetic scheme. The basic scheme, described by T. Reichstein and A. Groessner in 1934 (36), is still in use. The first step, in which D-glucose is reduced to D-sorbitol, is synthetic. Then, a biotransformation is effected by *Acetobacter suboxydans* to convert this to L-sorbose. The subsequent four steps are nonbiological but involve sufficiently straightforward chemistry that the overall yield from glucose to ascorbic acid is over 50%.

A similar example of a key microbial biotransformation came into commercial development because of a surging demand for a new medically relevant product. In 1949, Philip Hench of the Mayo Clinic in Rochester, Minn., announced the results of clinical trials in which the steroid cortisone had shown remarkable anti-inflammatory properties. But cortisone and other steroids were hard to come by. A synthetic route was developed from deoxycholic acid, which could be extracted from animals. However, the synthesis required 37 steps and proceeded with an overall yield of 0.15%. As a result, cortisone in 1950 sold for $200 a gram.

A group at the Upjohn Company decided to screen fungi for the ability to transform another starting compound, diosgenin, via a regiospecific and stereospecific reaction which could not be accomplished by organic chemists. The reaction, the 11-α-hydroxylation of the steroid ring structure, was found to be carried out cleanly by the filamentous fungi *Rhizopus stolonifer* and *Rhizopus arrhizus*. This formed the basis of a six-step synthesis, combining organic synthesis with the biotransformation step, which led the way to cheap production of cortisone, the price of which fell to $6 per gram initially and to $0.46 by 1980. Today, cortisone is widely sold as a non-prescription anti-inflammatory drug.

There are other examples of microbial and enzymatic transformations with industrial value which have historical significance. Since a number of these are still the primary methods of manufacturing the compound, they will be covered in chapter 10, which discusses industrial biotransformations.

Summary

Microbiology is a relatively young science, and there is much to learn from its brief history. The acetone-butanol fermentation developed in 1915 by Weizmann was a remarkable achievement. Subsequent decades saw the rise of microbes as model systems for the fundamental understanding of metabolism. Currently, microbiology is entering an important stage with respect to understanding and using the catalytic potential of microbes. Microbes have become important tools for chemical synthesis, a trend that will almost surely accelerate. Based on its past, the future of microbial catalysis looks bright.

References

1. **Beijerinck, M. W.** 1901. Enrichment culture studies with urea bacteria. *Zentbl. Bakteriol. Abt. 2* **7:**33–61.

2. **Bossert, I. D., and L. Y. Young.** 1986. Anaerobic oxidation of *p*-cresol by a denitrifying bacterium. *Appl. Environ. Microbiol.* **52:**1117–1122.

3. **Braus-Stromeyer, S. A., R. Hermann, A. M. Cook, and T. Leisinger.** 1993. Dichloromethane as the sole carbon source for an acetogenic mixed culture and isolation of a fermentative, dichloromethane-degrading bacterium. *Appl. Environ. Microbiol.* **59:**3790–3797.

*4. **Brock, T. D.** 1999. *Milestones in Microbiology.* ASM Press, Washington, D.C.

5. **Bryant, M. P., E. A. Wolin, M. J. Wolin, and R. S. Wolfe.** 1967. *Methanobacillus omelianskii,* a symbiotic association of two species of bacteria. *Arch. Mikrobiol.* **59:**20–31.

6. **Büchner, E.** 1897. Alcoholische Gärung ohne Hefezellen. *Ber. Dtsch. Chem. Ges.* **30:**117–124.

7. **Bull, A. T., and J. H. Slater (ed.).** 1982. *Microbial Interactions and Communities,* vol. 1. Academic Press, New York, N.Y.

8. **Bulloch, W.** 1938. *The History of Bacteriology.* Oxford University Press, Oxford, United Kingdom.

9. **Chakrabarty, A. M., C. F. Gunsalus, and I. C. Gunsalus.** 1968. Transduction and the clustering of genes in fluorescent *Pseudomonads. Proc. Natl. Acad. Sci. USA* **60:**168–175.

10. **Chakrabarty, A. M., G. Chou, and I. C. Gunsalus.** 1973. Genetic regulation of octane dissimilation plasmid in *Pseudomonas. Proc. Natl. Acad. Sci. USA* **70:** 1137–1140.

*11. **Clarke, P. H.** 1985. The scientific study of bacteria, 1780–1980, p. 1–37. *In* E. R. Leadbetter and J. S. Poindexter (ed.), *Bacteria in Nature,* vol. 1. Plenum Press, New York, N.Y.

12. **Cohn, F.** 1876. Studies on the biology of bacilli. *Beitr. Biol. Pflanzen* **2:**249–276.

13. **Dagley, S., W. C. Evans, and D. W. Ribbons.** 1960. New pathways in the oxidative metabolism of aromatic compounds by microorganisms. *Nature* **188:**560–566.

14. **DiMarco, A. A., T. A. Bobik, and R. S. Wolfe.** 1990. Unusual coenzymes of methanogenesis. *Annu. Rev. Biochem.* **59:**355–394.

15. **Dobell, C.** 1932. *Antonie van Leeuwenhoek and his "Little Animals."* Staples Press, London, England.

16. **Drummond, J. C.** 1957. *The Englishman's Food.* Jonathan Cape, London, England.

17. **Evans, W. C.** 1977. Biochemistry of the bacterial catabolism of aromatic compounds in anaerobic environments. *Nature* **270:**17–22.

18. **Fox, B. G., J. Shanklin, J. Ai, T. M. Loehr, and J. Sanders-Loehr.** 1994. Resonance Raman evidence for an Fe-O-Fe center in stearoyl-ACP desaturase. Primary sequence identity with other diiron-oxo proteins. *Biochemistry* **33:**12776–12786.

19. **Gibson, D. T., M. Hensley, H. Yoshioka, and T. J. Mabry.** 1970. Formation of (+)-*cis*-2,3-dihydroxy-1-methylcyclohexa-4,6-diene from toluene by *Pseudomonas putida. Biochemistry* **9:**1626–1630.

19a. **Hayaishi, O., M. Katagiri, and S. Rothberg.** 1955. Mechanism of the pyrocatechase reaction. *J. Am. Chem. Soc.* **77:**5450–5451.

20. **Hedegaard, J., and I. C. Gunsalus.** 1965. Mixed function oxidation. IV. An induced methylene hydroxylase in camphor oxidation. *J. Biol. Chem.* **240:**4038–4043.

21. **Hegeman, G.** 1985. The mineralization of organic materials under aerobic conditions, p. 97–112. *In* E. R. Leadbetter and J. S. Poindexter (ed.), *Bacteria in Nature,* vol. 1. Plenum Press, New York, N.Y.

22. **Hungate, R. E.** 1985. Anaerobic fermentations, p. 39–95. *In* E. R. Leadbetter and J. S. Poindexter (ed.), *Bacteria in Nature,* vol. 1. Plenum Press, New York, N.Y.

23. **Igarashi, N., H. Moriyama, T. Fujiwara, Y. Fukumori, and N. Tanaka.** 1997. The 2.8 A structure of hydroxylamine oxidoreductase from a nitrifying chemoautotrophic bacterium, *Nitrosomonas europaea. Nat. Struct. Biol.* **4:**276–284.

24. **Jacob, F., and J. Monod.** 1961. Genetic regulatory mechanisms in the synthesis of proteins. *J. Mol. Biol.* **3:**318–356.

25. **Kluyver, A. J.** 1957. Microbiology and industry, p. 165–185. *In* A. F. Kamp, J. W. M. La Riviere, and W. Verhoeven (ed.), *A. J. Kluyver: His Life and Work.* North-Holland, Amsterdam, The Netherlands.

26. **Lazcano, A., and S. L. Miller.** 1996. The origin and early evolution of life: prebiotic chemistry, the pre-RNA world, and time. *Cell* **85:**793–798.

27. **Lederberg, J.** 1952. Cell genetics and hereditary symbiosis. *Physiol. Rev.* **32:**403–430.

28. **Liborius, P.** 1886. Beiträge zur Kenntnis der Sauerstoffbedurfnisses der Bakterien. *Z. Hyg.* **1:**115–177.

29. **Lipmann, F.** 1941. Metabolic generation and utilization of phosphate bond energy. *Adv. Enzymol.* **1:**99–162.

30. **Mason, H. S., W. L. Fowlks, and E. Peterson.** 1955. Oxygen transfer and electron transport by the phenolase complex. *J. Am. Chem. Soc.* **77:**2914–2915.

31. **McTavish, H., J. A. Fuchs, and A. B. Hooper.** 1993. Sequence of the gene coding for ammonia monooxygenase in *Nitrosomonas europaea. J. Bacteriol.* **175:**2436–2444.

32. **Mitchell, P.** 1961. Coupling of phosphorylation to electron and hydrogen transfer by a chemiosmotic pump. *Nature* **191:**144–148.

33. **Pace, W. R.** 1997. A molecular view of microbial diversity and the biosphere. *Science* **276:**734–740.

34. **Pasteur, L.** 1860. Expériences relatives aux générations spontanées. *C. R. Acad. Sci.* **50:**303–675.

35. **Pasteur, L.** 1861. On the organized bodies which exist in the atmosphere; examination of the doctrine of spontaneous generation. *Ann. Sci. Nat. Ser. 4* **16:**5–98.

36. **Reichstein, T., and A. Groessner.** 1934. Eine ergiebige Synthese der 1-Ascorbinsaure (C-Vitamin). *Helv. Chim. Acta* **17:**311–328.

37. **Rheinwald, J. G., A. M. Chakrabarty, and I. C. Gunsalus.** 1973. A transmissible plasmid controlling camphor oxidation in *Pseudomonas putida. Proc. Natl. Acad. Sci. USA* **70:**885–889.

38. **Romine, M. F., L. C. Stillwell, K. K. Wong, S. J. Thurston, E. C. Sisk, C. Sensen, T. Gaasterland, J. K. Fredrickson, and J. D. Saffer.** 1999. Complete sequence of a 184-kilobase catabolic plasmid from *Sphingomonas aromaticivorans* F199. *J. Bacteriol.* **181:**1585–1602.

39. **Schloesing, J., and A. Muntz.** 1877. Sur la nitrification par les ferments organisés. *C. R. Acad. Sci.* **84:**301–303.

40. **Stanier, R. Y.** 1947. Simultaneous adaptation: a new technique for the study of metabolic pathways. *J. Bacteriol.* **54:**339–348.

41. **Stouthamer, A. H.** 1973. A theoretical study on the amount of ATP required for synthesis of microbial cell material. *Antonie Leeuwenhoek J. Microbiol. Serol.* **39:**545–565.

*42. **Thauer, R. K., K. Jungermann, and K. Decker.** 1977. Energy conversion in chemotrophic anaerobic bacteria. *Microbiol. Rev.* **41:**100–180.

43. **Tschech, A., and G. Fuchs.** 1987. Anaerobic degradation of phenol by pure cultures of newly isolated denitrifying pseudomonads. *Arch. Microbiol.* **148:**213–217.

44. **Tyndall, J.** 1881. *Floating Matter of the Air in Relation to Putrefaction and Infection*, 2nd ed. Longmans Green, London, England.

45. **Williams, P. A., and K. Murray.** 1974. Metabolism of benzoate and the methylbenzoates by *Pseudomonas putida* (arvilla) mt-2: evidence for the existence of a TOL plasmid. *J. Bacteriol.* **120:**416–423.

46. **Winogradsky, S.** 1889. Recherches physiologiques sur les sulfobactéries. *Ann. Inst. Pasteur* **3:**49–60.

47. **Winogradsky, S.** 1890. Sur les organismes de la nitrification. *C. R. Acad. Sci.* **110:**1013–1016.

48. **Wyndham, R. C., A. E. Cashore, C. H. Nakatsu, and M. C. Peel.** 1994. Catabolic transposons. *Biodegradation* **5:**323–342.

3

Identifying Novel Microbial Catalysis by Enrichment Culture and Screening

> It has been more than implied that results obtained from the breeding pen, the seed pan, the flower pot and the milk bottle do not apply to evolution in the "open," nature at large or to the wild types. To be consistent, this same objection should be extended to the spectroscope in the study of the evolution of the stars, to the use of the test tube and the balance by the chemist, of the galvometer by the physicist. All of these are unnatural instruments used to torture Nature's secrets from her.
>
> — *Thomas H. Morgan*

> . . . the growth of bacterial cultures does not constitute a specialized subject or branch of research: it is the basic method of microbiology.
>
> — *Jacques Monod*

Thomas Morgan was not a microbiologist. If he had been, he likely would have added to the examples above the technique of enrichment culture. Enrichment culture is an experimental method which has been used extensively by microbiologists to obtain bacteria in monotypic, or pure, culture. As implied in the statement above by Jacques Monod, enrichment culture has remained an important tool of the applied and environmental microbiologist because it rests on the central method of microbiology: bacterial cultivation. Some microbial ecologists have criticized its use on the grounds that it has not told us how microorganisms live in nature. But can one practically study a gram of soil containing 10^9 bacteria, with perhaps 10,000 different types, and hope to understand the entire mixture without any knowledge about the individuals that make up that mixture?

Enrichment culture has yielded key information on the metabolic activities and genes of thousands of microorganisms. As we enter the genomic age, we surely must realize that an organism will perform as allowed by its genes; the presence or absence of genes, in the context of information about their biological meaning, indicates in what ecological niche a bacterium will or will not reside. As Morgan would understand, the criticism is a philosophical stance; some people are intrinsically antireductionist. And while it is a virtue to look at the big picture, one needs to know something about the pieces of a jigsaw puzzle in order to put it together.

The overall value of the enrichment culture method has been substantiated by its constant use for over a century. This is dramatically illustrated in Table 3.1 showing some of the enrichment culture experiments published during 1998 and the first part of 1999. The reasons that investigators still use enrichment culture extensively are discussed below. Readers wishing to obtain more information on this topic may consult excellent reviews by Cook et al. (9) and Poindexter and Leadbetter (43).

Why Use Enrichment Culture?

In contrast to infection sites on animals, soil and water environments of macroscopic size are rarely, if ever, composed of pure cultures of bacteria. The lives of bacteria in the soils and waters of Earth are complex, and we understand very little of their ways, shrouded in the secrecy of their smallness and vast numbers. It has been argued that isolating soil bacteria and taking them into the laboratory does not provide insight into their true natures. However, the opposite is probably true. Pure-culture studies in the laboratory have been essential to show what and how microorganisms metabolize in natural environments.

Enrichment culture, yielding pure bacterial cultures, has often been considered a prerequisite for attempts to elucidate the underlying biochemistry by which a specific organic compound is biodegraded. This is due to the difficulties encountered when working with mixed cultures. For example, mixed-culture studies designed to identify the biochemical intermediates in catabolism are often plagued by the observation of compounds from competing pathways. Moreover, the metabolites yielded by mixed cultures often change over time as the composition of the microbial consortium changes. The complexity of the data and the significant possibility of lack of reproducibility argue against using mixed cultures for these types of studies.

A pure culture, by contrast, offers a range of techniques for revealing the molecular details of biodegradation. First, compounds structurally analogous to the starting material are putative metabolites of that one bacterium. Many biochemical intermediates do not accumulate, but different blocked mutants can be readily obtained to accumulate different metabolic intermediates in quantities sufficient for structural elucidation. Growth studies called "simultaneous adaptation" (49) can then be used to demonstrate that a series of compounds are likely being generated sequentially during metabolism. This is based on the observation that all the enzymes of a given catabolic pathway, as measured after growth on the starting compound, are often simultaneously induced.

The use of pure cultures has been considered almost mandatory for beginning to purify the enzymes involved in biodegradation. Thus, enrichment culturing has been instrumental in the identification of many novel enzymes that catalyze the metabolic transformations which constitute a part of the Earth's carbon cycle. In turn, the corresponding genes have been identified, first singly and now wholesale, as the result of genome-sequencing efforts, which have focused on prokaryotes initially due to their relatively small genome sizes. What would the state of our knowledge of biodegradation be without the isolation of pure cultures of bacteria for study? While it is hard to say precisely, there would likely be little biotech-

The Pervasiveness of Enrichment Culture

Table 3.1 is a partial list showing some of the compounds used in the enrichment culture method to obtain bacteria capable of degrading the compound. These studies used a single compound, often as a sole carbon source. The enrichment procedure was often followed up by one or more analytical chemical methods designed to demonstrate the course of biodegradation. Most commonly, this involved measuring the disappearance of the starting compound. In some cases, trace or significant levels of metabolites were detected, thus providing the first, and often important, clues to the pathway for the compound's catabolism.

There have been many more such studies than are indicated in Table 3.1. It is one measure of the pervasiveness of the technique that very little discussion of the details of enrichment culturing is provided in most publications. However, the details are important because the outcome is strongly influenced by the many variables of the selection process: the medium, the temperature, the choice of aerobic or anaerobic conditions, and the time elapsed between culture transfers. These details are discussed below.

Table 3.1 Substrates used in enrichment methods to obtain bacteria[a]

Class of compound	Compound (condition)	Reference
Alkane hydrocarbons	Methane	50
	Propane	50
	Pentane	13
Alkyl acids, tertiary	Dimethylmalonate	28
Alkene hydrocarbons	Isoprene	54
Alkyl amines	Diisopropylamine	15
	Taurine (anaerobic)	31
	Cyclohexylamine	22
Alkyl sulfides	Dimethyl sulfide (anaerobic)	32
Aromatic acids	Phthalates (anaerobic)	26
Aromatic amines	Aniline	52
Benzenoid monocyclic aromatic hydrocarbons	Benzene	56
	Benzene (anaerobic)	6
	Toluene	20, 56
	Toluene (anaerobic)	42
	Ethylbenzene	56
	Ethylbenzene (anaerobic)	42
	Xylenes (anaerobic)	17, 42
Benzenoid polycyclic aromatic hydrocarbons	Biphenyl	57
	Acenaphthene	48
	Phenanthrene	2
	Pyrene	36
Chlorinated alkenes	Tetrachloroethene (anaerobic)	23, 24
	Trichloroethene	4
Chlorobiphenyls	2,3,5,6-Tetrachlorobiphenyl	19
	2,3,5,6-Tetrachlorobiphenyl (anaerobic)	11
Halobenzoates	3-Chlorobenzoate (anaerobic)	16
N-Heterocyclic rings	Atrazine	29
	2-Methylpyridine	39
	5-Nitro-1,2,4-triazol-3-one	30
O-Heterocyclic rings	Carbosulfan	46
S-Heterocyclic rings	Dibenzthiophene	7
Nitriles	Adiponitrile	12
Nitroaliphatics	Nitroglycerin	1
Nitroaromatics	Nitrobenzene	41, 59
Organophosphorus pesticides	Methylparathion	44
Phenols	Phenol	58
	Phenol (anaerobic)	55
	Nonylphenol	53
	2,4-Dinitrophenol	5
	4-Chloro-2-hydroxyacetanilide	41
Phenylureas	Isoproton	45
Polymers	Poly-hexamethylene carbonate	51
Sulfonates	Cysteate (anaerobic)	35
Terpenes	*beta*-Myrcene	21
	alpha-Pinene	27, 47

[a]Published in 1998 or 1999 and indexed in Medline.

nology industry without an isolated *Escherichia coli* strain and the restriction enzymes derived from pure cultures of bacteria.

But why must enrichment culture be used? Why not plate directly on a selective or indicator agar medium? To answer this, we must think again of the complexity of many natural soil and water environments. For example, a soil rich in organic matter often contains 10^9 bacteria per gram of soil. If the compound being studied for biodegradation is an industrial chemical with no natural counterpart, there is a reasonable likelihood that the soil, if chosen at random, will have no or a very small number of bacteria capable of metabolizing the substance. So even if some are present, and they are plated at a typical dilution of 10^4 to 10^6, it is likely that no colonies will be obtained. Moreover, if a heavy soil suspension is made and plated without dilution, it is unlikely to obtain colonies that degrade the substance, perhaps because of antibiosis or other competing effects derived from some members of the large nondegrading population. These types of problems led to the development of enrichment culture, and the technique has served us well for 100 years.

The General Method

As pioneered by Beijerinck and Winogradsky (see chapter 2), enrichment culture techniques allow selective cultivation of one or more bacterial strains obtained from a complex mixture such as that found in most soils. The method typically relies on using a particular organic compound as the sole carbon source or, less frequently, as the nitrogen, sulfur, or phosphorus source.

Success in obtaining a pure culture(s) is usually taken as evidence that the environmental source contained bacterial strains with a certain metabolic activity. The converse cannot be assumed. That is, failure to culture an organism cannot be used as evidence that a particular soil lacks an organism capable of biodegrading a given substance. The choice of medium or the conditions used in the enrichment culture may not support the growth of microorganisms capable of metabolizing the compound of choice (Table 3.2). Moreover, the compound may be used in some other way, for example, as a final electron acceptor (e.g., with dimethyl sulfoxide or polychlorinated biphenyls). Alternatively, the compound of interest may be metabolized fortuitously in reactions that do not provide to the metabolizing

Table 3.2 Important conditions influencing the microbial types obtained from soil or water via enrichment culture

Condition	Property or organism selected for
Temperature	Psychrophile, mesophile, or thermophile
Heat-shock resistance[a]	Spore or cyst former
Oxygen tension	Anaerobe, microaerophile, or aerophile
Frequency of transfer	Fast- or slow-growing organism
Carbon source	Specific catabolic functions
Nitrogen source	Specific catabolic functions; ability to use specific N
Metals	Organisms needing metals in high concentration (Fe, W, Mo, or V)
pH	Acidophile, alkophile, or growth near neutrality

[a] Heat treatment before growth to kill most vegetative cells.

organism carbon, nitrogen, sulfur, phosphorus, or energy; the compound may be toxic at the concentrations used in the enrichment culture; or the compound may fail to induce the requisite enzyme(s). As an example of the last possibility, trichloroethylene was oxidized by *Burkholderia cepacia* G4 in medium prepared with water from a contaminated site but not with standard laboratory water (38). It was subsequently discovered that phenol was a contaminant in the site water and was responsible for inducing biosynthesis of the trichloroethylene-degrading enzyme (37).

When a researcher wishes to obtain a biodegradative bacterium from the environment, the best place to start is in the library. This is true even if there is no information available on the biodegradation of the organic compound of interest. In that case, the researcher needs to understand the chemical properties of the compound. What is its melting point? What is its boiling point? What is its solubility in water and other solvents? Is it completely stable under the experimental conditions to be used? What are its major reactions? Reduction? Oxidation? Hydrolysis? Are there clues about the metabolism of the compound from studies of animals or plants? Relevant information on chemical properties is now easier to obtain than in the past, with the recent advent of extensive free Internet resources on chemical compounds.

There are several very good reasons why physicochemical information is essential. First, the compound should be used correctly in the enrichment medium. For example, it is often a good practice to add water-insoluble solids at concentrations well above their solubility in the enrichment medium. The compound will continuously dissolve as it is being degraded in the liquid phase, and it is rarely toxic to the bacteria. In contrast, compounds that are liquids at room temperature and water insoluble must often be added judiciously to the medium. This is because the formation of two liquid phases in the enrichment medium will often prove to be very toxic to many bacteria. This problem can be overcome neatly with volatile, water-insoluble compounds by providing the compound as a vapor, with the liquid suspended in a glass bulb above the growth medium in an enrichment flask (Fig. 3.1). This provides the bacteria with a continuous supply of substrate without generating a two-phase system that might kill them by extracting the lipids from their cell walls. In this context, bacteria which grow on single-ring benzenoid compounds like toluene were considered rare in the environment when toluene was added directly to enrichment medium, but this was an artifact resulting from toxicity. The introduction of a vapor bulb technique, starting in the 1960s, led to the isolation of many bacteria that grow on toluene and some related aromatic hydrocarbons (8, 14).

A typical enrichment culture protocol is illustrated in Fig. 3.2. In one example, bacteria were obtained which used the herbicide atrazine, which contains five nitrogen atoms, as their sole source of nitrogen for growth (33, 34). The inorganic-salts medium used also contained sodium citrate and sucrose that could be used as carbon sources. To start the protocol, a soil was obtained in which atrazine had been continuously spilled over a period of years, attaining a concentration of several thousand parts per million in some spots. The site, in Little Falls, Minn., had been abandoned for several years, allowing the development of atrazine-degrading bacterial populations. Different samples of soil were taken and used to make soil suspensions for inoculating the enrichment medium in different flasks. The flasks were shaken in the presence of air, and aliquots were transferred into

Figure 3.1 Flask for growing bacteria, with a glass bulb containing a volatile organic compound. The compound, as indicated by the green spots, distributes throughout the gas and liquid phases.

Figure 3.2 Schematic representation of the enrichment culture method. A typical starting inoculum, such as soil, is put into liquid culture, with a specific compound serving as the selective substrate. The number of bacteria capable of using the compound, indicated in green, increases during the enrichment process.

fresh media periodically, after suitable growth was indicated by the presence of turbidity in the flasks. After numerous transfers, a loopful of the liquid medium containing bacteria was plated onto solid medium to obtain isolated colonies. Isolated colonies are typically reinoculated into fresh medium to determine if a pure culture can grow on the enrichment compound. In this example, advantage was taken of the relatively low water solubility of atrazine to incorporate it into the solid agar medium at a total concentration 15 times above its solubility limit in water (Fig. 3.3). The plates were opaque due to particulate atrazine. However, bacterial colonies which proved capable of growing on atrazine showed clear zones around the colonies. This allowed us to rule out the possibility that a bacterial colony growing on an atrazine plate might be fixing atmospheric nitrogen to meet its nitrogen needs. This approach has been used with other water-insoluble substances, such as phenanthrene (25), pyrene (18), and aliphatic polycarbonates (51).

Selection of Conditions and Medium

Often, the goal of an enrichment protocol is not merely to obtain any bacterium but one with a particular set of properties. This can be accomplished by incorporating the specific feature(s) into the enrichment culture protocol (Table 3.2). For example, the temperature can be elevated during growth to obtain thermophiles or it can be raised only in the soil suspension to obtain mesophiles that might be heat tolerant in a vegetative state, for example, due to spore formation or encystment. A very important consideration is whether one wants to select for anaerobic or aerobic bacteria. This parameter will often affect the choice of the medium, whether reducing agents and redox indicators are added, and the types of growth vessels the cells are cultivated in. Anaerobic enrichments are often carried out in thick-walled glass bottles with crimp-sealed septum tops. The gas phase above

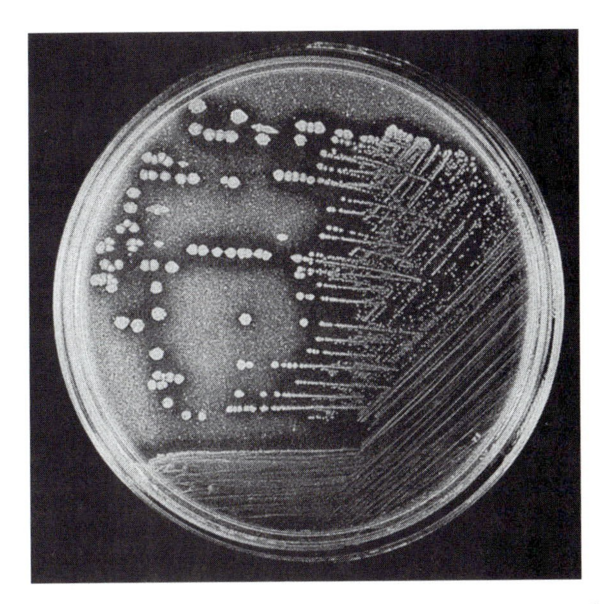

Figure 3.3 Petri plate of *Pseudomonas* sp. strain ADP showing clearing zones around colonies due to metabolism of the herbicide atrazine present in crystalline form on the plate.

the cultures is some inert gas lacking oxygen, typically purified nitrogen or argon. Since anaerobes often grow more slowly than aerobes, the time between transfers may be long.

The choice of enrichment medium is critical and deserves serious thought. For ideas, a general reference on bacterial cultivation methods may be consulted (3). Generally, a suitable minimal medium is chosen as the baseline, but even here decisions must be made. Different minimal media are thought to be best for different genera of bacteria, so the choice of minimal medium skews the enrichment toward obtaining certain bacterial types. It cannot be known a priori what genera of bacteria will be most active in degrading a particular compound in a given environment. In the absence of additional information, it is often prudent to use a relatively rich minimal medium containing many different metals to avoid limitations of inorganic compounds. It is generally wise to use a high buffer content. The pH chosen will be determined by the organisms which are desired; typically, a pH around neutrality is used.

Screening for Specific Biocatalytic Reactions

The specific reactions transforming the substrate provided in enrichment cultures or other bacterial growth experiments may or may not be of interest. If the goal is strictly focused on biodegradation, the pathway of the metabolism may not be important to the experimenter. However, in most cases, a knowledge of the pathway, the genes, and the enzymes is sought for the sake of learning the molecular basis of biodegradation or for purposes of applying one or more reactions in biotechnology.

With the growth of the organism and the disappearance of the starting compound established, the next step is often to look for any accumulating metabolites in the growth medium. The structures of metabolites and the stoichiometry of their formation can yield important clues as to the series of reactions leading to or from the metabolites. Increasingly, the reactions of an apparently new pathway are elucidated via molecular biological methods, typically cloning to obtain expression in an experimentally amenable genetic host, such as *E. coli*. Subclones which accumulate each of the stable metabolites of the pathway can be generated in this way, sometimes in gram quantities, thus providing clear evidence of the intermediates and the flow of metabolism.

In enrichment cultures conducted for obtaining a desired biocatalyst, a given pathway or reaction might be assumed; the goal is to obtain an enzyme of a known type which is highly active with a given substrate or under a specific set of conditions. In this case, the screening method is important, as it might be necessary to look at hundreds or thousands of enzymes yielded in the first round of screening (the enrichment step). The description of a wide range of screening methods is beyond the scope of this work, and excellent books are available for those seeking ideas for novel screening protocols (10, 40).

Another approach is to test existing strains to determine if they have the enzyme of interest and will show high activity with the substrate of interest. In this context, a limited taxonomic range of organisms is typically screened. The fungi screened include *Aspergillus*, *Fusarium*, *Trichoderma*, *Mucor*, and *Rhizomucor*. These genera are known to produce a host of hydrolytic biodegradative enzymes of importance in industry. The enzymes include cellulase, lipase, and rennilase. Bacteria known to produce useful hydrolytic enzymes are commonly found among *Bacillus* and *Pseudomonas* species. The microbial aspects of biodegradation and biocatalysis will be discussed in greater detail in the next chapter.

Summary

Enrichment culture has stood the test of time well: it has been used for over 100 years. But will the advent of widespread genomic sequencing of bacteria, including nonculturable bacteria, make enrichment culture obsolete? After all, enrichment culture does require the culturability of a bacterium. Will it not become increasingly feasible for sequence information and computers to tell us which bacteria can metabolize certain compounds without doing experiments?

The answer to these questions is no, at least for the near term. Genomics tells us only what already-isolated bacteria might have the potential to do. Even extensive sequencing of total DNA from a soil sample will only tell us about those genes whose functions have already been studied. It is likely that there is much more biocatalytic potential waiting to be discovered in microorganisms. Moreover, if an organism that degrades compound X at pH 2.0 at 90°C under anaerobic conditions is wanted, that capability will not be found by searching GenBank.

One possible new development will be the coupling of enrichment culture with genomic methods to pull out desirable genes without going through all the usual intermediate steps. For example, DNA from a late enrichment culture could be used to make artificial chromosomes in *E. coli*. These *E. coli* strains, with very large inserts of foreign DNA, could then be screened for various biodegradative phenotypes. When the desired function is identified, the recombinant genes are already in hand to be sequenced and exploited. But developments like this will not supplant enrichment culture. In fact, they will add to its usefulness. If Beijerinck and Winogradsky were alive today, they would no doubt be pleased.

References

1. **Accashian, J. V., R. T. Vinopal, B. J. Kim, and B. F. Smets.** 1998. Aerobic growth on nitroglycerin as the sole carbon, nitrogen, and energy source by a mixed bacterial culture. *Appl. Environ. Microbiol.* **64:**3300–3304.
2. **Aitken, M. D., W. T. Stringfellow, R. D. Nagel, C. Kazunga, and S. H. Chen.** 1998. Characteristics of phenanthrene-degrading bacteria isolated from soils contaminated with polycyclic aromatic hydrocarbons. *Can. J. Microbiol.* **44:**743–752.
3. **Atlas, R.** 1995. *Handbook of Media for Environmental Microbiology.* CRC Press, Boca Raton, Fla.
4. **Bielefeldt, A. R., and H. D. Stensel.** 1999. Biodegradation of aromatic compounds and TCE by a filamentous bacteria-dominated consortium. *Biodegradation* **10:**1–13.
5. **Blasco, R., E. Moore, V. Wray, D. Pieper, K. Timmis, and F. Castillo.** 1999. 3-Nitroadipate, a metabolic intermediate for mineralization of 2,4-dinitrophenol by a new strain of a *Rhodococcus* species. *J. Bacteriol.* **181:**149–152.
6. **Burland, S. M., and E. A. Edwards.** 1999. Anaerobic benzene biodegradation linked to nitrate reduction. *Appl. Environ. Microbiol.* **65:**529–533.
7. **Chang, J. H., S. K. Rhee, Y. K. Chang, and H. N. Chang.** 1998. Desulfurization of diesel oils by a newly isolated dibenzothiophene-degrading *Nocardia* sp. strain CYKS2. *Biotechnol. Prog.* **14:**851–855.
8. **Claus, D., and N. Walker.** 1964. The decomposition of toluene by soil bacteria. *J. Gen. Microbiol.* **36:**107–122.

*9. **Cook, A. M., H. Grossenbacher, and R. Hutter.** 1983. Isolation and cultivation of microbes with biodegradative potential. *Experientia* **39:**1191–1198.

10. **Cutler, H. G., and S. J. Cutler.** 1999. *Biologically Active Natural Products: Agrochemicals.* CRC Press, Boca Raton, Fla.

11. **Cutter, L., K. R. Sowers, and H. D. May.** 1998. Microbial dechlorination of 2,3,5,6-tetrachlorobiphenyl under anaerobic conditions in the absence of soil or sediment. *Appl. Environ. Microbiol.* **64:**2966–2969.

12. **Dhillon, J. K., and N. Shivaraman.** 1999. Biodegradation of cyanide compounds by a *Pseudomonas* species (S1). *Can. J. Microbiol.* **45:**201–208.

13. **Garnier, P. M., R. Auria, C. Augur, and S. Revah.** 1999. Cometabolic biodegradation of methyl *t*-butyl ether by *Pseudomonas aeruginosa* grown on pentane. *Appl. Microbiol. Biotechnol.* **51:**498–503.

14. **Gibson, D. T., J. R. Koch, and R. E. Kallio.** 1968. Oxidative degradation of aromatic hydrocarbons by microorganisms. I. Enzymatic formation of catechol from benzene. *Biochemistry* **7:**2653–2662.

15. **Gieg, L. M., D. L. Coy, and P. M. Fedorak.** 1999. Microbial mineralization of diisopropanolamine. *Can. J. Microbiol.* **45:**377–388.

16. **Haggblom, M. M., and L. Y. Young.** 1999. Anaerobic degradation of 3-halobenzoates by a denitrifying bacterium. *Arch. Microbiol.* **17:**230–236.

17. **Harms, G., K. Zengler, R. Rabus, F. Aeckersberg, D. Minz, R. Rossello-Mora, and F. Widdel.** 1999. Anaerobic oxidation of *o*-xylene, *m*-xylene, and homologous alkylbenzenes by new types of sulfate-reducing bacteria. *Appl. Environ. Microbiol.* **65:**999–1004.

18. **Heitkamp, M. A., W. Franklin, and C. E. Cerniglia.** 1988. Microbial metabolism of polycyclic aromatic hydrocarbons: isolation and characterization of a pyrene-degrading bacterium. *Appl. Environ. Microbiol.* **54:**2549–2555.

19. **Holoman, T. R., M. A. Elberson, L. A. Cutter, H. D. May, and K. R. Sowers.** 1998. Characterization of a defined 2,3,5,6-tetrachlorobiphenyl-*ortho*-dechlorinating microbial community by comparative sequence analysis of genes coding for 16S rRNA. *Appl. Environ. Microbiol.* **64:**3359–3367.

20. **Hubert, C., Y. Shen, and G. Voordouw.** 1999. Composition of toluene-degrading microbial communities from soil at different concentrations of toluene. *Appl. Environ. Microbiol.* **65:**3064–3070.

21. **Iurescia, S., A. M. Marconi, D. Tofani, A. Gambacorta, A. Paterno, C. Devirgillis, M. J. van der Werf, and E. Zennaro.** 1999. Identification and sequencing of beta-myrcene catabolism genes from *Pseudomonas* sp. strain M1. *Appl. Environ. Microbiol.* **65:**2871–2876.

22. **Iwaki, H., M. Shimizu, T. Tokuyama, and Y. Hasegawa.** 1999. Biodegradation of cyclohexylamine by *Brevibacterium oxydans* IH-35A. *Appl. Environ. Microbiol.* **65:**2232–2234.

23. **Kengen, S. W., C. G. Breidenbach, A. Felske, A. J. Stams, G. Schraa, and W. M. de Vos.** 1999. Reductive dechlorination of tetrachloroethene to *cis*-1,2-dichloroethene by a thermophilic anaerobic enrichment culture. *Appl. Environ. Microbiol.* **65:**2312–2316.

24. **Kennes, C., M. C. Veiga, and L. Bhatnagar.** 1998. Methanogenic and perchloroethylene-dechlorinating activity of anaerobic granular sludge. *Appl. Microbiol. Biotechnol.* **50:**484–488.

25. **Kiyohara, H., and K. Nagao.** 1978. The catabolism of phenanthrene and naphthalene by bacteria. *J. Gen. Microbiol.* **105:**69–75.

26. **Kleerebezem, R., P. L. W. Hulshoff, and G. Lettinga.** 1999. Anaerobic degradation of phthalate isomers by methanogenic consortia. *Appl. Environ. Microbiol.* **65:**1152–1160.

27. **Kleinheinz, G. T., S. T. Bagley, W. P. St. John, J. R. Rughani, and G. D. McGinnis.** 1999. Characterization of *alpha*-pinene-degrading microorganisms

and application to a bench-scale biofiltration. *Arch. Environ. Contam. Toxicol.* **37:**151–157.

28. **Kniemeyer, O., C. Probian, R. Rossello-Mora, and J. Harder.** 1999. Anaerobic mineralization of quaternary carbon atoms: isolation of denitrifying bacteria on dimethylmalonate. *Appl. Environ. Microbiol.* **65:**3319–3324.

29. **Kontchou, C. Y., and N. Gschwind.** 1999. Biodegradation of s-triazine compounds by a stable mixed bacterial community. *Exotoxicol. Environ. Saf.* **43:**47–56.

30. **Le Campion, L., A. Vandais, and J. Ouazzani.** 1999. Microbial remediation of NTO in aqueous industrial wastes. *FEMS Microbiol. Lett.* **176:**197–203.

31. **Lie, T. J., M. L. Clawson, W. Godchaux, and E. R. Leadbetter.** 1999. Sulfidogenesis from 2-aminoethanesulfonate (taurine) fermentation by a morphologically unusual sulfate-reducing bacterium, *Desulforhopalus singaporensis* sp. nov. *Appl. Environ. Microbiol.* **65:**3328–3334.

32. **Lomans, B. P., R. Maas, R. Luderer, H. J. Op den Camp, A. Pol, C. van der Drift, and G. D. Vogels.** 1999. Isolation and characterization of *Methanomethylovorans hollandica* gen. nov., sp. nov., isolated from freshwater sediment, a methylotrophic methanogen able to grow on dimethyl sulfide and methanethiol. *Appl. Environ. Microbiol.* **65:**3641–3650.

33. **Mandelbaum, R. T., D. L. Allan, and L. P. Wackett.** 1995. Isolation and characterization of a *Pseudomonas* sp. that mineralizes the *s*-triazine herbicide atrazine. *Appl. Environ. Microbiol.* **61:**1451–1457.

34. **Mandelbaum, R. T., L. P. Wackett, and D. L. Allan.** 1993. Mineralization of the *s*-triazine ring of atrazine by stable bacterial mixed cultures. *Appl. Environ. Microbiol.* **59:**1695–1701.

35. **Mikosch, C. A., K. Denger, E. M. Schafer, and A. M. Cook.** 1999. Anaerobic oxidation of cysteate: degradation via L-cysteate:2-oxoglutarate aminotransferase in *Paracoccus pantotrophus*. *Microbiology* **145:**1153–1160.

36. **Molina, M., R. Araujo, and R. E. Hodson.** 1999. Cross-induction of pyrene and phenanthrene in a *Mycobacterium* sp. isolated from polycyclic aromatic hydrocarbon contaminated river sediments. *Can. J. Microbiol.* **45:**520–529.

37. **Nelson, M. J. K., S. O. Montgomery, W. R. Mahaffey, and P. H. Pritchard.** 1987. Biodegradation of trichloroethylene and involvement of an aromatic biodegradative pathway. *Appl. Environ. Microbiol.* **53:**949–954.

38. **Nelson, M. J. K., S. O. Montgomery, E. J. O'Neill, and P. H. Pritchard.** 1986. Aerobic metabolism of trichloroethylene by a bacterial isolate. *Appl. Environ. Microbiol.* **52:**383–384.

39. **O'Loughlin, E. J., G. K. Sims, and S. J. Traina.** 1999. Biodegradation of 2-methyl, 2-ethyl, and 2-hydroxypyridine by an *Arthrobacter* sp. isolated from subsurface sediment. *Biodegradation* **10:**93–104.

40. **Omura, S.** 1992. *The Search for Bioactive Compounds from Microorganisms.* Springer-Verlag, New York, N.Y.

41. **Park, H. S., S. J. Lim, Y. K. Chang, A. G. Livingston, and H. S. Kim.** 1999. Degradation of chloronitrobenzenes by a coculture of *Pseudomonas putida* and a *Rhodococcus* sp. *Appl. Environ. Microbiol.* **65:**1083–1091.

42. **Phelps, C. D., and L. Y. Young.** 1999. Anaerobic biodegradation of BTEX and gasoline in various aquatic sediments. *Biodegradation* **10:**15–25.

*43. **Poindexter, J. S., and E. R. Leadbetter.** 1986. Enrichment cultures in bacterial ecology, p. 229–260. *In* J. S. Poindexter and E. R. Leadbetter (ed.), *Bacteria in Nature,* vol. 2. Plenum Press, New York, N.Y.

44. **Ramanathan, M. P., and D. Lalithakumari.** 1999. Complete mineralization of methylparathion by *Pseudomonas* sp. A3. *Appl. Biochem. Biotechnol.* **80:**1–12.

45. **Roberts, S. J., A. Walker, L. Cox, and S. J. Welch.** 1998. Isolation of isoproturon-degrading bacteria from treated soil via three different routes. *J. Appl. Microbiol.* **85:**309–316.

46. **Sahoo, A., N. Sethunathan, and P. K. Sahoo.** 1998. Microbial degradation of carbosulfan by carbosulfan- and carbofuran-retreated rice soil suspension. *J. Environ. Sci. Health B* **33:**369–379.

47. **Savithiry, N., D. Gage, W. Fu, and P. Oriel.** 1998. Degradation of pinene by *Bacillus pallidus* BR425. *Biodegradation* **9:**337–341.

48. **Selifonov, S. A., P. J. Chapman, S. B. Akkerman, J. E. Gurst, J. M. Bortiatynski, M. A. Nanny, and P. G. Hatcher.** 1998. Use of ^{13}C nuclear magnetic resonance to assess fossil fuel biodegradation: fate of [1-^{13}C]acenaphthene in creosote polycyclic aromatic compound mixtures degraded by bacteria. *Appl. Environ. Microbiol.* **64:**1447–1453.

49. **Stanier, R. Y.** 1947. Simultaneous adaptation: a new technique for the study of metabolic pathways. *J. Bacteriol.* **54:**339–348.

50. **Sukesan, S., and M. E. Watwood.** 1998. Effects of hydrocarbon enrichment on trichloroethylene biodegradation and microbial populations in finished compost. *J. Appl. Microbiol.* **85:**635–642.

51. **Suyama, T., H. Hosoya, and Y. Tokiawa.** 1998. Bacterial isolates degrading aliphatic polycarbonates. *FEMS Microbiol. Lett.* **161:**255–261.

52. **Tan, N. C., F. X. Prenafeta-Boldu, J. L. Opsteeg, G. Lettinga, and J. A. Field.** 1999. Biodegradation of azo dyes in cocultures of anaerobic granular sludge with aerobic aromatic amine degrading enrichment cultures. *Appl. Microbiol. Biotechnol.* **51:**865–871.

53. **Tanghe, T., W. Dhooge, and W. Verstraete.** 1999. Isolation of a bacterial strain able to degrade branched nonylphenol. *Appl. Environ. Microbiol.* **65:**746–751.

54. **van Hylckama Vlieg, J. E., J. Kingma, A. J. van den Wijngaard, and D. B. Janssen.** 1998. A glutathione *S*-transferase with activity towards *cis*-1,2-dichloroepoxyethane is involved in isoprene utilization by *Rhodococcus* sp. strain AD45. *Appl. Environ. Microbiol.* **64:**2800–2805.

55. **van Schie, P. M., and L. Y. Young.** 1998. Isolation and characterization of phenol-degrading denitrifying bacteria. *Appl. Environ. Microbiol.* **64:**2432–2438.

56. **Veiga, M. C., M. Fraga, L. Armor, and C. Kennes.** 1999. Biofilter performance and characterization of a biocatalyst degrading alkylbenzene gases. *Biodegradation* **10:**169–176.

57. **Wagner-Dobler, I., A. Bennasar, M. Vancanneyt, C. Strompl, I. Brummer, C. Eichner, I. Grammel, and E. R. Moore.** 1998. Microcosm enrichment of biphenyl-degrading microbial communities from soils and sediments. *Appl. Environ. Microbiol.* **64:**3014–3022.

58. **Watanabe, K., M. Teramoto, H. Futamata, and S. Harayama.** 1998. Molecular detection, isolation, and physiological characterization of functionally dominant phenol-degrading bacteria in activated sludge. *Appl. Environ. Microbiol.* **64:**4396–4402.

59. **Zhao, J. S., and O. P. Ward.** 1999. Microbial degradation of nitrobenzene and mono-nitrophenol by bacteria enriched from municipal activated sludge. *Can. J. Microbiol.* **45:**427–432.

4

Microbial Diversity: Catabolism of Organic Compounds Is Broadly Distributed

. . . we could imagine machines as small as a few microns across. . . .
— *Richard P. Feynman*

Machines a few micrometers across already exist: they are known as prokaryotes. From the flagellar motor to ribosomes and the intricate web of metabolism, microbes are machines that use some of the raw materials of their environment to make more machines.

The impact of microbial machines on the Earth will be discussed in the next several chapters. This chapter focuses on microbial diversity and the catabolism of nonintermediary chemical compounds by disparate genera of prokaryotes. Chapter 5 focuses on organic molecules and explores the breadth of organic functional groups that are produced in nature. Chapter 6 deals with physiological processes, such as enzyme reactions, compound uptake, and chemotaxis; perhaps we should think of these as the parts of the microbial machine.

The Importance of Microbial (Bio)diversity

There is currently a spotlight on biodiversity and its significance in maintaining global processes. Most of the biodiversity literature deals with animals, plants, and insects. The assumption is that these organisms are the most threatened and most important globally. This is a natural tendency. We think most about what we see and have a special fondness for *Homo sapiens* and its closest relatives. But a more objective review of current scientific knowledge suggests that microorganisms harbor the greatest biological diversity and play a more important role in maintaining global processes.

Microorganisms have been around since the start of life at least 3.6 billion years ago (36). Macroscopic organisms are thought to have proliferated largely with the accumulation of oxygen in the atmosphere, 1 billion years ago, or less than 30% of the time scale of microbial evolution. Microbes have thus had a longer time to evolve new genes and functions.

In addition, microorganisms reproduce, and thus evolve new diversity, faster than macroscopic organisms. One bacterium, dividing at the unspectacular rate, at least for bacteria, of once every 20 minutes, can theoretically give rise to 2^{144} cells in 48 h. This constitutes a mass exceeding that of the entire Earth. This assumes constant cell division and hence constant eating. Eating in this context is largely biodegradation, which ends when there is nothing more in the local environment to degrade. While bacterial numbers do not come close to 2^{144} cells, the number of bacteria attached to your body exceeds the entire human population of Earth.

How many prokaryotes are there on Earth? Recently, a global census of the prokaryotes was conducted by William Whitman and his coworkers (62). The estimates derive from the most recent data on microbial niches, the numbers of microbes in a representative sample of each niche, and the estimated volumes of that niche. The results of that monumental study are shown in Table 4.1. The overall estimate is that approximately 5×10^{30} prokaryotes reside on Earth.

Insects are often heralded as the champions of biodiversity, with perhaps 500,000 known species (64), but each individual is teeming with microorganisms. Termites cannot digest wood without the microflora in their guts, carrying out the biochemistry that produces acetate for the termite to absorb (see Fig. 1.3). The biochemistry is carried out by a very complex microbial ecosystem inside the gut. One termite species contains 1,000 or more bacterial species. The organic-matter-cycling capacity of termites, which is important globally, is based on the resident microbial flora. In termites, methanogenic bacteria occupy the terminal position of the anaerobic biodegradative food chain of cellulose digestion, and termites devoid of bacteria cannot degrade cellulose. This is a common phenomenon. A cow does not digest the cellulose of grass directly but is dependent on the microorganisms in its rumen. The microbial ecosystem of a cow's rumen has been studied for over 20 years; new species are constantly being discovered, but the surface has only been scratched in characterizing the biodiversity inside this one mammal. Now consider that 4×10^{24} prokaryotes reside in or on domestic animals on Earth (Table 4.1).

With many documented examples of animal, plant, and insect dependence on microorganisms, could many of those life forms exist without them? The definitive experiments would be difficult to do, but the answer is probably not. What about the converse? Could microorganisms exist without eukaryotes? That experiment has already been done, and the answer is, resoundingly, yes. For the first billion or more years of life, microorganisms flourished on their own. This early life yielded fossils observed

Table 4.1 Niches and numbers of prokaryotes on Earth[a]

Environment	No. of prokaryotes
Ocean subsurface (below 200 m)	3.55×10^{30}
Terrestrial subsurface (below 8 m)	$(0.25–2.5) \times 10^{30}$
Soil	0.26×10^{30}
Open ocean (above 200 m)	0.12×10^{30}
Domestic animals	4×10^{24}
Atmosphere (bottom 3 km)	5×10^{19}

[a]Data from reference 62 by permission of W. Whitman.

today as 100-m-high stromatolithic belts and microfossils of trichomic microorganisms from 3.5 billion years ago (36).

Fungi in Biocatalysis and Biodegradation

As described in chapter 2, the fungal biodegradation of plant material has been observed for a long time. The process became the subject of intense scientific scrutiny during World War II. The American forces fighting in the South Pacific islands found their cotton clothing, knapsacks, and tents deteriorating at an alarming rate (28). The cause was clearly fungal growth, but this did not suggest an obvious solution.

To obtain a solution, a research program was begun at a U.S. Army center to cultivate the fungi and try to find methods to control their growth. In all, more than 10,000 strains of fungi, obtained for their ability to hydrolyze cellulose, were studied. One outcome of the screening process was the identification of a fungus with strong cellulose-degrading activity. The fungus, *Trichoderma viride,* was isolated from a rotting piece of equipment obtained from New Guinea. Even today, *T. viride* strains are among the most proficient secretors of extracellular cellulases known.

Fungi are prominent in many environmental biodegradation processes and also in industrial biocatalysis. With regard to the latter, the yeast *Saccharomyces* has been most important in ethanol beverage fermentation. Later, filamentous fungi, notably *Aspergillus niger,* were developed for fermentation to produce citric acid. Still later, fungal hydroxylation reactions were exploited to produce high-value products, such as 11-α-hydroxyprogesterone used in the biosynthesis of cortisone (46). Fungi are secretors of biodegradative enzymes in nature, and this has formed the basis of industrial processes to extract enzymes for biotechnology. The enzymes used industrially are lipases, cellulases, and proteases.

Four of the five major phyla of fungi are commonly used in industry (32). Of the ascomycota, *Saccharomyces* and *Schizosaccharomyces* are the major fungal genera used in alcoholic-beverage fermentations. The latter is preferred for the African beer called pombe, made from millet, and the Oriental beer arrack, made from molasses, rice, or cocoa palm sap. Of the basidiomycota, the edible mushroom *Agaricus bisporus* is heavily cultivated in Europe and the United States. Of the zygomycota, *Mucor* and *Rhizopus* are highly significant in industry. *Rhizopus* strains, as previously noted, are important in citric acid production. *Mucor* strains make a significant number of important lipases and catalyze the hydroxylation of a wide range of chemical compounds. Some of the hydroxylation reactions observed in the *Zygomycota* (Table 4.2) are catalyzed by *Mucor* species. The deuteromycota includes some of the most celebrated industrial fungi, the various *Penicillium* spp., which produce numerous antimicrobial secondary products. In addition to the antibacterial β-lactam compounds, *Penicillium griseofulvum* produces the clinically important antifungal agent griseofulvan.

Fungi are particularly important industrially because of their ability to catalyze novel hydroxylation reactions. In one large survey of fungi for the ability to hydroxylate the polycyclic aromatic hydrocarbon naphthalene, 55 species were observed to have some activity (11). The ratio of 1-naphthol to 2-naphthol produced was, in virtually all cases, suggestive of naphthalene 1,2-epoxide as the initial oxidation product. This implicated a monooxygenase in the reaction. In the cases investigated, a cytochrome P450 monooxygenase had been shown to catalyze the epoxidation of aromatic

Prokaryotes versus Eukaryotes: Fundamental Differences in Biodegradation

In a broad sense, virtually all living things participate in biodegradation. That is, as part of the need to derive chemical energy to make ATP or to produce metabolic intermediates, virtually all free-living organisms have the ability to catabolize some set of organic molecules. The metabolism we think of as biodegradation deals with catabolism of nonintermediary compounds, the much greater diversity of biomolecules generated by specific plants, bacteria, and now organic chemists. While intermediary chemical compounds number in the thousands, nonintermediary chemical compounds number in the millions.

Is there a fundamental difference in the metabolism of nonintermediary compounds by prokaryotes and eukaryotes? In fact, some important distinctions can be made. The metabolism of nonintermediary compounds by mammals has been best studied because of the health issues underlying this research. Many drugs are metabolized by the same enzyme systems that metabolize industrial chemicals. The enzymes involved in this metabolism are thought to have evolved to handle the many toxic natural compounds produced by plants and fungi that early foraging humans were probably exposed to in very large quantities. In modern life, humans eat a much smaller number of plants, all of which are well known to have nutritive value and to be nontoxic. The same enzyme systems that metabolize exotic natural plant products also metabolize pharmaceutical compounds and environmental pollutants that enter our bodies through ingestion, injection, or passive transport through mucous membranes. Such compounds are sometimes called xenobiotic, meaning foreign to metabolism. We believe that the term *xenobiotic* has frequently been misused in the biodegradation literature, and we prefer to use the term *nonintermediary* to describe the metabolism of compounds which are not common to most living things. This point will be discussed further in chapter 5 on chemical compounds.

The metabolic strategy for the mammalian metabolism of nonintermediary chemicals is fundamentally different from that of bacteria (Fig. 4.1). Mammals typically metabolize hydrophobic, potentially toxic chemicals by rendering them more hydrophilic and then excreting them from the body, often in the urine. This is thus purely detoxification and not energy-linked metabolism. Understanding this strategy has been important in pharmacology, where it is necessary to predict the lifetime of a drug in the body in order to determine the medical dosage and frequency of administration. A key set of enzymes involved in this metabolism is the cytochrome P450 monooxygenases, which use molecular oxygen to hydroxylate or epoxidize hydrophobic compounds. This might increase the water solubility of the compound somewhat. Further reactions, catalyzed by enzymes such as epoxide hydrolases, glutathione S-transferases, and UDP-glucoronyl transferases, further increase the water solubility of nonintermediary compounds and aid in their mobilization into the urine. It has been found that some fungi metabolize nonintermediary compounds in a manner analogous to that of mammals. For example, one study demonstrated that over 70 different fungi metabolize the polycyclic hydrocarbon naphthalene via an apparent epoxidation reaction identical to the reaction catalyzed by mammals. This has further suggested a prokaryote-versus-eukaryote dichotomy in the way nonintermediary compounds are metabolized (Fig. 4.1). It was startling to discover that metabolism of the potent carcinogen benzo[*a*]pyrene by rat enzymes and the fungus *Cunninghamella elegans* even produced the same major diastereomer of the diol epoxide product (10).

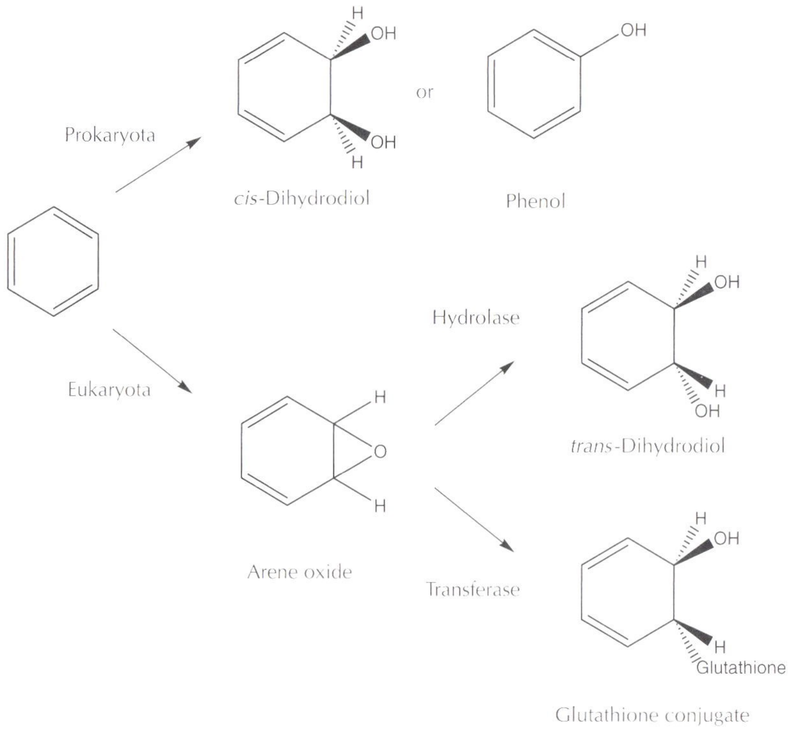

Prokaryota
cis-Dihydrodiol
or
Phenol
Eukaryota
Arene oxide
Hydrolase
trans-Dihydrodiol
Transferase
Glutathione
Glutathione conjugate

Figure 4.1 Divergence in the metabolism of aromatic hydrocarbons by prokaryotes and eukaryotes.

Table 4.2 Phyla and classes of fungi active in hydroxylation reactions[a]

Phylum and class	No. of species
Ascomycota	
Euascomycetes	28
Hemiascomycetes	5
Basidiomycota	
Gasteromycetes	0
Hymenomycetes	32
Uereodinomycetes	0
Ustilagomycetes	0
Deuteromycota	
Deuteromycetes	18
Coelomycetes	19
Hyphomycetes	222
Chytridiomycota	
Chytridiomycetes	0
Hyphochtriomycetes	0
Oomycetes	1
Zygomycota	
Trichomycetes	0
Zygomycetes	103

[a]Data from reference 32 with permission from Elsevier Science.

hydrocarbons. Fungi of the genera *Mucor, Fusarium, Rhizopus,* and *Cunninghamella* were observed to have substantially higher activity than many of the others. In a larger survey of hydroxylation reactions more generally, fungi of the classes *Hyphomycetes* and *Zygomycetes* were most heavily represented (Table 4.2). Many screening programs look for novel fungal hydroxylation reactions of steroidal substrates because of the medicinal value of this class of compounds. Some examples of the steroid transformation reactions, and the fungi carrying out the transformations, are shown in Table 4.3.

In addition to in vivo transformations, fungi are an important source of industrial enzymes. Some major fungal enzymes used in biocatalytic processes are listed in Table 4.4.

Distribution in the Prokaryotic World of Biodegradative and Novel Biocatalytic Capabilities

Previous sections discussed the role that prokaryotes and fungi play collectively in biodegradation and how their extensive biocatalytic potential derives from a long evolutionary history. But microbial diversity also implies that individual bacteria and fungi are metabolically unique. In this context, it is useful to think of microbes as metabolic machines, dependent on gathering chemicals from their environment to obtain carbon, other elements, and energy to compete favorably against other microbes. There is likely an optimum genome size in prokaryotes, somewhere between 2 and 10 Mb (see chapter 11 on genomics), which is limited by the time and energy required to replicate the DNA. More is not always better, and thus bacteria carry a restricted set of genetic baggage. This mandates a strategy of met-

Table 4.3 Fungal hydroxylation of steroid substrates[a]

Substrate	Position of oxidation	Fungus
Androst-4-ene-3,17-dione	1-α	*Penicillium* sp.
Androst-4-ene-3,17-dione	1-β	*Xylaria* sp.
Androstane-7,17-dione	3-α	*Diaporthe celastrinia*
17β-Hydroxyandrostan-11-one	3-β	*Wojnowicia graminis*
Progesterone	11-α	*Rhizopus* sp.
11-Deoxycortisone	11-β	*Curvularia lunata*
17β-Hydroxy-ester-4-ene-3-one	12-β	*Collectrichum derridis*

[a]Data from reference 28 with permission from Gordon and Breach Publishers (© Overseas Publishers Association).

abolic specialization, and it also raises an important question: is there a clustering of metabolic diversity directed against nonintermediary compounds within a relatively small subset of prokaryotes?

This is not an easy question if we seek a broad answer dealing with biodegradation and biotransformation in the soil and water of planet Earth. As stated in a previous chapter, most prokaryotes have never been cultivated, and some microbial phyla, identified by 16S rRNA sequencing, are not represented by a single isolated strain. Thus, the potential for currently uncultivatable prokaryotes to harbor diverse catabolic functions has not been adequately addressed. While some suggest that whole-genome sequencing of the unculturable organisms will answer the question, this is only half true. Yes, genome sequencing can identify putative catabolic genes with sequences resembling those of catabolic genes from cultivated organisms. No, it will not identify the functions of genes with sequences which are not significantly homologous with those of known biodegradative genes. Moreover, there will be many ambiguities even when new genes are found to be homologous to known genes. This will be discussed in greater detail in chapter 11.

We will only address here the correlation between taxonomy and novel biocatalytic potential with respect to those microorganisms which have been cultivated and whose metabolisms have been studied in some detail. There are additional caveats in this context as well. For example, certain genera of bacteria have been obtained with high frequency because of the nature of enrichment culturing techniques (see chapter 3), which are often done under conditions selecting for fast-growing organisms with few nu-

Table 4.4 Industrially important enzymes produced by fungi

Enzyme	Fungal species
Lipase	*Candida cylindracea*
Alcohol dehydrogenase	*Mortierella isabellina*
	Aspergillus niger
	Thermoanaerobium brockii
Monooxygenase (used in vivo)	*Rhizopus arrhizus*
	Aspergillus niger
	Cunninghamella echinutata
Cellulase	*Trichoderma reesei*
Racemase	*Exophiala wilhansii*

tritional requirements. Moreover, there is a bias in taxonomy in that some genera are much broader than others, as indicated by criteria such as percent G+C content. This is a historical artifact; at one time, virtually any aerobic, nonfermenting, gram-negative soil bacterium with a broad capability to metabolize exotic compounds was denoted *Pseudomonas* sp. That is being partly rectified with the reassignment of many such strains, based on 16S rRNA analysis, to other genera, such as *Sphingomonas*, *Burkholderia*, and *Comamonas* (Table 4.5).

Still, many of the biodegrading organisms described are found among the alpha, beta, and gamma subdivisions of the division *Proteobacteria* and the high- and low-G+C gram-positive bacteria. However, these groups represent a significant percentage of the known bacterial strains. Since the assignment of biodegradation activity to specific strains requires their isolation and characterization, it is no wonder that biodegradation is most established in the major groups that have been isolated to date. Moreover, many of the proteobacteria and high-G+C gram-positive bacteria were isolated in enrichment cultures specifically for the purpose of studying biodegradation, with organic compounds as the sole source of carbon. Most of those enrichments have been done under conditions conducive to the isolation of members of the alpha, beta, and gamma subdivisions of *Proteobacteria* or the high-G+C gram-positive bacteria.

With these caveats in mind, we can look at compiled information on biodegradative organisms. One tool to do this is the microorganism index found in the University of Minnesota Biocatalysis/Biodegradation Database (UM-BBD) (http://umbbd.ahc.umn.edu/search/micro.html). The list is made up largely of prokaryotes, but some fungi are also included. The output from August 1999 is shown in Table 4.6. That compilation, derived from the published literature, was obtained after novel biodegradative reactions were selected and annotated. As part of the annotation, the microorganisms known in the literature (i.e., in a paper indexed by Medline) to catalyze that reaction were indicated. The list was then compiled alphabetically with the microorganism's genus, species (when known), and strain (when indicated) names. The list is not biased toward the current analysis,

Table 4.5 Reclassification of bacterial strains previously classified as *Pseudomonas* species

Former name	Name following reclassification
P. acidovorans	*Comamonas acidovorans*
P. aminovorans	*Aminobacter aminovorans*
P. avenae	*Acidovorax avenae*
P. cepacia	*Burkholderia cepacia*
P. diminuta	*Brevundimonas diminuta*
P. flava	*Hydrogenophaga flava*
P. luteola	*Chryseomonas luteola*
P. maltophilia	*Stenotrophomonas maltophilia*
P. marina	*Deleya marina*
P. mesophilica	*Methylobacterium mesophilicum*
P. mixta	*Telluria mixta*
P. oryzihabitans	*Flavimonas oryzihabitans*
P. paucimobilis	*Sphingomonas paucimobilis*

Table 4.6 Microbial genera known to biodegrade organic compounds, represented in the UM-BBD on 15 August 1999

Genus or group	Compound
Acinetobacter	Cyclohexanol
Actinomycetes	2,4,6-Trinitrotoluene
Aeromonas	Phenanthrene
Agrobacterium	Glyphosate
Alcaligenes	2-Aminobenzenesulfonate
	2,4-Dichlorobenzoate
	2,4-Dichlorophenoxyacetic acid
	Toluene-4-sulfonate
Ancylobacter	1,3-Dichloro-2-propanol
	2,4-Dichlorobenzoate
	1,2-Dichloroethane
	2,4-Dichlorophenoxyacetic acid
	Glyphosate
	4-Nitrophenol
	Parathion
	Tyrosine
Arthrobacter	2-Aminobenzoate
	1,3-Dichloro-2-propanol
	2,4-Dichlorobenzoate
	2,4-Dichlorophenoxyacetic acid
	Fluorene
	Glyphosate
	Methyl *tert*-butyl ether
	Nicotine
	4-Nitrophenol
	Parathion
	Phenanthrene
	Tyrosine
Aspergillus	2-Aminobenzoate
	Phenanthrene
Azoarcus	Benzoate
	Toluene
Azotobacter	2,4-Dichlorophenoxyacetic acid
	Thiocyanate
Beijerinckia	Xylene
Brevibacterium	Dibenzofuran
Burkholderia	3-Chloroacrylic acid
	2,4-Dichlorophenoxyacetic acid
	Pentachlorophenol
	Phthalate
	Toluene
	1,2,4-Trichlorobenzene
	Trichloroethylene
	2,4,5-Trichlorophenoxyacetic acid
	o-Xylene
Chelatobacter	Nitrilotriacetate
Clostridium	Phenol
Comamonas	3-Methylquinoline
	Nitrobenzene
	Phthalate
	Toluene-4-sulfonate
Corallinus	Atrazine
Corynebacterium	1,3-Dichloro-2-propanol
	2,4-Dichlorobenzoate
Cunninghamella	Phenanthrene
Dehalobacter	Tetrachloroethene
Dehalococcoides	Tetrachloroethene
Dehalospirillum	Tetrachloroethene
Desulfitobacterium	Tetrachloroethene
Dunaliella	DDT[a]
Enterobacter	DDT
	Glyphosate
	Pentaerythritol tetranitrate
Escherichia	Carbazole
	Methionine
	Organomercury
	3-Phenylpropionate
	Threonine
Eubacterium	Gallate
Exophiala	Styrene
Flavobacterium	Bromoxynil
	2,4-Dichlorophenoxyacetic acid
	Glyphosate
	Parathion
	Pentachlorophenol
Fusarium	2-Nitropropane
Hydrogenophaga	4-Carboxy-4′-sulfoazobenzene
Hyphomicrobium	Dichloromethane
Klebsiella	Acetylene
	Benzonitrile
	Bromoxynil
Methanosarcina	Tetrachloroethene

(continued next page)

Table 4.6 *continued*

Genus or group	Compound
Methylobacterium	Dichloromethane
	Methyl *tert*-butyl ether
Methylophilus	Dichloromethane
Methylosinus	Trichloroethylene
Methylosulfonomonas	Methanesulfonic acid
Moraxella	Naphthalenesulfonate
	4-Nitrophenol
Mycobacterium	Methyl *tert*-butyl ether
Neurospora	2-Nitropropane
Nitrosomonas	Dimethyl ether
	Methyl fluoride
Nocardia	Methyl *tert*-butyl ether
	Methyl ethyl ketone
Paracoccus denitrificans	Thiocyanate
Phanerochaete	Phenanthrene
Pleurotus	Phenanthrene
Proteobacteria	Methanesulfonic acid
	Nitrilotriacetate
Proteus	DDT
Pseudomonas	Acrylonitrile
	2-Aminobenzoic acid
	1-Aminocyclopropane-1-carboxylate
	Atrazine
	Biphenyl
	Bromoxynil
	(+)-Camphor
	Caprolactam
	Chlorobenzene
Pseudomonas (*continued*)	4-Chlorobiphenyl
	m-Cresol
	DDT
	Dibenzothiophene
	2,4-Dichlorobenzoate
	Dodecyl sulfate
	Ethylbenzene
	Mandelate
	Methyl *tert*-butyl ether
	Naphthalene
	Naphthalenesulfonate
	Nitrobenzene
	Nitroglycerin
	n-Octane
	Orcinol
	Organomercury
	Parathion
	Phenanthrene
	Styrene
	Thiocyanate
	Toluene
	Toluene-4-sulfonate
	Trichloroethylene
	2,4,6-Trinitrotoluene
	Tyrosine
	Xylene
Rhodobacter	Dimethyl sulfoxide
Rhodococcus	Acetylene
	Atrazine
	Benzonitrile
	Bromoxynil
	Cyclohexanol
Rhodococcus (*continued*)	Dibenzothiophene
	Methyl *tert*-butyl ether
	Styrene
	Tetrahydrofuran
Rhodopseudomonas	Benzoate
Sphingomonas	Dibenzo-*p*-dioxin
	Dibenzofuran
	gamma-1,2,3,4,5,6-Hexachlorocyclohexane
	Xylene
Sporomusa	Tetrachloroethene
Staphylococcus	Dibenzofuran
	Fluorene
	2,4,6-Trinitrotoluene
Streptomyces	Atrazine
	Phenanthrene
Syncephalastrum	Phenanthrene
Synechococcus	DDT
	Phenanthrene
Terrabacter	Dibenzofuran
Tetrahymena	Dimethyl sulfoxide
Thauera	Benzoate
	Phenol
	Toluene
Thiobacillus	Thiocyanate
Williopsis	2-Nitropropane
Xanthobacter	1,2-Dichloroethane
	2-Nitropropane
	Propylene

[a]DDT, 1,1,1-trichloro-2,2-bis-(4-chlorophenyl)ethane.

as the compilation was made without consideration for the organism but rather for the utility of the reaction for adding biochemical novelty to the database.

As of August 1999, the genus with the largest number of entries in the UM-BBD was *Pseudomonas,* with 66 different strains acting on 47 different substrates. This represented approximately 50% of the starting compounds for pathways contained in the UM-BBD. The compounds metabolized represent aliphatic and aromatic hydrocarbons; aliphatic and aromatic acids; aliphatic and aromatic compounds with chloro, cyano, nitro, and phosphoryl groups substituted; and cycloaliphatic compounds. The genus with the second largest set of substrates metabolized is *Arthrobacter,* with 12. The major substrates for the *Arthrobacter* strains listed are aromatic and N-heterocyclic ring substrates. Following that, *Rhodococcus* and *Burkholderia* are most highly represented in the UM-BBD.

We can now use the database entries, over 250 different biodegradative strains, to see whether biodegradation reactions are catalyzed most heavily by genera in certain subdivisions of bacteria. Table 4.7 shows the prokaryotic divisions:

1. High-G+C gram-positive bacteria
2. Low-G+C gram-positive bacteria
3. The alpha, beta, gamma, delta, and epsilon subdivisions of *Proteobacteria*
4. The *Cytophagales*–green sulfur group
5. The green non-sulfur group

The UM-BBD contains a significant number of prokaryotes from the following groups:

1. High-G+C gram-positive bacteria
2. Low-G+C gram-positive bacteria
3. The alpha, beta, and gamma subdivisions of *Proteobacteria*

While there is not an even distribution, there are, with the exception of the epsilon subdivision of *Proteobacteria,* biodegradative bacteria in all the groups in Table 4.7. A significant number of reactions (greater than five) are catalyzed by one or more strains from the genera *Arthrobacter, Rhodococcus, Clostridium, Sphingomonas, Burkholderia, Escherichia, Pseudomonas,* and *Flavobacterium.* These genera cut across the taxonomic lines shown and indicate that no genus has a monopoly on biodegradative metabolism.

Of course, vast taxonomic territory is not covered in Tables 4.6 and 4.7 (see Fig. 2.1, the taxonomic tree of life, for comparison). More than 99% of bacterial genera are thought to be yet unisolated, and many taxa of isolated bacteria are largely unstudied. For example, the divisions *Planctomycetes* and OP3 to OP10 (Fig. 4.2) have few isolated strains (44). Taxonomic and distributional data about these strains are largely derived from 16S rRNA sequences following PCR amplification of DNA extracted from soils. Without isolated strains, it is difficult or impossible to determine if those divisions of bacteria contribute significantly to catabolic diversity and biodegradation.

Another indicator that biodegradative activity is more widespread than previously thought resides in recent scientific literature. With our increased ability to culture organisms or identify an organism's metabolism without isolating it (see, for example, reference 50), biodegradative capabilities in new branches of the tree of life are quickly emerging (Table 4.8). This in-

Table 4.7 Taxonomic distribution of bacteria for which biodegradation reactions were depicted on the UM-BBD as of 15 August 1999[a]

		Proteobacteria					
High- G + C gram positive	**Low- G + C gram positive**	α	β	γ	∂/ε	***Cytophagales*–green sulfur**	**Green non-sulfur**
Arthrobacter	*Bacillus*	*Agrobacterium*	*Achromobacter*	*Acinetobacter*	*Desulfovibrio*	*Flavobacterium*	*Dehalococcoides*
Brevibacterium	*Clostridium*	*Ancylobacter*	*Alcaligenes*	*Aeromonas*			
Clavibacter	*Desulfitobacterium*	*Brevundimonas*	*Azoarcus*	*Azotobacter*			
Corynebacterium	*Eubacterium*	*Chelatobacter*	*Burkholderia*	*Enterobacter*			
Dehalobacter	*Staphylococcus*	*Hyphomicrobium*	*Comamonas*	*Escherichia*			
Nocardia		*Methylobacterium*	*Hydrogenophaga*	*Klebsiella*			
Rhodococcus		*Paracoccus*	*Ralstonia*	*Methylobacter*			
Streptomyces		*Rhodobacter*	*Thauera*	*Methylococcus*			
Terrabacter		*Sphingomonas*	*Thiobacillus*	*Moraxella*			
			Thauera	*Pseudomonas*			

[a]Greater than five strains in the genera highlighted in green catalyze biodegradative reactions.

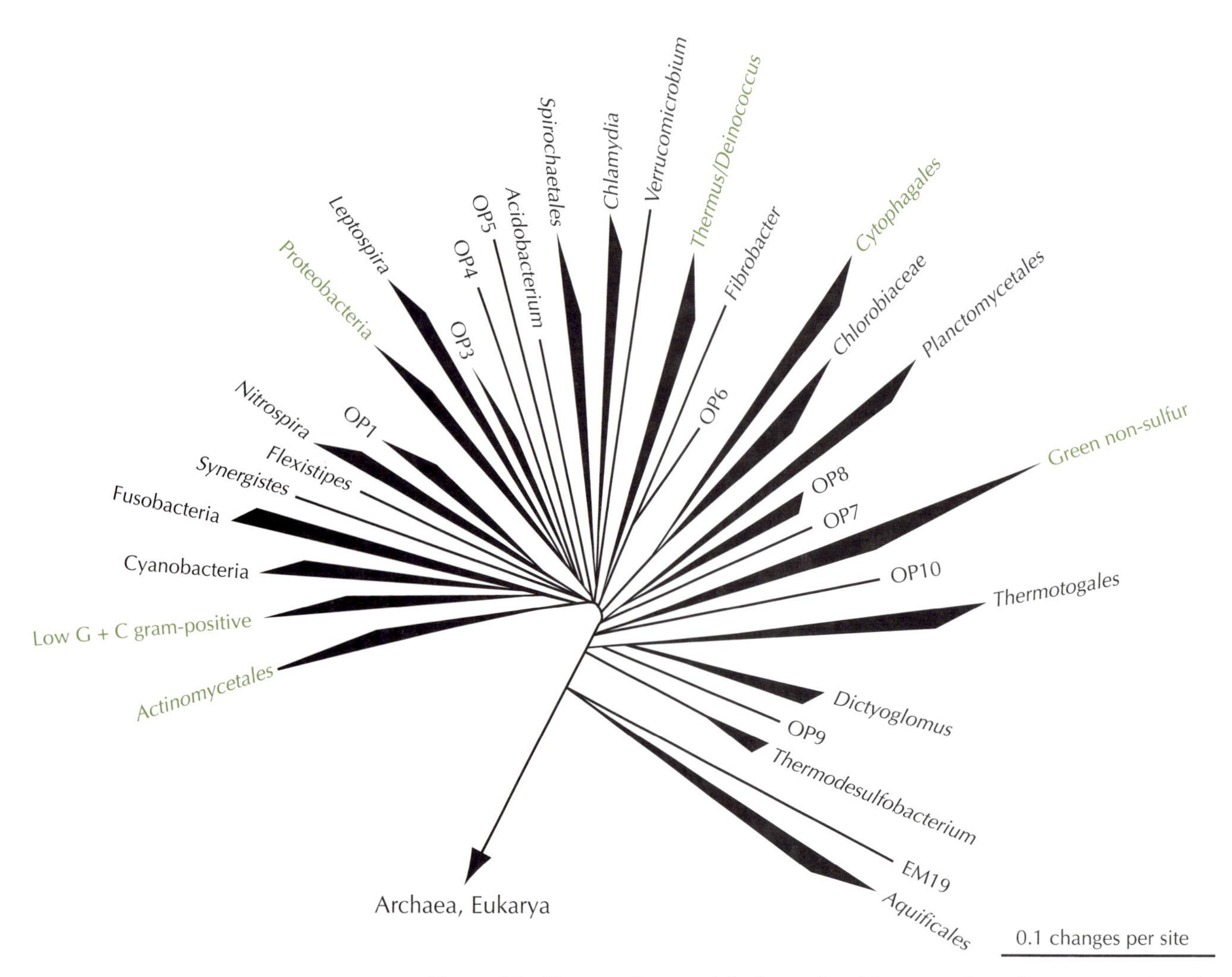

Figure 4.2 Taxonomic tree of the bacteria with groups heavily represented in the UM-BBD highlighted in green. (From reference 44 with permission.)

cludes archaea capable of metabolizing aromatic acids, a member of the genus *Fusobacterium* metabolizing multi-ring phenols, and a member of the *Desulfuromonas* group which metabolizes the unactivated hydrocarbon benzene. Most recently, uncultivated *Syntrophus* spp., acting in concert with archaeal methanogens, were implicated in the metabolism of hexadecane to methane in strictly anaerobic sediments (66).

It is interesting that most of the novel organisms identified in biodegradation have been reported only in 1998 or 1999. This likely reflects their very recent cultivation or identification in consortia. Moreover, not all of those studies had as their goal the elucidation of biodegradation reactions; researchers in the biodegradation community more typically isolate bacteria under standard aerobic conditions that yield aerobic proteobacteria. More recently, the increased laboratory cultivation of prokaryotes under anaerobic conditions has been expanding our perception of the diverse catabolic activities of these organisms.

These observations strongly suggest that much novel biodegradative metabolism remains to be discovered. In one example, using trichloroben-

Table 4.8 Prokaryotes from taxonomic groups underrepresented in the biodegradation literature, recently reported to catalyze biodegradation reactions

Organism	Taxonomic group	Substrate	Reference
Archaea			
Haloferax	*Halobacteriaceae*	3-Phenylpropionate	24
		Gentisate	25
Thermosphaera aggregans	*Crenarcaeota*	Lactose	14
Bacteria			
Haloanaerobium praevalens	*Haloanaerobiales*	Nitrobenzene	43
		o-Nitrophenol	
		m-Nitrophenol	
		p-Nitrophenol	
		Nitroanilines	
		2,4-Dinitrophenol	
		2,4-Dinitroaniline	
Sporohalobacter marismortui	*Haloanaerobiales*	Nitrobenzene	43
		o-Nitrophenol	
		m-Nitrophenol	
		p-Nitrophenol	
		Nitroanilines	
		2,4-Dinitrophenol	
		2,4-Dinitroaniline	
Borrelia burgdorferi	*Spirochaetales*	Benzamides	22
Thermus	*Thermus/Deinococcus*	Starch	33
Bacteroides	*Cytophagales*	Rhamnosidase	
Bacteroides fragilis	*Cytophagales*	Alkylhydroperoxides	49
Fibrobacter intestinalis	*Fibrobacter/Acidobacter*	Cellobiose	38
Syntrophus	*∂-Proteobacteria*	Alkanes	66
Fervidobacterium pennavorans	*Thermotogales*	Pullulan	4
		Starch	
		Amylopectin	
Thermotoga maritima	*Thermotogales*	Starch	40
		Cellulose	5
		Xylan	13
Ruminococcus	*Clostridiaceae*	Cellulose	31
Caldicellulosiruptor	*Syntrophomonas/Thermoanaerobacter*	Xylan	41
Thermoanaerobacterium	*Syntrophomonas/Thermoanaerobacter*	Xylan	41
Dictyoglomus thermophilum	*Syntrophomonas/Thermoanaerobacter*	Xylan	41
Thermoanaerobacter	*Syntrophomonas/Thermoanaerobacter*	Cyclodextrins	1
Desulfitobacterium dehalogenans	*Heliobacterium*	PCBs	63
		Chlorophenols	59
Desulfitobacterium hafniense	*Heliobacterium*	3-Chloro-4-hydroxy-phenylacetate	15
Sporotomaculum hydroxybenzoicum	*Heliobacterium*	3-Hydroxybenzoate	8
Syntrophus gentianae	*∂-Proteobacteria*	Benzoate	52
Syntrophus sp.	*∂-Proteobacteria*	Hexadecane	66
Desulfobacula toluolica	*∂-Proteobacteria*	*p*-Toluidine	47
Geobacter spp.	*∂-Proteobacteria*	Benzene	50
Pelobacter acetylenicus	*∂-Proteobacteria*	Acetylene	39
Pelobacter acidigallici	*∂-Proteobacteria*	Pyrogallol	48
Desulfovibrio	*∂-Proteobacteria*	2,4,6-Tribromophenol	7

zene as the substrate, 51 different taxonomic classes were identified by 16S rRNA analysis, only one of which could be defined as closely related to a known dechlorinating *Dehalobacter* sp. (60). This and related studies, such as those shown in Table 4.8, are consistent with the idea that biodegradative capabilities are pervasive in the prokaryotic world.

Specialized Biodegradation by Microorganisms with Specialized Metabolisms

In general, one cannot say that certain bacteria are predominant in metabolizing certain classes of compounds: for example, aromatic hydrocarbons, chlorinated aliphatic compounds, or heterocyclic ring compounds. One notable exception to this is the class of single-carbon (C_1) pollutants and the C_1-metabolizing bacteria known as methanotrophs and methylotrophs. These bacteria are largely classified in the division *Proteobacteria*. They grow on methane or methanol as their sole carbon sources, respectively. Methanotrophs metabolize methane to methanol and then, through a series of further two-electron oxidation reactions, to carbon dioxide to generate ATP via electron transport coupled to a membrane ATPase. Carbon is fixed at the level of formaldehyde, an intermediate whose level must be carefully regulated. This is a specialized form of metabolism, and these organisms often fail to grow on many other more complex organic substrates. Despite this, their metabolism gives them the unique ability to metabolize certain substituted C_1 compounds, some of which are natural products and some of which are largely or exclusively of anthropogenic origin (Fig. 4.3). For example, chloromethane is a natural (and also industrial) product made in quantities of thousands of tons by soil fungi (30). Dichloromethane is principally synthesized by humans and is heavily used as an industrial solvent. In a number of cases, one additional enzyme can bring about a transformation of the C_1 compound to funnel the carbon into the C_1 oxidative cycle. This will be discussed in greater depth in chapter 8 dealing with metabolic logic.

Aerobic versus Anaerobic Microorganisms in Biodegradation

Aerobic and anaerobic bacteria do not divide along strictly phylogenic lines, but because this is an important distinction with respect to biodegradation and industrial biotransformations, it is covered as a defined topic here. The presence or absence of oxygen often dictates the type of biodegradative pathway and the types and number of bacteria involved in the biodegradation of a particular compound. In practical biodegradation, or industrial biotransformations, the choice of fostering aerobic or anaerobic conditions is often a crucial one.

In general, we know more about biodegradation under aerobic conditions. This reflects the greater knowledge accumulated about aerobic and facultatively aerobic microorganisms. Historically, organisms such as *Pseudomonas* sp. or *Escherichia coli* that can be grown overnight with simple equipment to yield high cell densities were preferred for many studies. Anaerobic biotransformation studies sometimes present other difficulties, too. Anaerobic enrichment cultures may initially show very long lag phases, perhaps 6 months or 1 year, before significant biodegradation occurs. Upon repeated transfers, the lag phase often shortens continually. Still, many years may be required to achieve significantly high rates of biodeg-

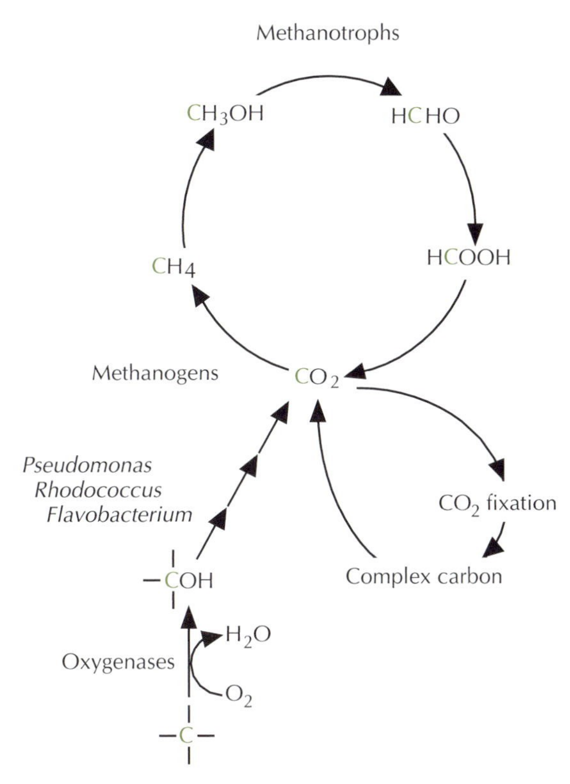

Figure 4.3 Part of the carbon cycle showing the metabolism of single-carbon compounds and reactions which feed into the cycle.

radation, and they may never approach the rates of comparable aerobic biodegradation. In most cases, a definitive explanation for the lag phase phenomenon is lacking.

Anaerobic biodegradation studies often involve complex, undefined bacterial consortia. This is due partly to the slower cultivation cycles and also to the complexity of anaerobic metabolic interactions that make pure-culture isolation sometimes difficult or impossible. A classic example is that of the anaerobic ethanol-degrading organism *Methanobacterium omelianskii,* described as a pure culture (3) but later shown to be a tight obligate assemblage of an ethanol-oxidizing bacterium and an acetate-utilizing methanogen (9).

By contrast, it is often easier to obtain aerobic biodegrading bacteria in pure culture. This is reflective of higher growth rates and a lesser metabolic interdependence. The latter is often due to oxidative metabolism of the organic compound to carbon dioxide and water, a process that is usually highly exergonic. Aerobic processes typically yield more energy, generate a commensurately greater number of ATP equivalents, and generate more biomass per unit of compound transformed (16, 57). Thus, compounds with a significant energy yield upon oxygenative metabolic combustion to carbon dioxide and water are often rapidly biodegraded under aerobic conditions. This is most notable with hydrocarbons, for example, fossil fuel components, such as octane, other alkanes, benzene, toluene, and xylenes. These compounds are more rapidly degraded in most well-aerated soil environments and will be metabolized more rapidly in the laboratory by aerobic cultures. A telling comparison can be made with hexadecane, a linear alkane found in fossil fuels. Aerobic cultures which demonstrably metabolize and grow on hexadecane overnight are known (54). By contrast, an anaerobic culture of hexadecane-metabolizing bacteria has been described, but it took over 1 year after the lag phase for the culture to grow (66). This was proposed to be indicative of a very slow methanogenic anaerobic transformation of alkanes occurring in sediments in nature.

Representative Microorganisms with Broad Catabolic Abilities

While it is certainly important to look at catabolic metabolism by many diverse microorganisms, in-depth study of a few is also important to learn about physiological processes in detail. In that context, we have chosen a few microorganisms which have been relatively well studied for more detailed treatment. The examples given below are by no means representative of all the well-studied microorganisms which have contributed to our understanding of biodegradation and microbial biocatalysis. The choices reflect our biases and are partly due to our experience with them and our access to unpublished information about them. Other well-characterized aerobic microorganisms with broad catabolic activities are *Pseudomonas* sp. strain B3, *Pseudomonas* sp. strain CBS3, *Pseudomonas* sp. strain NCIB 9816, *Pseudomonas chlororaphis* B23, *Pseudomonas putida* mt-2, *Sphingomonas* sp. strain RW1, and *Burkholderia cepacia* AC1100. Among the best-studied anaerobes are *Desulfomonile tiedjei* DCB-1, *Thauera aromatica*, *Desulfitobacterium dehalogenans*, *Acetobacterium woodii*, *Rhodopseudomonas palustris*, *Dehalobacter restrictus*, and *Dehalococcoides ethenogenes* 195.

Many of the best-studied prokaryotes are aerobic proteobacteria and the high-G+C gram-positive bacteria. These, along with the fungus *Cunninghamella elegans*, are the examples discussed below. It is important to point out that some of the extensive catabolic activities of this class are based on the broad substrate specificities of a few oxygenases and related enzymes which handle the oxygenated intermediates. This is perhaps best illustrated with *P. putida* F1 because the substrate specificity of toluene dioxygenase has been extensively investigated.

Pseudomonas putida F1

P. putida F1 was initially isolated from soil at the edge of Boneyard Creek in Urbana, Ill., with ethylbenzene as the enrichment substrate. It was isolated and first studied with the idea that bacteria would likely initiate metabolism of aromatic hydrocarbons in a manner similar to that of mammalian systems (see Fig. 4.1), based on a single report in the literature (37). This proved not to be the case, and a series of definitive studies showed that a *cis*-dihydrodiol was the first intermediate and that it derived from the action of a dioxygenase enzyme (26). Since that time, scores of researchers have reported the bacterial formation of *cis*-diols from aromatic ring compounds, establishing a major theme in microbial metabolic logic for this large class of compounds.

Like other members of the genus *Pseudomonas*, *P. putida* F1 can grow on a defined minimal medium with a large number of compounds as its sole source of carbon and energy. These are shown in Table 4.9. The table includes compounds experimentally demonstrated to support growth of *P. putida* F1, those compounds known to support growth of all *P. putida* strains (45), and the known substrates for toluene dioxygenase. In 1997, Eaton

Table 4.9 Compounds known or proposed to be oxidized by *P. putida* F1[a]

Known substrates		
1,1-Dichloro-1-propene	3-Methylphenol	Glucose
1,1-Dichloroethene	3-Nitrotoluene	Indan
1,2-Dichlorobenzene	4-Bromotoluene	Indene
1,2-Dihydronaphthalene	4-Chlorobiphenyl	Indole
1,2-Dimethylbenzene	4-Chlorostyrene	L-Arginine
1,2-Methylenedioxybiphenyl	4-Chlorotoluene	L-Glutamate
1,3-Dibromobenzene	4-Fluorotoluene	*m*-Bromobenzotrifluoride
1,3-Dichlorobenzene	4-Methylphenol	Methylphenyl sulfide
1,3-Dimethylbenzene	4-Nitrotoluene	Methyl *p*-nitrophenyl sulfide
1,4-Dichlorobenzene	Acetate	Methyl *p*-tolyl sulfide
1,4-Dimethylbenzene	Acetophenone	*o*-Iodotoluene
1-Bromo-2,3-difluorobenzene	Anisole	*p*-Cumate
1-Bromo-4-iodobenzene	Benzene	*p*-Cymene
1-Chloro-2-methyl-propene	Benzenenitrile	Perdeuteriobenzene
1-Fluoro-4-iodobenzene	Benzoate	Phenetole
2-(2-Bromoethyl)bromobenzene	Benzocyclohept-1-ene	Phenol
2,3-Dichloro-1-propene	Biphenyl	Phenylethanol
2,3-Dimethoxybiphenyl	Bromobenzene	*p*-Hydroxybenzoate
2-Acetoxyethylbenzene	Chlorobenzene	*p*-Iodotoluene
2-Azidoethylbenzene	*cis*-1,2-Dichloroethene	*p*-Methoxyphenylmethyl sulfide
2-Bromoethylbenzene	*cis*-1,4-Dichloro-2-butene	Propoxybenzene
2-Bromostyrene	*cis*-1-Bromo-1-propene	Protocatechuate
2-Chlorobiphenyl	*cis*-1-Chloro-1-propene	Pyruvate
2-Chlorostyrene	*cis*-2-Chloro-2-butene	Styrene
2-Cyanoethylbenzene	*cis*-β-Bromostyrene	Succinate
2-Indanone	*cis*-Dibromoethene	Toluene
2-Isothiocyanatoethylbenzene	Citrate	*trans*-1,4-Dichloro-2-butene
2-Methoxybiphenyl	Cyanobenzene	*trans*-1-Bromo-1-propene
2-Methoxynaphthalene	D-Glucose	*trans*-1-Chloro-1-propene
2-Methylphenol	*cis*-β-Bromostyrene	*trans*-2-Chloro-2-butene
2-Nitrotoluene	Ethylbenzene	*trans*-Cinnamonitrile
2-Thiocyanatoethylbenzene	Ethylphenyl sulfide	*trans*-Dibromoethene
3,4-Dichloro-1-butene	Ethynylbenzene	Trichloroethylene
3-Chlorobiphenyl	Fluorobenzene	Trifluoromethoxybenzene
3-Chlorostyrene	Fumarate	Trifluorotoluene
Proposed substrates		
2-Ketogluconate	L-Alanine	Propionate
Betaine	L-Aspartate	Putrescine
Butylamine	L-Histidine	Quinate
Butyrate	L-Isoleucine	Saccharate
Caprate	L-Leucine	Sarcosine
Caproate	L-Malate	Spermine
Caprylate	L-Ornithine	Tryptamine
D-Fructose	L-Phenylalanine	Valerate
Gluconate	L-Proline	α-Aminovalerate
Glutarate	L-Tyrosine	α-Ketoglutarate
Glycerol	L-Valine	β-Alanine
Heptanoate	Malonate	β-Hydroxybutyrate
Isovalerate	*n*-Butanol	δ-Aminovalerate
Lactate	Pelargonate	

[a]Known substrates are those experimentally demonstrated with *P. putida* F1. Proposed substrates are those known to be substrates for *P. putida* strains as defined in *Bergey's Manual of Systematic Bacteriology*.

reported the involvement of a cymene dioxygenase in the metabolism of cymene, or *p*-isopropyltoluene, by *P. putida* F1 (23). The genes are encoded by the *cym* operon, located just upstream of the *cmt* operon, which encodes the further catabolism of *p*-cumate. Both of these operons are located upstream of the *tod* operon encoding toluene oxidation.

The presence of multiple dioxygenases in *Pseudomonas* spp. and related strains dramatically expands the range of substrates capable of being catabolized. The dioxygenases which act on aromatic hydrocarbons are dramatically broad in their substrate specificities. On the order of 100 substrates are known to be oxidized by the well-studied toluene dioxygenase from *P. putida* F1. These have been compiled on the UM-BBD (http://umbbd.ahc.umn.edu/tol/tdo.html). Toluene dioxygenases in other bacterial isolates have been described, for example, in *P. putida* UV4 and JT107. The relationship of the sequences of these toluene dioxygenases to that from *P. putida* F1 is not known. In this context, the only substrates listed in Table 4.9 are those known to react with the $todABC_1C_2$ gene products of *P. putida* F1.

The oxidation of compounds to carbon dioxide requires the concerted action of numerous enzymes. The enzymes which metabolize dioxygenase-generated products also have relatively broad substrate specificities, thus allowing a large number of hydrocarbons to be used as carbon and energy sources. If a single dioxygenase and associated enzymes can confer a significantly larger substrate range, how many oxygenases may be found in organisms related to *P. putida* F1, many of which are known to contain catabolic oxygenases? More answers will begin to come with the completion of genome-sequencing projects for *P. putida, Pseudomonas aeruginosa,* and related genera. Consider, however, that a single catabolic plasmid, recently isolated from *Sphingomonas aromaticivorans* F199 and sequenced, contains gene modules encoding the multicomponent dioxygenase subunits which could theoretically combine to make 49 different four-component oxygenase systems. This strengthens the idea that broad-specificity dioxygenases contribute to the celebrated metabolic diversity of *Pseudomonas* and related genera. This will be discussed in greater detail in chapter 8, dealing with metabolic logic.

Sphingomonas yanoikuyae B1

Sphingomonas yanoikuyae B1, previously known as *Beijerinckia* sp. strain B1, was originally isolated for its ability to use biphenyl as a carbon source. It was the first bacterium for which definitive metabolites were shown for multi-ring polycyclic compounds, such as benzo[*a*]pyrene and benzo[*a*]anthracene (27). These were demonstrated with *S. yanoikuyae* B8/36, which had a mutation in an aromatic hydrocarbon diol dehydrogenase gene and thus accumulated *cis*-dihydrodiols. In this context, the strain was instrumental in demonstrating that *cis*-dihydrodiols were common bacterial oxidation products of polycyclic, as well as monocyclic, aromatic hydrocarbons.

S. yanoikuyae B1 is remarkable as a bacterium capable of oxidizing many aromatic compounds as its source of carbon and energy. As such, it is representative of the genus *Sphingomonas,* a group of gram-negative bacteria generally known for their diverse catabolic activities. The sphingomonads have been recognized as a group only since around 1990, with a number of their first representatives having been previously assigned to the genus *Pseudomonas.* A signature feature of *Sphingomonas* spp. is a sphingolipid composed of 18- to 21-carbon nonbranched and cyclopropane fatty acyl side chains attached to a dihydrosphingosine moiety which is part of a ceramide glycolipid in the cell wall (34). The corresponding component of the cell wall in *Pseudomonas* spp. and other gram-negative bacteria is the well-known lipopolysaccharide. Many *Sphingomonas* spp. are flagellated, like pseudomonads, but contain carotenoid pigments, unlike *Pseudomonas* spp. For an excellent treatment of *Sphingomonas* spp. and their diverse catabolic activities, the reader may wish to consult a special issue of the *Journal of Industrial Microbiology and Biotechnology* (35) which is completely devoted to the topic.

Rhodococcus spp.

Rhodococcus strains are well represented in the UM-BBD and the biodegradation and biocatalysis literature. The range of compounds they catabolize is diverse, from small gaseous compounds to the fuel additive methyl *tert*-butyl ether, terpenes, and the herbicide atrazine. Some of the greatest interest in the members of this group derives from their potential and current use in biotechnological processes, a few examples of which are given below. While *Rhodococcus erythropolis* has been implicated in some of these transformations, in other cases the species is not known. Thus, the reactions discussed below are not necessarily all catalyzed by the same bacterial strain.

One of the largest-volume biocatalytic industrial processes is the conversion of acrylonitrile to acrylamide. The world's annual acrylamide requirement is 400,000,000 lb. The conventional chemical synthesis, based on a copper catalyst, is plagued by the high cost of catalyst removal and undesirable purity of the product. A bacterial enzyme, nitrile hydratase, circumvents these problems (see Fig. 1.2). Moreover, it has a very high turnover and is thus suitable for very high-yield production of acrylamide. Figure 4.4 shows a laboratory fed-batch experiment in which *Pseudomonas chlororaphis* B23 produces 400 g of acrylamide per liter in 7 h. Nitto Chemical Company in Japan has developed a commercial process that is based completely on a microbiological catalyst.

Most of what is known about the structure and mechanism of nitrile hydratase derives from study of the enzyme from *Rhodococcus* spp. Nitrile hydratase is a metalloenzyme, typically with iron(III), or some forms Co(II), being required for catalytic activity. It is now known from X-ray crystallography (65) that the active site of the ferric enzyme has a unique ligand structure. It has been proposed that two cysteines which coordinate the iron atom are oxidized to sulfenic and sulfinic acid residues. Moreover, nitric oxide (NO) is found bound to the enzyme under conditions of darkness but is absent after the enzyme undergoes photoactivation. This unusual property is thought to provide a control mechanism for the activity of the enzyme. In any case, nitrile hydratase is useful as an industrial catalyst because it has a high specific activity, determined to be 1 mmol per min per mg of enzyme, and requires only water as a cosubstrate.

Rhodococcus strains have also been investigated for biocatalytic potential with respect to another very high-volume industrial transformation, the desulfurization of fossil fuels. It is very desirable to remove sulfur from heterocyclic rings. The combustion of fossil fuels, which contain variable but often substantial levels of sulfur, leads to sulfur dioxide release. When this goes into the atmosphere, it causes acid rain downwind from the combustion source, a major environmental concern with fuel combustion.

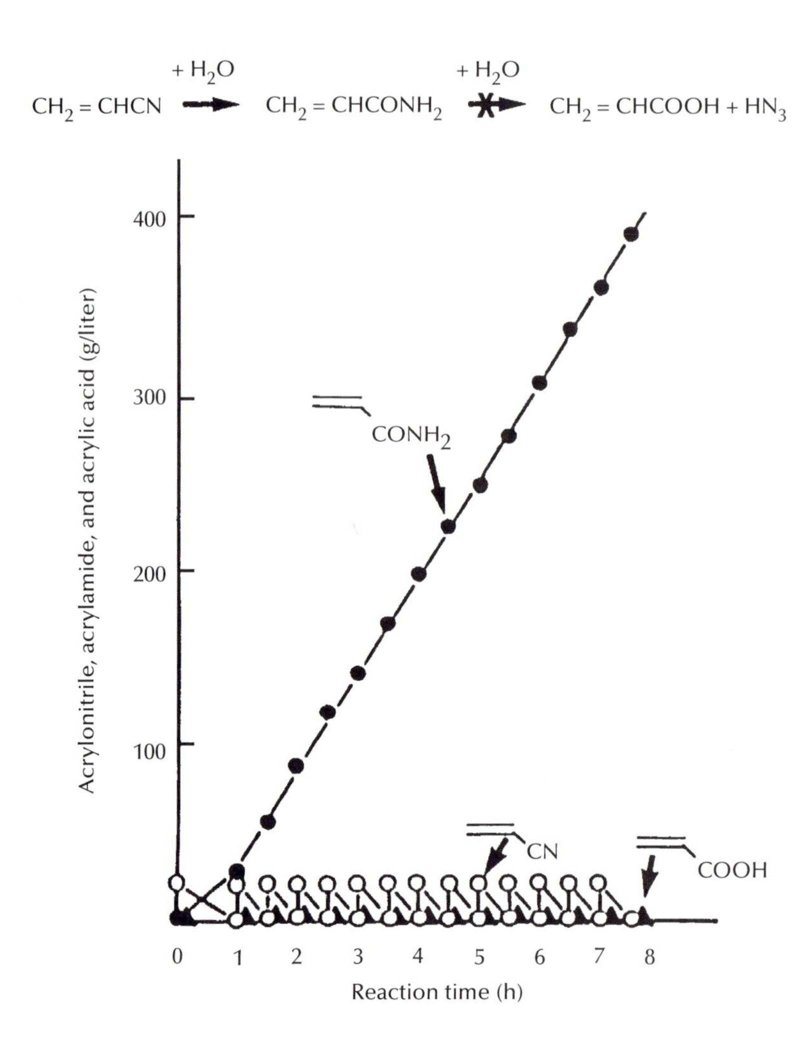

Figure 4.4 Continuous acrylamide production by *P. chlororaphis* B23 in a fed-batch reactor. (From reference 42 with kind permission from Kluwer Academic Publishers.)

Chemical catalysts designed to scrub the sulfur at various points in the process suffer from being very expensive and showing a substantial loss of sulfur-removing activity over a short time. This opens the way for a biotechnological process, although a very large-scale process would be required to address the current problem.

It is also important for economic reasons to remove organic sulfur without substantial oxidation of the carbon atoms surrounding the sulfur, which would decrease the thermal units derived from burning the fuel later. The ideal catalyst must be fast and efficient at removing sulfur and leave the surrounding organic structure largely intact.

Against this backdrop, several *Rhodococcus* strains have been identified which are capable of snipping out the sulfur from sulfur-heterocyclic rings, which are the major organosulfur structures in fossil fuels (29). The structure of dibenzothiophene and the biochemical pathway for its desulfurization are shown in Fig. 4.5. The desulfurization is accomplished by a series of oxidation reactions at the sulfur, followed by adjacent oxidations that lead to the removal of sulfur as sulfite. The *Rhodococcus* strains with these enzymes can use dibenzothiophene as the sole sulfur source. This physio-

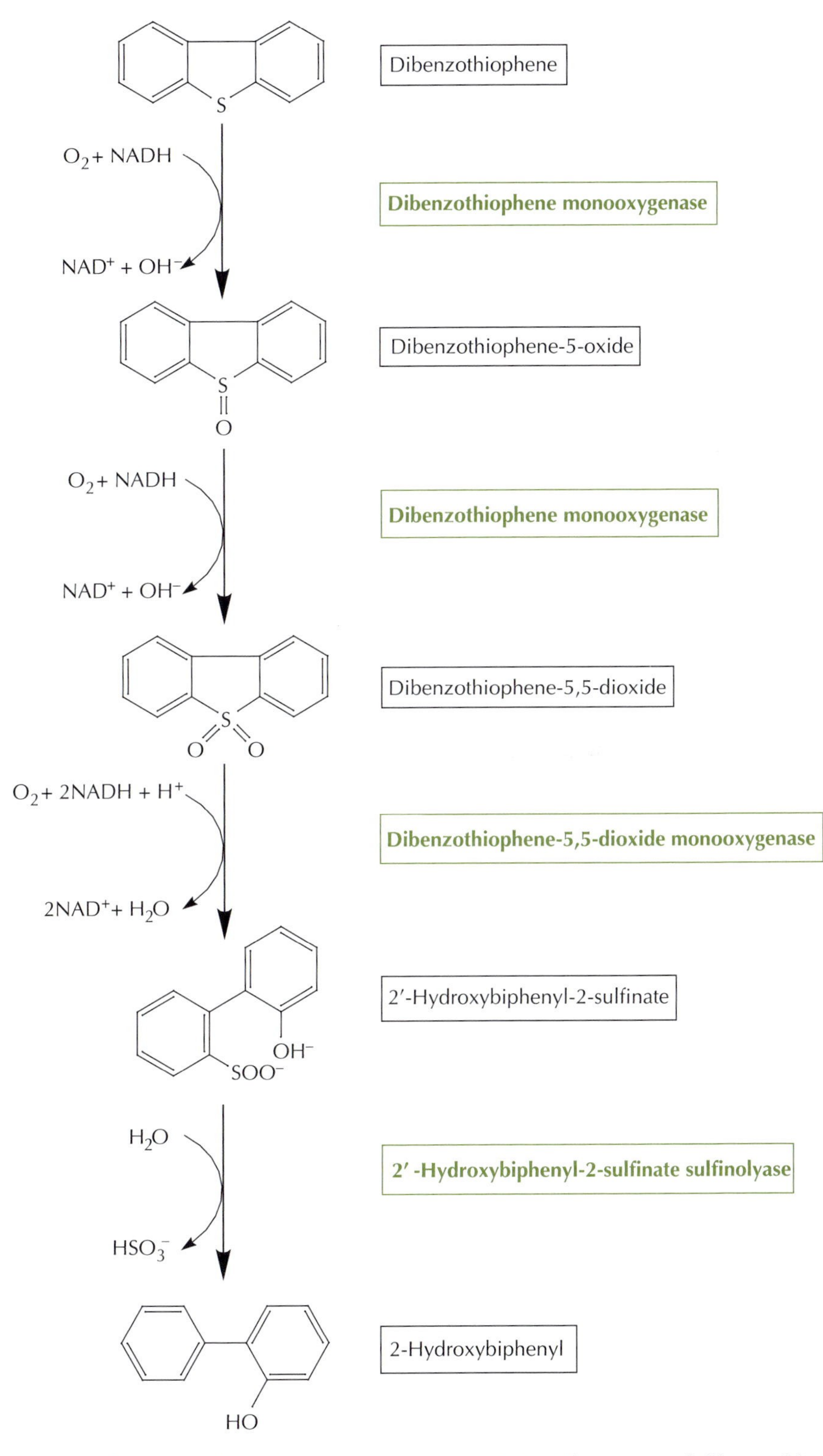

Figure 4.5 Metabolic pathway for the microbial desulfurization of dibenzothiophene. Substrate and enzyme (green) names are shown in boxes.

logical purpose underlies the low expression of the enzymes in native strains; microbes have relatively low needs for sulfur compared to the need for carbon. This is currently being circumvented by cloning and expressing the genes for fuel desulfurization in *Pseudomonas* spp., where the production of the enzymes can be uncoupled from natural physiological regulation.

Another recent application of *Rhodococcus* spp. has been in the production of a fine chemical, a derivative of 1-amino-2-hydroxyindan, in a stereospecific fashion (Fig. 4.6). This compound is a key structural fragment of a new anti-HIV drug, indinavir, that is currently marketed by Merck. To meet demand and keep the price down, a biological source of the compound, or its direct precursor, would be valuable. *Rhodococcus* sp. strain I24 (58) is one of a number of strains that oxidize indene to a *cis*-1,2-dihydrodiol, which can be transformed by chemical synthesis to the target compound.

Cunninghamella elegans

C. elegans is a fungus in the class *Zygomycetes*. It should not be confused with *Caenorhabditis elegans,* the worm studied in great detail for insights into its developmental cycle. The fungus *C. elegans* was isolated for its ability to grow on crude oil as a sole source of carbon and energy (12). The organism was observed to utilize the alkane fraction for growth. Other alkane-utilizing filamentous fungi and yeasts have been identified. However, *C. elegans* distinguishes itself from the others by its remarkable ability to oxidize an enormous array of aromatic ring compounds. While many fungi have the ability to oxidize naphthalene to α-naphthol (11), *C. elegans* is probably the most active fungal species known for the metabolism of polycyclic aromatic hydrocarbons. The organism oxidizes from two- to five-ring polycyclic aromatic hydrocarbons and many substituted aromatic compounds, which include important pharmaceutical ring compounds. In all of the many reactions examined, metabolism appears to be initiated by monooxygenation reactions of the aromatic ring or a substituent. It is not known if this virtuoso metabolism of aromatic compounds is due to a single monooxygenase or multiple monooxygenases. Cell-free monooxygenase activity was demonstrated with naphthalene and was shown to be due to a cytochrome P450 type of monooxygenase(s). However, monooxygenase activity in the microsomal cell fraction was not successfully fractionated further to resolve the enzymological basis of the many different reactivities.

In addition to monooxygenase activity, which yields initial aromatic ring epoxides (Fig. 4.1), *C. elegans* contains enzyme activities which process epoxides and phenols to more water-soluble metabolites, and in this property, its activities resemble the metabolic transformations carried out by mammals (61). The startling observation was made that *C. elegans* metabolism of the strong carcinogen benzo[*a*]pyrene generates a series of metabolites that includes a single enantiomer of the 7,8-diol-9,10-epoxide of benzo[*a*]pyrene, which was shown to be the ultimate metabolite generating the carcinogenic state in rats (10). This, in turn, was consistent with the proposition that *C. elegans* could be used as a microbial model for the metabolism of carcinogens to learn about possible pathways of metabolic activation of these compounds (55).

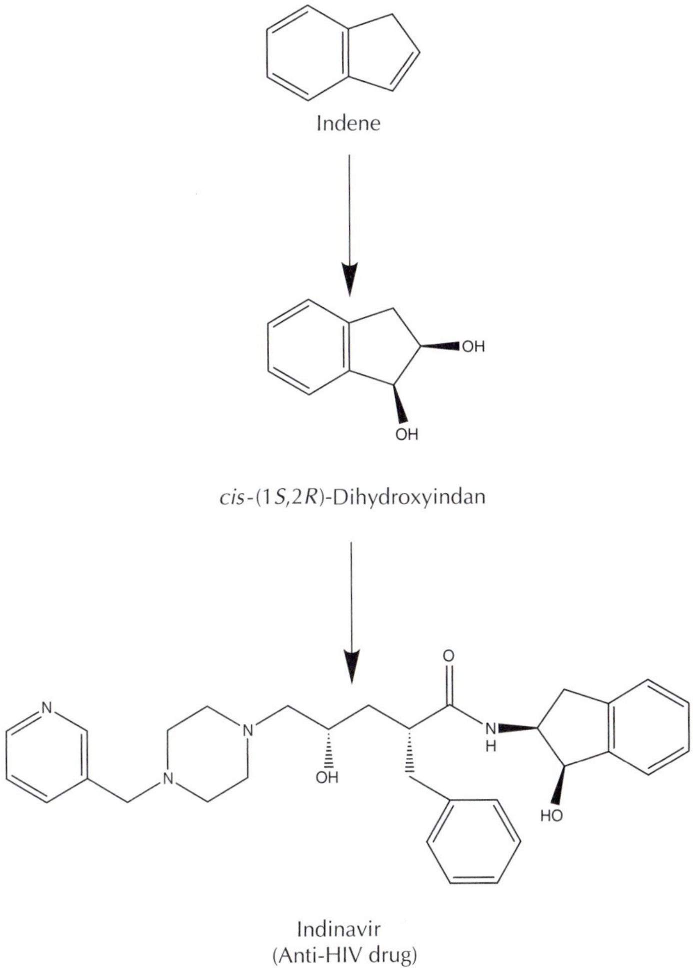

Indene

cis-(1*S*,2*R*)-Dihydroxyindan

Indinavir
(Anti-HIV drug)

Figure 4.6 Dioxygenase-catalyzed oxidation of indene to yield *cis*-(1*S*,2*R*)-dihydroxyindan, useful for the production of the anti-HIV drug indinavir.

Microbial Consortia in Biodegradation

It is sometimes stated, largely without proof, that most biodegradation in nature occurs via consortial metabolism, that is, by sequential metabolism in which part of the pathway is found in one microorganism and part in others. This is a difficult conjecture to prove or disprove and impossible to consider adequately without knowing the genetic and enzymological details of biodegradation derived from the thousands of studies conducted with microbial pure cultures.

It is important to consider studies of consortium-based systems conducted in the laboratory. What has been learned? First, stable mixed cultures are sometimes obtained from either batch enrichment cultures or continuous-culture techniques. These have often been found to be due to metabolite excretion causing cross feeding from one bacterial strain to another. For example, one partially characterized system involved a stable four-member community growing on the insecticide parathion. An organism identified as *Pseudomonas stutzeri* hydrolyzed parathion to *p*-nitrophenol, which yields no usable carbon or energy for that organism (17). The growth medium was found to contain *p*-nitrophenol, which was metabolized by a *P. aeruginosa* strain. The metabolites and lysis products of that strain fed the *P. stutzeri* strain and additional strains, one of which was a coryneform bacterium.

More details of a consortial catabolic system were obtained in a continuous-culture study using the herbicide Dalapon, or 2,2-dichloropropionic acid, as the growth-limiting carbon source. Pyruvate, the product of Dalapon dehalogenation, was also supplied at a growth-limiting concentration. Inoculation of such a chemostat with soil led to the prolonged establishment of a six-member microbial community consisting of several *Pseudomonas* strains, a *Flavobacterium* sp., and a fungus, *Trichoderma viride,* which adhered to the glass walls of the vessel and so likely avoided washout even with a lower growth rate (53). One of the most fascinating observations of that study was a sudden increase in the density of the population, which was accompanied by a great increase in Dalapon degradation activity by the *P. putida* strain. This changed both the overall growth rate and the population dynamics of the culture. The precise molecular events underlying the change were not revealed.

More recently, bacterial growth on the herbicide atrazine has been shown to be dependent on the genes *atzA, atzB,* and *atzC,* which encode hydrolytic enzymes which transform atrazine to cyanuric acid (6, 20, 51). Few bacteria catabolize atrazine, while many catabolize cyanuric acid, so these three genes will confer an atrazine-degrading phenotype on many soil bacteria. One atrazine-degrading bacterium, *Pseudomonas* sp. strain ADP, was shown to contain all three genes, *atzABC,* on a broad-host-range 96-kb plasmid (21). Several genera of bacteria, recently isolated from different regions around the world, contain virtually identical *atzABC* genes and grow on atrazine as their sole nitrogen source (19).

David Crowley and his coworkers obtained, by enrichment culture, a stable atrazine-catabolizing consortium which consisted of at least four distinct bacteria (2). None of the individual bacteria readily emerged as a pure culture growing on atrazine. This stable consortium was subsequently probed by PCR to learn the distribution of atrazine catabolism genes in the members of the consortium. Both genetic and biochemical data established that a *Clavibacter* sp. metabolized atrazine to *N*-ethylammelide (Fig. 4.7), which in turn was metabolized by *Pseudomonas* sp. strain CN (18). Another strain was able to grow on cyanuric acid. The role of the other uncharacterized bacterium or bacteria is proposed to be that of metabolizing the *N*-alkyl side chains which are liberated by the AtzB- and AtzC-catalyzed reactions (Fig. 4.7).

Global Biodegradation and the Supraorganism Concept

This chapter focuses on individual organisms that have been studied for their ability to biodegrade specific compounds, since that is how we have learned much about the underlying molecular events involved. However, when one wishes to consider the fate of compounds in the environment, to understand the evolution of catabolic pathways, or to follow gene transfer and dissemination, it is wise to look at the aggregate potential of all of the bacteria in a given environment. That is, it is often useful to consider the bacterial flora of the world as a "supraorganism" that metabolizes collectively, shares biodegradative genes, and evolves collectively to biodegrade

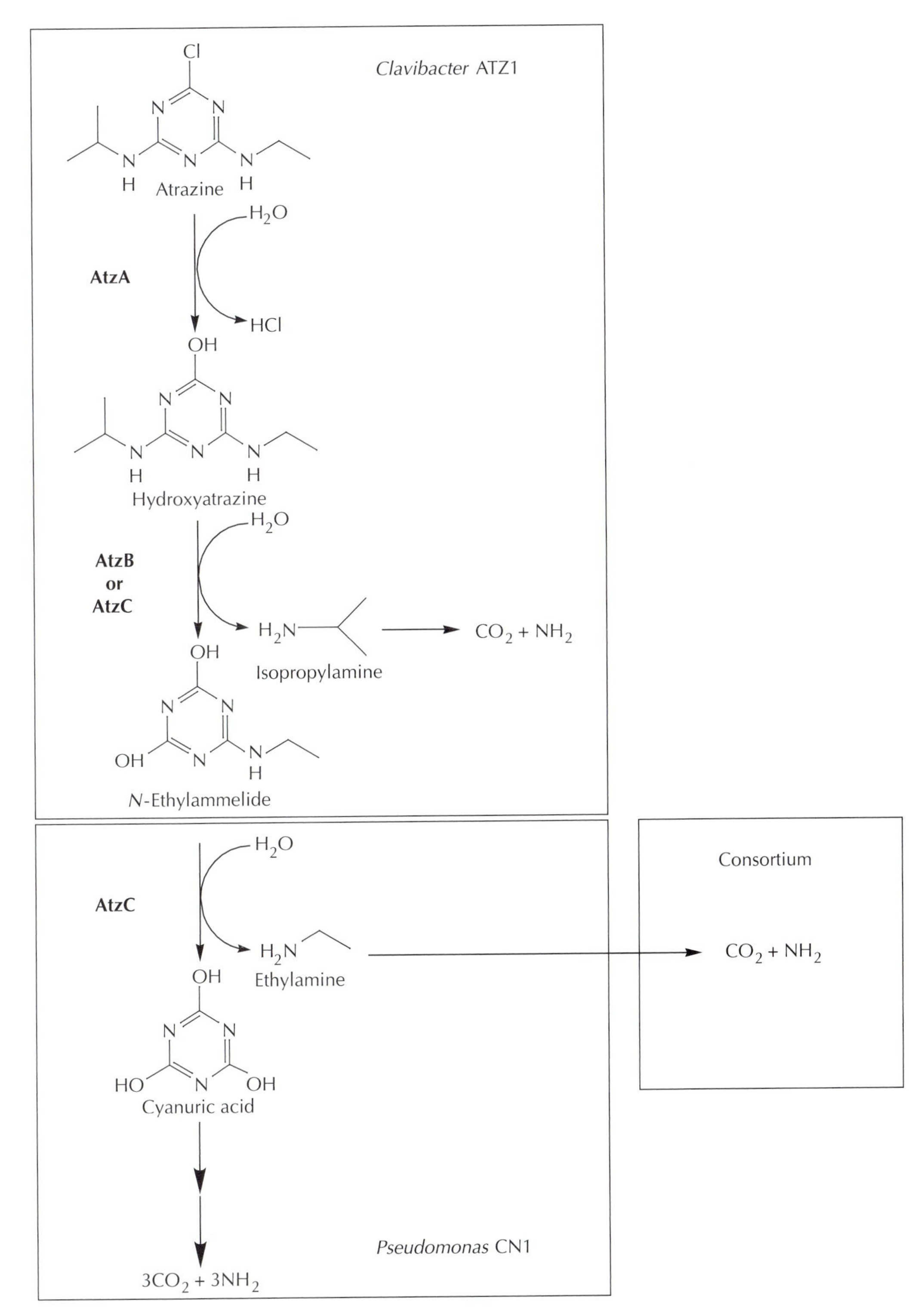

Figure 4.7 Consortial metabolism of atrazine, with reactions catalyzed by different bacteria in separate boxes.

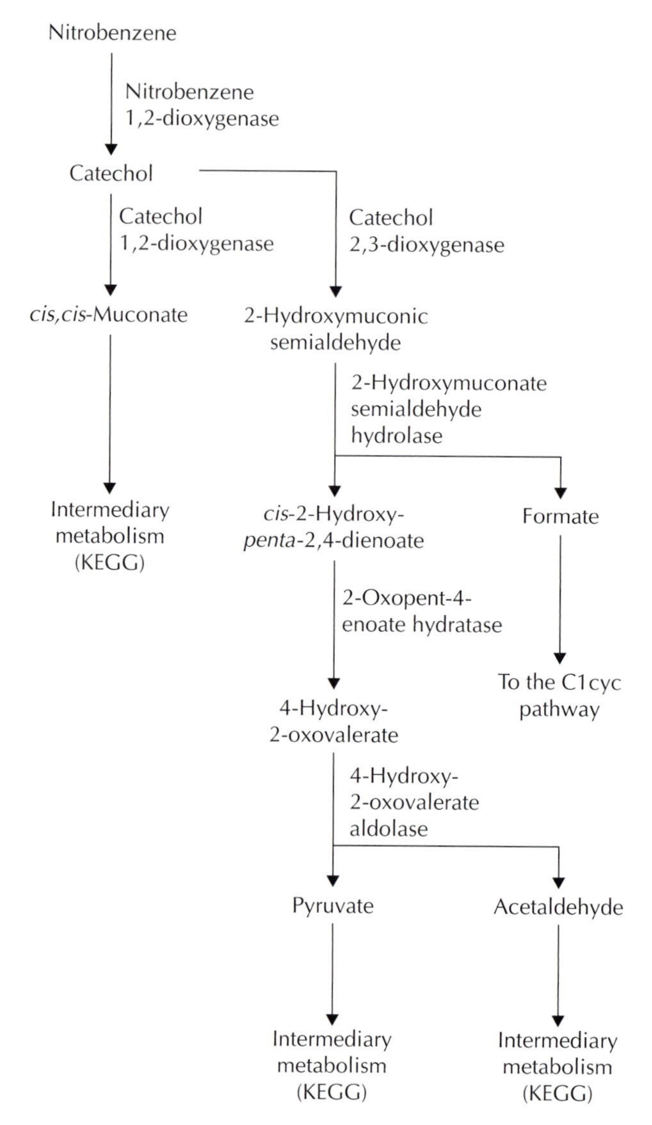

Figure 4.8 Output of the UM-BBD "create a pathway" function starting with nitrobenzene.

new compounds that enter an environmental niche. This view of the prokaryotic world as a supraorganism has been expressed previously in a different context (56), but it is particularly useful when thinking about biodegradation in the environment.

There is another value to thinking of biodegradation and biocatalysis reactions in this context, and that has to do with the way the information is best organized in a database format. The UM-BBD offers its information on the World Wide Web to scientists interested in applied biodegradation, basic aspects of biodegradation, metabolic engineering for biocatalysis, or other biotechnological applications. This information is best organized with a focus on individual reactions rather than on individual bacteria. Certain reactions [for example, mercury(II) reduction to mercury(0)] are known to be catalyzed by dozens of different bacteria. Moreover, initial reactions in some catabolic pathways are catalyzed by one known bacterium and subsequent reactions in the pathway are known to be catalyzed by different bacteria. This represents a "virtual world" of consortial metabolism that may or may not exist in the real world (but it represents the extent of known information).

This is an attractive way to view microbial catabolism. For example, it provides ideas as to what organisms and genes could complete a given pathway, both in nature and in engineered environments. A particular pathway, not known to exist in any natural organism, may thus be generated by genetic-engineering techniques. In this context, the reactions in the UM-BBD, linked by the CORE (Compound, Organism, Reaction, and Enzyme) object-oriented database structure, can be stitched together in novel combinations by the "create a pathway" function. An example is shown in Fig. 4.8. The interconnected pathways shown do not exist on any preconfigured page in the database. Rather, the user decides to retrieve all reactions stored in the database which connect to the product of a given reaction. This can yield a multibranched upside-down tree of catabolic metabolism, some of which may exist only on the user's computer screen. Linkage of the reactions may or may not be thermodynamically or practically feasible; for example, anaerobic and aerobic reactions may be alternately linked in a pathway. However, this puts reactions together in new ways and may provide thought-provoking new arrangements of reactions. In a sense, this is what nature does continually through the process of evolution, but in that case, we don't see the failures.

Summary

Historically, there has been a focus on a limited number of prokaryotes and fungi with respect to the biotransformation of diverse organic compounds. This has been due partly to methodological limitations and partly to people choosing to study organisms that others have studied. Now, anaerobic methods are well developed and the extraordinary diversity of microbes is more fully appreciated. It is becoming increasingly clear that metabolic diversity is distributed throughout many branches on the tree of life. Continuing studies in catabolic metabolism, augmented by whole-genome sequencing of diverse microbes, will add to our understanding of the catalytic potential of microbes.

References

1. **Alcalde, M., F. J. Plou, C. Andersen, M. T. Martin, S. Pedersen, and A. Ballesteros.** 1999. Chemical modification of lysine side chains of cyclodextrin glycosyltransferase from *Thermoanaerobacter* causes a shift from cyclodextrin glycosyltransferase to alpha-amylase specificity. *FEBS Lett.* **445:**333–337.

2. **Alvey, S., and D. E. Crowley.** 1996. Survival and activity of an atrazine-mineralizing bacterial consortium in rhizosphere soil. *Environ. Sci. Technol.* **30:**1596–1603.

3. **Barker, H. A.** 1940. Studies on the methane fermentation: IV. The isolation and culture of *Methanobacterium omelianskii*. *Antonie Leeuwenhoek* **6:**201–220.

4. **Bertoldo, C., F. Duffner, P. L. Jorgensen, and G. Antranikian.** 1999. Pullulanase type I from *Fervidobacterium pennavorans* Ven5: cloning, sequencing, and expression of the gene and biochemical characterization of the recombinant enzyme. *Appl. Environ. Microbiol.* **65:**2084–2091.

5. **Bok, J. D., D. A. Yernool, and D. E. Eveleigh.** 1998. Purification, characterization, and molecular analysis of thermostable cellulases CelA and CelB from *Thermotoga neapolitana*. *Appl. Environ. Microbiol.* **64:**4774–4781.

6. **Boundy-Mills, K. L., M. L. de Souza, R. T. Mandelbaum, L. P. Wackett, and M. J. Sadowsky.** 1997. The *atzB* gene of *Pseudomonas* sp. strain ADP encodes the second enzyme of a novel atrazine degradation pathway. *Appl. Environ. Microbiol.* **63:**916–923.

7. **Boyle, A. W., C. D. Phelps, and L. Y. Young.** 1999. Isolation from estuarine sediments of a *Desulfovibrio* strain which can grow on lactate coupled to the reductive dehalogenation of 2,4,6-tribromophenol. *Appl. Environ. Microbiol.* **65:**1133–1140.

8. **Brauman, A., J. A. Muller, J. L. Garcia, A. Brune, and B. Schink.** 1998. Fermentative degradation of 3-hydroxybenzoate in pure culture by a novel strictly anaerobic bacterium, *Sporotomaculum hydroxybenzoicum* gen. nov., sp. nov. *Int. J. Syst. Bacteriol.* **1:**215–221.

9. **Bryant, M. P., E. A. Wolin, M. J. Wolin, and R. S. Wolfe.** 1967. *Methanobacillus omelianskii,* a symbiotic association of two species of bacteria. *Arch. Mikrobiol.* **59:**20–31.

10. **Cerniglia, C. E., and D. T. Gibson.** 1980. Fungal oxidation of (+/−)-9,10-dihydroxy-9,10-dihydrobenzo[a]pyrene: formation of diastereomeric benzo[a]pyrene 9,10-diol 7,8-epoxides. *Proc. Natl. Acad. Sci. USA* **77:**4554–4558.

11. **Cerniglia, C. E., R. L. Hebert, P. J. Szaniszlo, and D. T. Gibson.** 1978. Fungal transformation of naphthalene. *Arch. Microbiol.* **117:**135–143.

12. **Cerniglia, C. E., and J. J. Perry.** 1972. Crude oil degradation by microorganisms isolated from the marine environment. *Z. Allg. Mikrobiol.* **13:**299–306.

13. **Chen, C. C., and J. Westpheling.** 1998. Partial characterization of the *Streptomyces lividans xlnB* promoter and its use for expression of a thermostable xylanase from *Thermotoga maritima. Appl. Environ. Microbiol.* **64:**4217–4225.

14. **Chi, Y. I., L. A. Martinez-Cruz, J. Jancarik, R. V. Swanson, D. E. Robertson, and S. H. Kim.** 1999. Crystal structure of the *beta*-glycosidase from the hyperthermophile *Thermosphaera aggregans:* insights into its activity and thermostability. *FEBS Lett.* **445:**375–383.

15. **Christiansen, N., B. K. Ahring, G. Wohlfarth, and G. Diekert.** 1998. Purification and characterization of the 3-chloro-4-hydroxy-phenylacetate reductive dehalogenase of *Desulfitobacterium hafniense. FEBS Lett.* **436:**159–162.

16. **Dagley, S.** 1975. A biochemical approach to some problems of environmental pollution, p. 81–138. *Essays in Biochemistry,* vol. 11. Academic Press, London, United Kingdom.

17. **Daughton, C. G., and D. P. Hsieh.** 1977. Parathion utilization by bacterial symbionts in a chemostat. *Appl. Environ. Microbiol.* **34:**175–184.

18. **de Souza, M. L., D. Newcombe, S. Alvey, D. E. Crowley, A. Hay, M. J. Sadowsky, and L. P. Wackett.** 1998. Molecular basis of a bacterial consortium: interspecies catabolism of atrazine. *Appl. Environ. Microbiol.* **64:**178–184.

19. **de Souza, M. L., J. Seffernick, B. Martinez, M. J. Sadowsky, and L. P. Wackett.** 1998. The atrazine catabolism genes *atzABC* are widespread and highly conserved. *J. Bacteriol.* **180:**1951–1954.

20. **de Souza, M. L., M. J. Sadowsky, and L. P. Wackett.** 1996. Atrazine chlorohydrolase from *Pseudomonas* sp. strain ADP: gene sequence, enzyme purification, and protein characterization. *J. Bacteriol.* **178:**4894–4900.

21. **de Souza, M. L., L. P. Wackett, and M. J. Sadowsky.** 1998. The *atzABC* genes encoding atrazine catabolism are located on a self-transmissible plasmid in *Pseudomonas* sp. strain ADP. *Appl. Environ. Microbiol.* **64:**2323–2326.

22. **Dettori, G., R. Grillo, P. Cattani, A. Calderaro, C. Chezzi, J. Milner, K. Truelove, and R. Sellwood.** 1995. Comparative study of the enzyme activities of *Borrelia burgdorferi* and other non-intestinal and intestinal spirochaetes. *New Microbiol.* **18:**13–26.

23. **Eaton, R. W.** 1997. *p*-Cymene catabolic pathway in *Pseudomonas putida* F1: cloning and characterization of DNA encoding conversion of *p*-cymene to *p*-cumate. *J. Bacteriol.* **179:**3171–3180.

24. **Fu, W., and P. Oriel.** 1999. Degradation of 3-phenylpropionic acid by *Haloferax* sp. D1227. *Extremophiles* **3:**45–53.

25. **Fu, W., and P. Oriel.** 1998. Gentisate 1,2-dioxygenase from *Haloferax* sp. D1227. *Extremophiles* **2:**439–446.

26. **Gibson, D. T., M. Hensley, H. Yoshioka, and T. J. Mabry.** 1970. Formation of (+)-*cis*-2,3-dihydroxy-1-methylcyclohexa-4,6-diene from toluene by *Pseudomonas putida. Biochemistry* **9:**1626–1630.

27. **Gibson, D. T., V. Mahadevan, D. Jerina, H. Yagi, and H. Yeh.** 1975. Oxidation of the carcinogens benzo(a)pyrene and benzo(a)anthracene. *Science* **189:**295–297.

*28. **Glazer, A. N., and H. Nikaido.** 1995. *Microbial Biotechnology: Fundamentals of Applied Microbiology.* W. H. Freeman and Company, New York, N.Y.

29. **Gray, K. A., O. S. Pogrebinsky, G. T. Mrachko, L. Xi, D. J. Monticello, and C. H. Squires.** 1996. Molecular mechanisms of biocatalytic desulfurization of fossil fuels. *Nat. Biotechnol.* **14:**1705–1709.

30. **Harper, D. B.** 1994. Biosynthesis of halogenated methanes. *Biochem. Soc. Trans.* **22:**1007–1011.

31. **Julliand, V., A. de Vaux, L. Millet, and G. Fonty.** 1999. Identification of *Ruminococcus flavefaciens* as the predominant cellulolytic bacterial species of the equine cecum. *Appl. Environ. Microbiol.* **65:**3738–3741.

32. **Kieslich, K.** 1998. General introduction to biocatalysis and screening, p. 3–11. *In* K. Kieslich, C. P. van der Beek, J. A. M. de Bont, and W. J. J. van den Tweel (ed.), *New Frontiers in Screening for Microbial Biocatalysis.* Elsevier Science, Amsterdam, The Netherlands.

33. **Kim, T. J., M. J. Kim, B. C. Kim, J. C. Kim, T. K. Cheong, J. W. Kim, and K. H. Park.** 1999. Modes of action of acarbose hydrolysis and transglycosylation catalyzed by a thermostable maltogenic amylase, the gene for which was cloned from a *Thermus* strain. *Appl. Environ. Microbiol.* **65:**1644–1651.

34. **Laskin, A. I., and D. C. White.** 1999. Preface to special issue on *Sphingomonas. J. Ind. Microbiol. Biotechnol.* **23:**231.

*35. **Laskin, A. I., and D. C. White (ed.).** 1999. Special issue on the genus *Sphingomonas. J. Ind. Microbiol. Biotechnol.* **23:**231–445.

*36. **Lazcano, A., and S. L. Miller.** 1996. The origin and early evolution of life: prebiotic chemistry, the pre-RNA world, and time. *Cell* **85:**793–798.

37. **Marr, E. K., and R. W. Stone.** 1961. Bacterial oxidation of benzene. *J. Bacteriol.* **81:**425–430.

38. **Matheron, C., A. M. Delort, G. Gaudet, and E. Forano.** 1998. In vivo ^{13}C NMR study of glucose and cellobiose metabolism by four cellulolytic strains of the genus *Fibrobacter. Biodegradation* **9:**451–461.

39. **Meckenstock, R. U., R. Krieger, S. Ensign, P. M. Kroneck, and B. Schink.** 1999. Acetylene hydratase of *Pelobacter acetylenicus.* Molecular and spectroscopic properties of the tungsten iron-sulfur enzyme. *Eur. J. Biochem.* **264:**176–182.

40. **Meissner, H., and W. Liebl.** 1998. *Thermotoga maritima* maltosyltransferase, a novel type of maltodextrin glycosyltransferase acting on starch and malto-oligosaccharides. *Eur. J. Biochem.* **258:**1050–1058.

41. **Morris, D. D., M. D. Gibbs, M. Ford, J. Thomas, and P. L. Bergquist.** 1999. Family 10 and 11 xylanase genes from *Caldicellulosiruptor* sp. strain Rt69B.1. *Extremophiles* **3:**103–111.

42. **Nagasawa, T., and H. Yamada.** 1990. Bioconversion of nitriles to amides and acids, p. 277–318. *In* D. A. Abramowicz (ed.), *Biocatalysis.* Van Nostrand Reinhold, New York, N.Y.

43. **Oren, A., P. Gurevich, and Y. Henis.** 1991. Reduction of nitrosubstituted aromatic compounds by the halophilic anaerobic eubacteria *Haloanaerobium prevalens* and *Sporohalobacter marismortui. Appl. Environ. Microbiol.* **57:**3367–3370.

44. **Pace, N. R.** 1997. A molecular view of microbial diversity and the biosphere. *Science* **276:**734–740.

45. **Palleroni, N.** 1984. Family I. *Pseudomonaceae,* p. 140–220. *In* N. R. Kreig (ed.), *Bergey's Manual of Systematic Bacteriology,* vol. 1. Williams and Wilkins, Baltimore, Md.

46. **Peterson, D. H., and H. C. Murray.** 1952. Microbiological oxygenation of steroids at carbon 11. *J. Am. Chem. Soc.* **74:**1871–1872.

47. **Raber, T., T. Gorontzy, M. Kleinschmidt, K. Steinbach, and K. H. Blotevogel.** 1998. Anaerobic degradation and transformation of *p*-toluidine by the sulfate-reducing bacterium *Desulfobacula toluolica. Curr. Microbiol.* **37:**172–176.

48. **Reichenbecher, W., and B. Schink.** 1999. Towards the reaction mechanism of pyrogallol-phloroglucinol transhydroxylase of *Pelobacter acidigallici. Biochim. Biophys. Acta* **1430:**245–253.

49. **Rocha, E. R., and C. J. Smith.** 1999. Role of the alkyl hydroperoxide reductase (*ahpCF*) gene in oxidative stress defense of the obligate anaerobe *Bacteroides fragilis. J. Bacteriol.* **181:**5701–5710.

50. **Rooney-Varga, J. N., R. T. Anderson, J. L. Fraga, D. Ringelberg, and D. R. Lovley.** 1999. Microbial communities associated with anaerobic benzene degradation in a petroleum-contaminated aquifer. *Appl. Environ. Microbiol.* **65:**3056–3063.

51. **Sadowsky, M. J., M. L. de Souza, Z. Tong, and L. P. Wackett.** 1998. AtzC is a member of the amidohydrolase protein superfamily and is homologous to other atrazine-metabolizing enzymes. *J. Bacteriol.* **180:**152–158.

52. **Schocke, L., and B. Schink.** 1998. Membrane-bound proton-translocating pyrophosphatase of *Syntrophus gentianae,* a syntrophically benzoate-degrading fermenting bacterium. *Eur. J. Biochem.* **256:**589–594.

*53. **Senior, E., A. T. Bull, and J. H. Slater.** 1976. Enzyme evolution in a microbial community growing on the herbicide Dalapon. *Nature (London)* **263:**476–479.

54. **Singer, M. E., S. M. Tyler, and W. R. Finnerty.** 1985. Growth of *Acinetobacter* sp. strain HO1-N on *n*-hexadecanol: physiological and ultrastructural characteristics. *J. Bacteriol.* **162:**162–169.

55. **Smith, R. V., and J. P. Rosazza.** 1974. Microbial models of mammalian metabolism. Aromatic hydroxylation. *Arch. Biochem. Biophys.* **161:**551–558.

56. **Sonea, S., and M. Panisset.** 1983. *A New Bacteriology.* Jones and Bartlett Publishers, Inc., Boston, Mass.

57. **Thauer, R. K., K. Jungermann, and K. Decker.** 1977. Energy conversion in chemotrophic anaerobic bacteria. *Microbiol. Rev.* **41:**100–180.

58. **Treadway, S. L., K. S. Yanagimachi, E. Lankenau, P. A. Lessard, G. Stephanopoulos, and A. J. Sinskey.** 1999. Isolation and characterization of indene bioconversion genes from *Rhodococcus* strain I24. *Appl. Microbiol. Biotechnol.* **51:**786–793.

59. **van de Pas, B. A., H. Smidt, W. R. Hagen, J. van der Oost, G. Schraa, A. J. Stams, and W. de Vos.** 1999. Purification and molecular characterization of *ortho*-chlorophenol reductive dehalogenase, a key enzyme of halorespiration in *Desulfitobacterium dehalogenans. J. Biol. Chem.* **274:**20287–20292.

60. **von Wintzingerode, F., B. Selent, W. Hegemann, and U. B. Gobel.** 1999. Phylogenetic analysis of an anaerobic, trichlorobenzene-transforming microbial consortium. *Appl. Environ. Microbiol.* **65:**283–286.

61. **Wackett, L. P., and D. T. Gibson.** 1982. The metabolism of xenobiotic compounds by enzymes in cell extracts of the fungus *Cunninghamella elegans. Biochem. J.* **205:**117–122.

*62. **Whitman, W. B., D. C. Coleman, and W. J. Wiebe.** 1998. Prokaryotes: the unseen majority. *Proc. Natl. Acad. Sci. USA* **95:**6578–6583.

63. **Wiegel, J., X. Zhang, and Q. Wu.** 1999. Anaerobic dehalogenation of hydroxylated polychlorinated biphenyls by *Desulfitobacterium dehalogenans*. *Appl. Environ. Microbiol.* **65:**2217–2221.

64. **Wilson, E. O.** 1992. *The Diversity of Life*. Harvard University Press, Cambridge, Mass.

65. **Yamada, H., and M. Kobayashi.** 1996. Nitrile hydratase and its application to industrial production of acrylamide. *Biosci. Biotechnol. Biochem.* **60:**1391–1400.

66. **Zengler, K., H. H. Richnow, R. Rossello-Mora, W. Michaelis, and F. Widdel.** 1999. Methane formation from long-chain alkanes by anaerobic microorganisms. *Nature* **401:**266–269.

5

Organic Functional Group Diversity: the Unity of Biochemistry Is Dwarfed by Its Diversity

Science attempts to confront the possible with the actual, and by so doing must inevitably renounce a unified world concept.

— *Francois Jacob as paraphrased by Miriam Rothschild*

. . . it is part of the adventure of science to try to find a limitation in all directions. . . .

— *Richard P. Feynman*

The variety of structures that carbon can form is due in part to its unparalleled ability to form single, double and triple bonds with another carbon atom or with other atoms.

— *P. W. Atkins*

The compounds to be considered in the context of biodegradation and biocatalysis are the known organic molecules, an ever-expanding set of over 10 million compounds. Most natural products and industrial chemicals are thought to be subjected to biodegradation at some time and some place on Earth. Yet, the vast majority of known compounds have not been studied with respect to microbial metabolism. This knowledge gap will be discussed further in chapter 9, which deals with predicting biodegradation. In this chapter, we will consider how to classify organic compounds by their organic functional groups.

Because the number of compounds subjected to microbial metabolism is enormous, it is imperative to consider substructures of larger chemical entities and how they are catabolized. We will not typically focus on individual atoms but on small clusters of atoms which constitute a definable chemical functional group with a definable set of chemical reactivities. In this way, the compounds of biodegradation can be defined systematically. Then, we can examine what is known about the biodegradation of individual organic functional groups. It is important to specify both what is known and what is not.

We will also examine how compounds are often subdivided in the scientific literature that deals with biodegradation and suggest that this clustering is incongruent with current biological and biochemical knowledge.

For example, one common fallacy is to categorize many industrial chemicals as xenobiotic, or foreign to the biosphere. This presupposes that most industrial compounds have functional groups which do not occur naturally in biological systems. This supposition is proven to be false if evidence of most industrial chemical functional groups is found in the biological world. In this chapter, we present evidence from the natural-product literature to support the idea that many functional groups typically referred to as xenobiotic are in fact found in the natural biological world, exclusive of organic synthesis.

The corollary of this widespread natural distribution of "exotic" chemicals is that biodegradation potential is almost certainly much greater than what has been discovered. Not only is there enormous microbial taxonomic diversity (discussed in chapter 4), there is likely to be a corresponding catalytic diversity in microbes still waiting to be discovered. To start the present discussion, we will consider biodegradation globally by defining it as a key component of the Earth's carbon cycle.

Microbial Global Cycling of the Elements

> In all this time we have been looking at the forest, we have seen only half of what is going on, exactly half. . . . for each bit of growth, there must be the same amount of decay. . . .
>
> — *Mr. Feynman to his son, Richard P. Feynman*

In its simplest conceptual form, the Earth's carbon cycle consists of carbon dioxide, organic molecules, and their interconversion (Fig. 5.1). In many textbooks, the biological catalysts driving the carbon cycle are depicted as

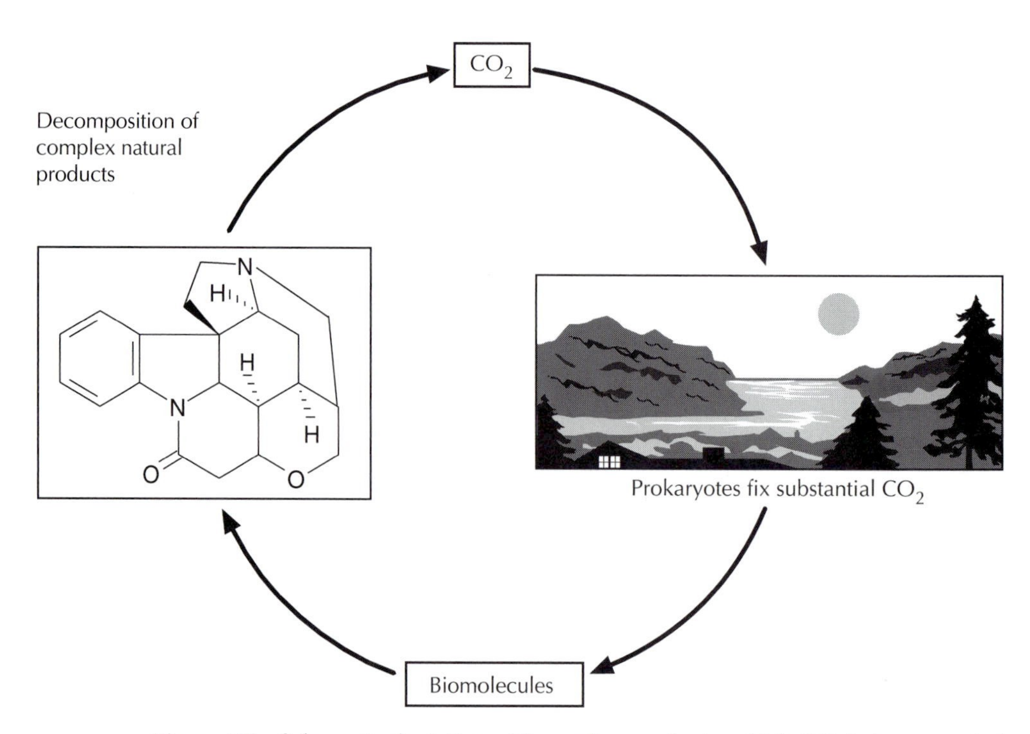

Figure 5.1 Schematic depiction of the carbon cycle, in which CO_2 is incorporated into organic compounds by plants and microbial autotrophs and complex organic compounds are oxidized back to CO_2, largely by prokaryotes.

plants; large land animals, such as humans; and macroscopic sea animals. While all of these organisms do impact the biospheric carbon cycle, such depictions belie the more important role of microorganisms in cycling carbon. Clearly, microorganisms decompose the greatest diversity of organic compounds to yield CO_2 and fix a significant fraction of CO_2 deposited in the atmosphere and hydrosphere (Fig. 5.1). Complex organic matter is in turn recycled to CO_2 via microbial oxidation reactions. Microbial oxidative metabolism is collectively diverse because biological systems and natural diagenic processes make millions of unique organic compounds.

The carbon structures made by biological systems present an imposing challenge for microbial catalysts. For example, in the C_1 (single-carbon) cycle, CO_2 is reduced to methane, CH_4, by anaerobic methanogenic bacteria (Fig. 5.2). Methane is very unreactive; the strength of the C-H bond is 117 kcal/mol. An impressive catalyst is required to break the C-H bond without combusting it via a flame to carbon dioxide and water. Methanotrophic bacteria accomplish this feat and transform methane to methanol stoichiometrically. Methanol is further oxidized in three more discrete enzyme-catalyzed two-electron steps to yield CO_2. In the process, chemical energy

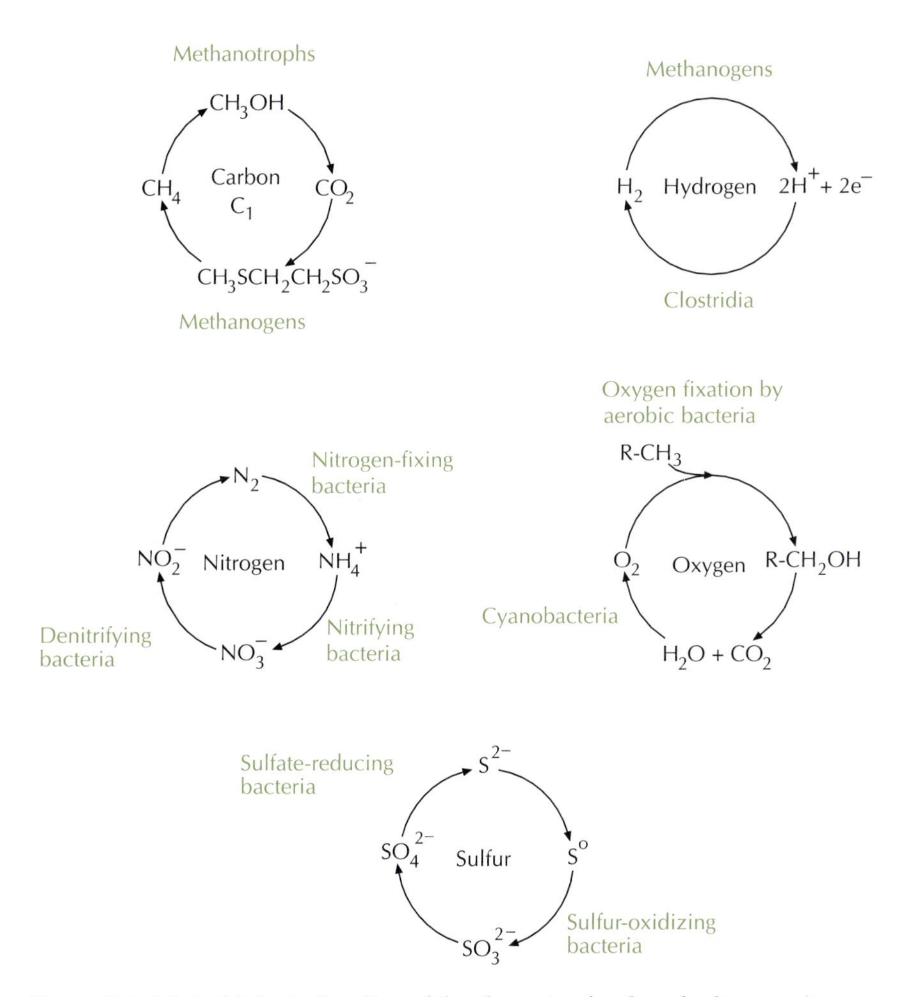

Figure 5.2 Major biological cycling of the elements of carbon, hydrogen, nitrogen, oxygen, and sulfur is catalyzed largely by microorganisms. Some representative microorganisms catalyzing certain transformations are shown.

is trapped in the form of ATP. By contrast, the combustion of methane in a flame releases virtually all the chemical energy as heat.

More complex organic molecules also present a great challenge to microbial metabolism. For example, the major structural polymers of plants are cellulose and lignin. Lignin is more heterogeneous because it derives from radical-based biosynthetic reactions. Lignin biodegradation is typically the rate-determining step for the metabolic recycling of woody plant material. This is because its biodegradation is relatively slow and it is metabolized by only a limited number of microorganisms. Ligninolytic, or lignin-degrading, fungi and bacteria have been mentioned previously in chapters 1 and 4. The biodegradation process is fundamentally a radical depolymerization. The reactions release organic-acid products that can be further metabolized by a wider range of bacterial species, thus recycling the carbon to CO_2.

Elements do not, of course, cycle in isolation. For example, the carbon and oxygen cycles intersect at key points (Fig. 5.2). Oxygen atoms are incorporated into organic molecules via direct insertion of one or both oxygen atoms from dioxygen, gaseous O_2 (Fig. 5.2), or metabolism involving dehydrogenation followed by hydration, as discussed in chapter 2. The first type of reaction is catalyzed by monooxygenases and dioxygenases, respectively. This will be discussed further in chapter 8, dealing with metabolic logic.

Facts and Fallacies: Natural Products versus Synthetic Chemicals and Their Biodegradation

A common practice in correlating the types of organic molecule with their ease of biodegradation is to define them as (i) natural products or (ii) industrial chemicals. The logic underlying that division is that microbes have been exposed to natural products for millions of years and thus have evolved enzymes to catabolize them for obtaining carbon and energy. This argument has some merit. Most natural products do not accumulate in the environment and thus are likely readily biodegraded. Consider that plants are known to make thousands of different compounds that fall into some of the following broad categories: (i) monoterpenes, (ii) sesquiterpenes, (iii) diterpenoids, and (iv) flavonoids (Table 5.1). Some plants make up to several percent of their leaf mass with such compounds, and the leaves are biodegradable.

Let's consider another class of molecules that long predate the evolutionary entry of humans on Earth. Aromatic hydrocarbons (Fig. 5.3) are formed naturally by diagenesis, or slow burning. Organic matter of diverse origins slowly rearranges its carbon-carbon and carbon-hydrogen bonds to form the most thermodynamically stable structures, resonance-stabilized aromatic rings (3). These compounds are abundant in liquid petroleum and are found extensively cross-linked in coal. Modern civilization combusts these compounds in factories and motor vehicles to use the energy for mechanical work or heat. Long before the advent of human societies, microorganisms learned to combust these same compounds to carbon dioxide with the capture of carbon and chemical energy. It is well known that fossil fuel spills in soil are among the most amenable of chemical wastes to biological cleanup. These treatments usually consist of stimulating bacteria already present at the site, since the capacity for catabolism of alkanes and aromatic-ring compounds is a very widespread microbial phenotype in virtually all well-aerated soils.

Table 5.1 Major classes of secondary plant compounds[a]

Class	Approx. no. of structures	Distribution	Physiological activity
Nitrogen compounds			
Alkaloids	6,500	Widely in angiosperms, especially in root, leaf, and fruit	Many toxic and bitter tasting
Amines	100	Widely in angiosperms, often in flowers	Many repellent smelling; some hallucinogenic
Amino acids (nonprotein)	400	Especially in seeds of legumes but relatively widespread	Many toxic
Cyanogenic glycosides	30	Sporadic, especially in fruit and leaf	Poisonous (as HCN)
Glucosinolates	75	*Cruciferae* and 10 other families	Acrid and bitter (as isothiocyanates)
Terpenoids			
Monoterpenes	1,000	Widely in essential oils	Pleasant smells
Sesquiterpenes	1,500	Mainly in *Compositae* but increasingly in other angiosperms	Some bitter and toxic; also allergenic
Diterpenoids	2,000	Widely, especially in latex and plant resins	Some toxic
Saponins	600	In over 70 plant families	Hemolyze blood cells
Limonoids	100	Mainly in *Rutaceae*, *Meliaceae*, and *Simaroubaceae*	Bitter tasting
Cucurbitacins	50	Mainly in *Cucuribitaceae*	Bitter tasting and toxic
Cardenolids	150	Especially common in *Apocyanaceae*, *Asclepiadaceae*, and *Scrophulariaceae*	Toxic and bitter
Carotenoids	500	Universal in leaf; often in flower and fruit	Colored
Phenolics			
Simple phenols	200	Universal in leaf; often in other tissues as well	Antimicrobial
Flavonoids	4,000	Universal in angiosperms, gymnosperms, and ferns	Often colored
Quinones	800	Widely, especially *Rhamnaceae*	Colored
Other			
Polyacetylenes	650	Mainly in *Compositae* and *Umbelliferae*	Some toxic

[a]Data from reference 14 by permission of the publisher, Academic Press.

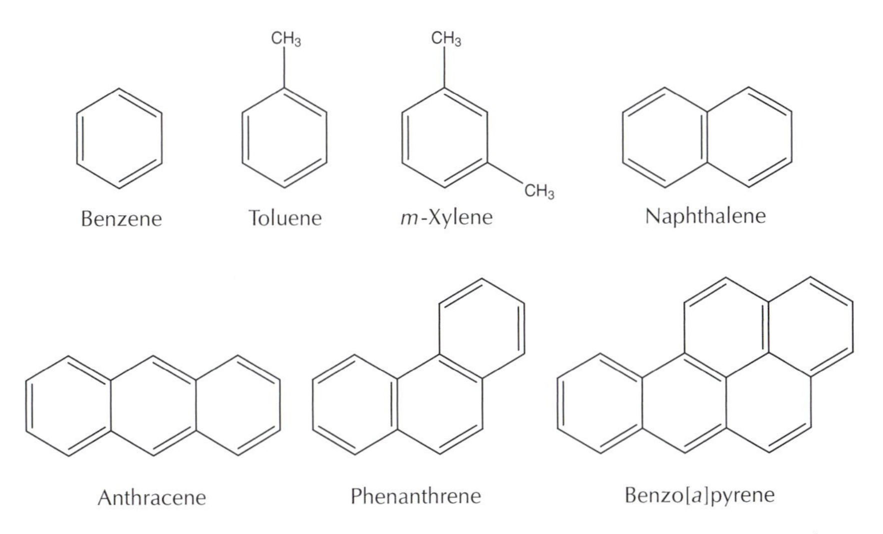

Figure 5.3 Thousands of aromatic hydrocarbons are made naturally; several representative structures are shown.

These examples and other similar observations might lead one to embrace the notion that it is useful to break compounds into the natural and synthetic groups of compounds to best think about their susceptibilities to microbial metabolism. But every classification system gives rise to a particular way to think about a subject, and this one generates a presupposition that there is some fundamental difference, in general, between compounds made in nature and those made in a factory. This idea should be examined more closely.

The biodegradation literature largely deals with the catabolism of industrial chemicals, because the distribution of those chemicals is much more frequently regulated by government agencies. Hence, research on the biodegradation of synthetic chemicals is readily funded by government and industry sources. Research on the biodegradation of natural plant products, by contrast, is not pursued nearly so vigorously. Thus, while the biodegradation of common biological molecules, like amino acids, is known to be fast, there is a dearth of information on the biodegradation of complex natural products, of which over 100,000 are biosynthesized by plants alone. It is largely assumed that these compounds are very biodegradable because they are products of nature. However, this has not been well studied biochemically.

Another bias that creeps into the biodegradation literature is the frequent assumption that industrial compounds are harmful to humans. The introductory sections of many biodegradation research articles justify the experiments on the grounds that it is important to understand the environmental fate of the compound(s). This is true in the sense that government regulatory agencies increasingly demand environmental fate data. However, the fates of such compounds, especially at trace levels, may not be important for ecosystem or human health reasons. In this context, a review of the biodegradation literature indicates that statements about the carcinogenicity or toxicity of the compound(s) under study are often not supported by a critical evaluation of the relevant toxicological research.

An Organic Functional Group Classification

The classification of organic compounds used in this book will not be based on the origin of the molecule. Instead, the compound's structure alone will determine its class. As pointed out previously, over 10 million distinct organic chemical structures are described in the literature. With such complexity of structures, it is necessary to follow Albert Einstein's edict, "Make things as simple as possible, but not simpler."

The focus here will be on each compound's substructure, that is, a fragment which possesses cohesive and defining chemical properties. This is commonly referred to as an organic functional group. Organic functional groups are defined in many beginning and advanced textbooks dealing with organic chemistry as a way to understand the type of chemical reactions given compounds undergo. This same logic will be applied here to microbial biocatalytic reactions.

Many organic functional groups are acted upon by the individual enzymes of biodegradation. A given enzyme typically transforms one organic functional group in isolation, for example, oxidizing an alcohol to an aldehyde or hydrolyzing an amide to a carboxylic acid and an amine. The rest of the structure helps determine if the molecule is bound productively by the enzyme, but is often unmodified in the reaction. In this context, if a

The Use of the Term *Xenobiotic* with Respect to Biodegradation Should Be Discontinued

It is worth considering the historic roots of the term xenobiotic. Its common scientific usage in the 1940s and 1950s differentiated the mammalian metabolism of drugs and environmental pollutants from intermediary metabolism, as discussed in chapter 4. There are many differences. Intermediary metabolism transforms compounds that are commonly biosynthesized by biological systems, for example, amino acids, nucleotides, and lipids, via enzymes that are typically specific to a given metabolic step. In contrast, mammalian xenobiotic metabolism transforms compounds that are often not biosynthesized by mammals. Moreover, the metabolism is carried out via highly nonspecific enzymes, like cytochrome P450 monooxygenases and glutathione transferases (see Fig. 4.1). Individual isozymes of those enzyme classes have been shown to transform dozens, if not hundreds, of substrates. The reactions typically functionalize a drug or pollutant compound with the effect of removing it from the body, a process that is often known as detoxification.

More recently, the idea has emerged that these systems evolved to handle toxic chemicals in plants that were ingested by insects and animals (14). Plants produce many toxins to protect themselves against ingestion, and insects and animals have evolved enzymatic protective mechanisms against them. This concept is now well established, with many examples of cytochrome P450- and glutathione transferase-catalyzed reactions in which the substrates are toxic natural plant products. Humans, too, likely suffered from the toxic effects of the many plants they ingested while foraging for food prior to the invention of agriculture.

Are these natural products also xenobiotic chemicals? Well, to humans they are. They are compounds not biosynthesized by animals in most cases, and they can be very toxic. For example, fluoroacetate, produced by 34 species of *Gastrolobium* and *Oxylobium* plants, is toxic to most mammals at a concentration of 1 μg per kg of body weight, making it one of the most toxic organic chemicals known. Is fluoroacetate a xenobiotic chemical to the bacteria known to catabolize it? It cannot be considered to be foreign because of its overall chemical structure and its structural similarity to acetate and glycollate, the latter being the enzymatic hydrolysis product. Moreover, fluoroacetate is produced by biological systems, and it is metabolized by bacteria, which use it as a sole source of carbon and energy. Thus, the term xenobiotic, or foreign to biological systems, would be inappropriate to apply to fluoroacetate.

Fluoroacetate is a natural product; what about compounds common to nature and industry? Chloromethane is produced by the chemical industry and is considered to be a xenobiotic compound (Fig. 5.4). It is a reactive alkylating agent and thus tightly regulated. Fifteen years ago, it was discovered that chloromethane is biosynthesized by soil fungi in massive quantities, 10^{10} lb per year (15). In fact, its biosynthetic production dwarfs its manufacture by the chemical industry, at 5×10^7 lb per year. Is chloromethane xenobiotic, or foreign to biological systems? Not surprisingly, numerous soil prokaryotes are capable of catabolizing chloromethane as their sole source of carbon and energy (44).

The examples cited above are not rare. There are thousands of natural products that have both biosynthetic and human synthetic origins. In one striking recent review article, over 1,000 naturally produced organohalogen compounds were described (12). Some of them are halogenated solvents and halophenol toxins that are made as antibiotics by marine algae to kill organisms in their immediate environment.

Another similar class of compounds is the aromatic hydrocarbons, thousands of which are products of nature and are used industrially. Are they xenobiotic? Polycyclic aromatic hydrocarbons are found in Martian meteorites (28) and interstellar space. Everywhere in the universe where life might exist, and on many other planets as well, aromatic hydrocarbons might be present.

In light of these observations, we will not refer to the substrates of microbial biodegradation as being xenobiotic. The additional point is made below that biological systems produce a wider array of organic functional groups than is generally appreciated.

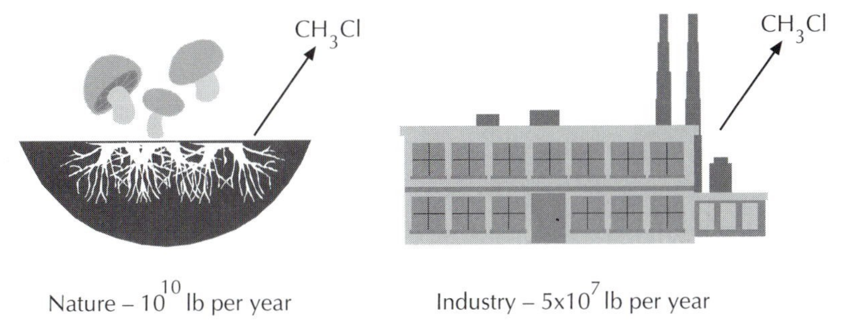

Figure 5.4 Chloromethane is made principally by microorganisms; 0.5% of the world output is made industrially.

common functional group, such as a primary alkyl alcohol, is present on a nonpolymeric environmental chemical, a bacterial enzyme likely exists somewhere in nature which can bind the substrate and oxidize the alcohol functional group. In this way, we can use organic functional groups as the elements for describing an unbiased, formal compilation of organic compounds in order to depict their biotransformations.

We recognize that what a functional group is connected to can influence the course of its metabolism. For example, a chlorine atom bonded to a carbon atom differs substantially depending on whether the carbon atom is part of an alkane, alkene, or aromatic-ring system. Moreover, the number of chlorine atoms connected to a single carbon atom will alter its metabolic fate. For example, chloromethane is metabolized via different biochemical mechanisms than carbon tetrachloride. These features highlight the richness of organic and biological chemistry. Despite this, it is still advantageous to focus on functional groups and describe their individual metabolisms as a way to begin to simplify the highly complex array of molecules and reactions that nature has evolved. Chapter 8 on metabolic logic deals largely with the metabolism of single organic functional groups or combinations of at most two groups bonded together. After that chapter, it is possible to consider combinations of functional groups and deduce rules for their metabolism when connected. This is the treatment we will follow in chapter 9, which deals with predicting microbial biocatalytic pathways.

Organic Functional Groups Found in Nature

It is not a new idea to formally classify biodegradation by using organic functional groups as the elements, but this treatment has often been biased by the idea that certain functional groups are xenophores and hence resist biodegradation (1). A common functional group said to be a xenophore is a halogen atom bonded to carbon. This is because thousands of new organohalogen compounds were synthesized, and some were commercialized, during the 1930s, 1940s, and 1950s. Subsequent to that, some organohalide pesticides and solvents were found to be toxic to mammals and birds. Thus, the biodegradation of compounds containing one or more carbon-halogen bonds became of great interest.

However, the approach of calling a halogen substituent a xenophore is too simplistic. As noted above, over 1,000 halogenated natural products are known (12). When biodegradation information is available for some of the heavily biosynthesized compounds, for example, chloromethane, the data indicate that they are readily biodegraded. It is true that the presence of many halogen substituents on a molecule may slow its biodegradation. However, the presence of multiple methyl groups on an alkyl carbon or an aromatic ring will also slow its biodegradation, and a methyl group is not considered to be a xenophore.

In Table 5.2, we show that many organic functional groups described in the biodegradation literature as being xenobiotic are in fact biosynthesized by one or more biological systems. The organisms producing some of the rarer functional groups are plants or members of the bacterial genus *Streptomyces*. The diversity of chemical groups provided by nature is clearly impressive. Of those shown in Table 5.2 alone, only about one-half are associated with the compounds of intermediary metabolism. Compounds containing the other organic groups, when function has been assigned, often serve the organism in signaling or protection from predators. Nature is rife with chemical warfare, and the best weapons are chemical bullets, compounds that react with sensitive targets in the enemy. For example, strategically generated isothiocyanates or ketenes are as toxic to sensitive organisms as military nerve gases are to humans. As indicated above, fluoroacetate is exceedingly toxic. It is a Trojan horse toxin, chemically very similar to acetate, and it reacts like acetate to yield fluorocitrate, which shuts down energy metabolism in many cells. Other fluorinated compounds, like fluorohistidine, are made by *Streptomyces* to induce potential competitors to incorporate it into proteins, with lethal effects.

In warfare, the development of a weapon invites the victim to develop an antiweapon system. In nature, toxic chemicals are combated with enzymes that transform the molecule into a nontoxic analog. For example, wyerone acid, a plant-produced fungal toxin containing an alkyne functional group, is detoxified by the fungus *Botrytis cinerea,* which is pathogenic for the plant (14). The detoxification is attributed to enzymatic reduction of the alkyne. So, just as novel groups are biosynthesized, corresponding enzymes evolve to transform them and thus provide a defense against the toxin. With wyerone acid, the enzymatic transformation is the four-electron reduction of the alkyne functional group into the much more common alkane functional group. In another example, bacteria have been identified which metabolize the alkyne functional group via a hydratase which yields a ketone functional group (36). Over 1,000 acetylinic natural products have been identified, and thus, it is not surprising that multiple biochemical mechanisms exist for their metabolism.

Ring Compounds Found in Nature

Table 5.2 depicts several ring structures, but they represent only a small fraction of the total number found in nature. Those shown in Table 5.2 include (i) monocyclic benzenoid rings, (ii) polycyclic benzenoid rings, (iii) hydrocarbon ring assemblies, (iv) cyclic aliphatic rings, (v) bicyclic aliphatic rings, (vi) tricyclic aliphatic rings, and (vii) small heterocyclic rings (azetidine, aziridine, epoxide, oxetane, and endoperoxide rings).

The last group, heterocyclic rings, can be expanded to include an immense class of molecules. Biological organic heterocyclic-ring compounds commonly contain nitrogen, oxygen, or sulfur atoms, but still, an enormous variety of structures are attainable. For example, the nitrogen heterocyclic-ring structures found in some natural products are shown in Fig. 5.5. The figure shows core ring structures which form one part of much larger molecules, the names of which are given. The complete structures are not shown for clarity and to conserve space. The reader can consult various chemical-structure indexes, *The Merck Index,* or other resources if the entire structure is needed.

Some of the nitrogen ring structures shown in Fig. 5.5 are fairly common in biological systems, for example, the imidazole ring found in the amino acid histidine. Others, such as the 1,2,4-triazine ring, are much less commonly known, being found in the antibiotic fervenulin, which is produced by a relatively small number of bacteria. Only single-ring heterocyclic structures are shown, but fused, multi-ring heterocyclic nitrogen compounds are reasonably common. For example, chemical structures not adequately covered here include those where the heteroatom is at a bridgehead, that is, shared between different fused rings. This structural configuration is common in the alkaloid family, for which over 10,000 different

Table 5.2 Organic functional groups found in natural products

Name of functional group	Structure	Approx. no. in nature	Example	Function or source	Reference
Acetal	$-C(OR)_2$ (with single bond below C)	Many	Cellulose	Plants	
Acid anhydride, mixed	$-C(=O)-O-P(=O)(OH)-O^-$	Several	Glyceroylphosphate	Energy metabolism	
Alcohol	$R-OH$	Many	Glycerol	Many	
Aldehyde	$-C(H)=O$	Many	Glyceraldehyde 3-phosphate	Many	
Alkane, primary	$-CH_3$	Many	Crude oil	Diagenesis	
Alkane, secondary	$-CH_2-C$	Many	Crude oil	Diagenesis	
Alkane, tertiary	$-C(C)(H)-C$	Many	Crude oil	Diagenesis	
Alkane, quaternary	$-C(C)(C)-C$	Many	Crude oil	Diagenesis	
Alkene	$-C=C-$	Many	Unsaturated fatty acids	Many	
Alkyne	$-C\equiv C-$	More than 1,000	Wyerone acid	Phytoalexin	14
Allene	$-C=C=C-$	Many	Fucoxanthin	Carotenoid	23
Amide	$-C(=O)-NH_2$	Many	Proteins	Many	
Amidine	$-C(=NH_2^+)-NH_2$	Many	Purine biosynthon	Metabolic intermediate	
Amine, primary	$-C-NH_3^+$	Many	Amino sugar	Cell surface	
Amine, secondary	$-NH_2^+$ (with single bond above N)	Many	Alkaloids	Toxins	
Amine, tertiary	$-NH^+$ (with single bonds above and below N)	Many	Alkaloids	Toxins	

Amine, quaternary	$-N^{+}-$	Many	Alkaloids	Toxins	
Antimony, organo	—C—Sb—	Several	Stibinolipids	Algae	43
Arsenic, organo	—C—As-C—	At least several	Arsenobetaine	Algae	5
Azetidine	HN—CH_2 / —C—C— (H H)	At least several	Azetidine-2-carboxylic acid	*Streptomyces*	10
Aziridine	NH / —C—C— (H H)	At least several	Mitomycin A	Antibiotic	47
Azo	—C—N = N—C—	Several	Azoserine	*Streptomyces*	11
Azoxy	O^- / —C—N = N —C—	Few	Elaiomycin	Antibiotic	29
Benzenoid ring		Many	Phenylalanine	Amino acid	
Bicycloaliphatic ring		Many	Camphor	Plant terpenes	14
Boron, organo	$-C-B^{-}-$	One	Boromycin	Antibiotic	24
Bromine, organo	—C—Br	Many	Bromoform	Marine algae	13
Carboxylate	O ‖ $-C-O^{-}$	Many	Formic acid	Ant irritant	14
Chlorine, organo	—C—Cl	Hundreds	Chloromethane	Soil fungi	15
Cobalt, organo	—C—Co	At least one	Methyl-B_{12}	Methanogens	
Cyanate	—O—C≡N	One	Cyanate		2
Cyano	—C≡N	At least dozens	Cyanogenic glycosides	Toxins	14
Cyanohydrin	OH / —C—C≡N (H)	Many	Mandelonitrile	Insect defense	8

(continued next page)

Table 5.2 *continued*

Name of functional group	Structure	Approx. no. in nature	Example	Function or source	Reference
Cycloalkane		Many	Crude oil	Diagenesis	
Diazo	—C = N^+ = N^-	Several	Thrazarine	Antibiotic	18
Diselenide	—C—Se-Se-C—	Several	tRNA	Anaerobic bacteria	40
Disulfide	—C—S—S—C—	Thousands	Glutathione disulfide	Redox balance	
Endoperoxide	O—O (each O bonded to a C; both C joined through C_n)	Many	Ascaridole	Anthelminthic from chenopodium oil	14
Epoxide	—C—C— (both C bridged by O)	Many	Jurocimene 2	Insect hormone	14
Ester, carboxylate	—C(=O)—O—R	Many	Fatty acyl esters	Many	
Ether	R—O—R	Many	Ether phospholipids	Methanogens	
Fluorine, organo	—C—F	Moderate	Fluorothreonine	Antibiotic	34
Fused polycyclic aromatic		Many	Anthracene	Diagenesis	
Germanium, organo	—C—Ge— (Ge with bonds up and down)	Several	Dimethylgermanium	Natural waters	26
Hemiacetal	—C(OH)—O—R	Many	Glucose	Sugars, all organisms	
Hydrazide	—C(=O)—NH—NH_2	Several	XK-90	Antibiotic	42
Hydrazine	—C—NH—NH_2	Numerous	4-(Hydroxymethyl)-phenylhydrazine	*Agaricus*	45
Hydrazo	—C—NH—NH—C—	Several	Spinamycin	Antifungal	29
Hydrocarbon ring assembly		Several	Biphenyl	Diagenesis	

Hydroperoxide	$-C(=O)-O-OH$	Several	Linolic acid hydroperoxide	Many	35
Hydroxylamino	$-C-NH-OH$	At least several	Hydroxylaminobenzene	Nitrobenzene metabolite	30
Hydroxamate	$-C(=O)-NH-OH$	Many	Pseudobactins	Metal chelation	20
Imine	$R-C=N-$	Many	Formylglycinamidine	Biosynthesis	
Iodo	$-C-I$	At least several	Thyroxin	Hormone	
Isothiocyanate	$R-N=C=S$	At least several	Allyl isothiocyanate	Toxins, cabbage	14
Ketene	$=C=C=O$	Few	Epoxidation of acetylenes	Plant protection (transiently generated)	
Ketone	$-C=O$	Many	Pyruvate	Intermediary metabolism	
Lactone	$(H_2C)_n$–C=O ring closed via H_2C-O	Many	Many	Intermediary metabolism	
Lead, organo	$-C-Pb-$	At least one	Tetramethyl-lead	Bacteria	37
Nickel, organo	$-C-Ni$	At least one	Methyl-Ni-F430	Methanogenesis	
Nitrate ester	$-C-O-NO_2$	Several	Nitrated intermediary compounds	Nitration of intermediary compounds via nitric oxide	46
Nitro	$-C-NO_2$	At least several	Nitroglycosides	*Astragalus*	14
Nitroso	$-C-NO$	At least several	Viridomycins	*Streptomyces*	
N-Nitroso	$-C-N-NO$	Several	*N*-Nitrosodimethylamine	Mammals	1
Oxetane	(four-membered ring with O)	Several	Taxol	*Taxus brevifolia*	41
Oximino	$=N-OH$	Several	Indoleacetaldoxine	Plants	19
Peroxyester	$-C(=O)-O-O-C-$	Some	Epiplakinic acid methyl ester	Sea sponge	17

(continued next page)

Table 5.2 *continued*

Name of functional group	Structure	Approx. no. in nature	Example	Function or source	Reference
Phosphate, organo	$-C-O-P(=O)(O^-)-O^-$	Many	Glucose 6-phosphate	Bacteria	
Phosphinate	$-C-P(=O)(O^-)-C-$	At least one	Phosphinothricin	Antibiotic	38
Phosphoanhydride	$-C-O-P(=O)(O^-)-O-P(=O)(O^-)-OH$	Many	Adenosine diphosphate	Bacteria	
Phosphonamide	$-C-N(H)-P(=O)(O^-)-O-$	At least several	Phosphocreatine	Muscle tissue	
Phosphonate	$-C-P(O^-)(OH)=O$	Many	Phospholipids	*Tetrahymena*	16
S-Selanylsulfide	$-C-S-Se^-$	One	Carbon monooxide dehydrogenase	Bacteria	6
Selenoether	$-C-Se-C-$	Few	Selenocystathionine	Metabolic intermediate	39
Selenol	$-C-Se^-$	Many	Selenocysteine	Bacteria, humans	39
Selenourea	$-N-C(=Se)-N-$	Few	Selenouridine	tRNA	39
Spiro rings	C_n(C, C)C(C, C)$C_{n'}$ (spiro)	At least several	Sanglifehrins	*Streptomyces*	7
Sulfate ester	$-C-O-SO_3^-$	Many	Chondroitin sulfate	Connective tissue	
Sulfenic acid	$-C-S-OH$	Several	Cysteine sulfenic acids	Bacterial enzymes	4

Sulfinate	O ‖ —C—S—O^-	At least one	Cysteine sulfinic acids	Nitrile hydratase	31
Sulfone	O ‖ —C—S—C— ‖ O	Many	Dibenzothiophene sulfone	Metabolic intermediate	33
Sulfonate	O ‖ —C—S—O^- ‖ O	Many	Taurine	Bacteria	
Sulfonium cation	C— | —C—S^+—C—	Many	*S*-Adenosylmethionine	Many	
Sulfoxide	O ‖ —C—S—C—	Many	Methionine sulfoxide	Bacteria	
Telluroether	—C—Te-C—	One	Dimethyltelluride	*Penicillium*	9
Thioamide	S ‖ —C—NH_2	Several	Sulfinemycin	Antibiotic	25
Thiolester	O ‖ —C—C—S—	Many	Acyl-coenzyme A	Many	
Thiol	—C—SH	Many	Cysteine	Proteins	
Tin, organo	—C—Sn	At least one	Trimethyltin	Bacteria	48
Tricycloaliphatic ring	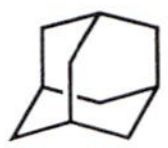	At least a few	Adamantane	Petroleum	
Trisulfide	—C—S—S—S—C—	Several	Methylallyl trisulfide	Garlic	27
Thiosulfinate	O ‖ —C—S—S—	At least one	Methylmethanothiosulfinate	Cabbage	22
Thiosulfonate	O ‖ —C—S—S— ‖ O	At least one	Methylmethanothiosulfonate	Cabbage	22

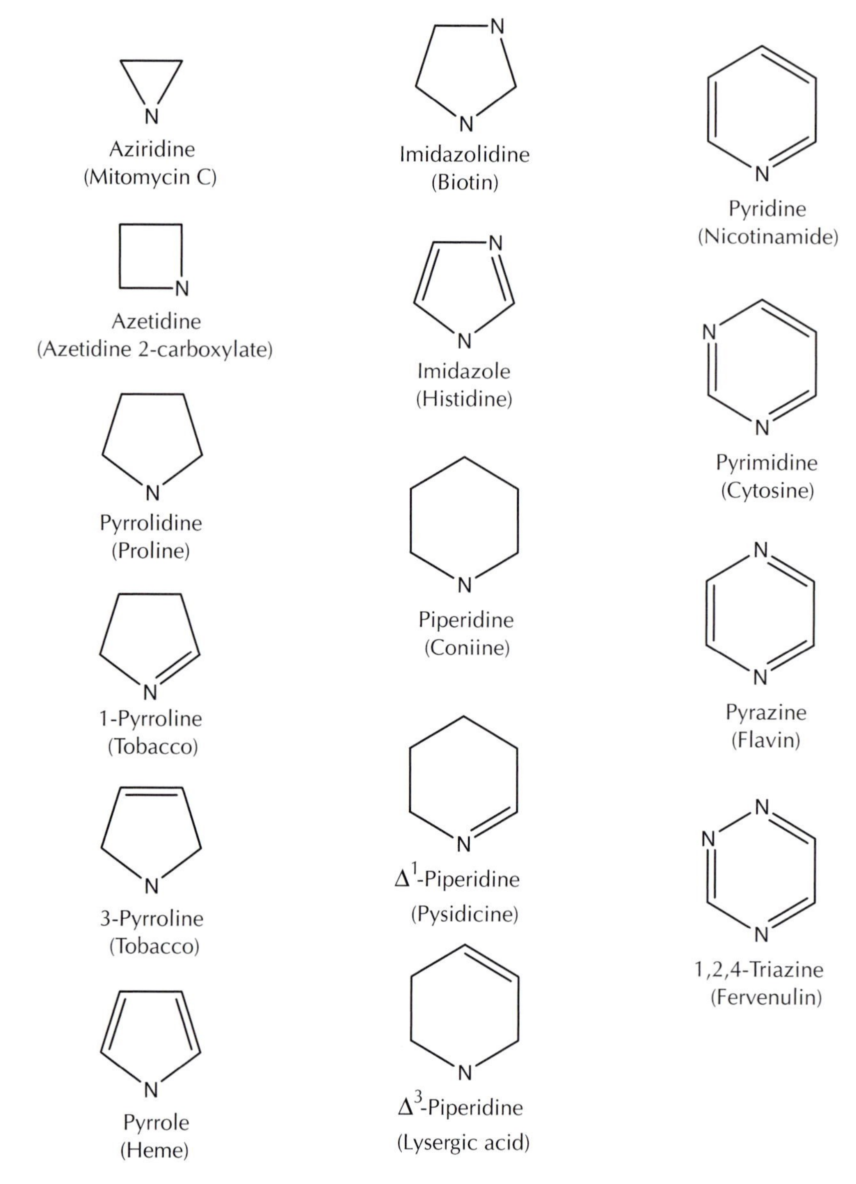

Figure 5.5 Some representative nitrogen heterocycles found in natural products.

structures are known (Table 5.1). The diversity of naturally occurring nitrogen heterocyclic-ring structures is truly staggering.

Figure 5.6 shows a corresponding set of naturally occurring oxygen heterocyclic-ring structures. Generally, oxygen-containing rings are not as widespread in intermediary compounds as the nitrogen heterocyclic compounds. Nonetheless, some plants produce significant quantities of compounds containing oxygen heterocyclic rings as pigments (cyanin red) and toxin protectants (psoralens). As for the nitrogen heterocycles, only the basic ring structure is shown; some of the natural products listed are much larger molecules comprising many additional organic functional groups.

Figure 5.7 shows naturally occurring sulfur and mixed-heteroatom ring structures. Some are found in intermediary metabolism and are thus widespread throughout the biosphere, for example, tetrahydrothiophene, found

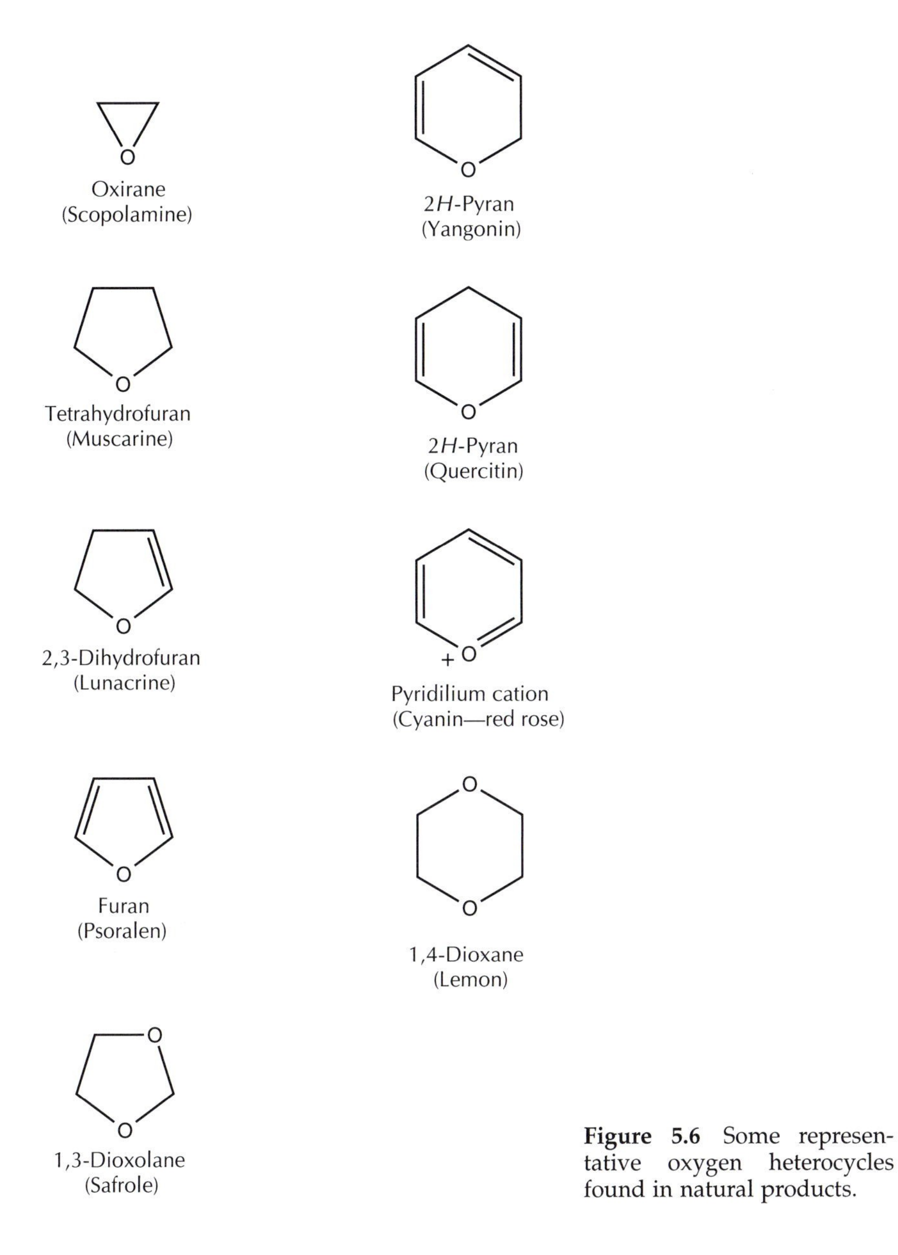

Figure 5.6 Some representative oxygen heterocycles found in natural products.

in the vitamin biotin, and the thiazole ring found in the vitamin thiamine. Others are found in more specialized metabolism, for example, the isoxazolidine ring found in the antibiotic cycloserine. Other mixed heterocyclic-ring structures have been identified in combustion mixtures, for example, in roasted meats. Note that natural combustion processes occurred long before humans walked the Earth. It is well documented that forest fires have occurred cyclically, caused primarily by lightning strikes, for millions of years. Thus, natural combustion to generate complex ring structures has continually occurred on the Earth's surface and has generated structures of the types shown in Fig. 5.7.

The occurrence of natural cycloaliphatic-ring compounds is shown in Table 5.2. Compounds such as steroids with five- and six-member cycloaliphatic rings are well known, but the diversity of cycloaliphatic-ring sizes found in natural products is quite impressive. Some of those ring

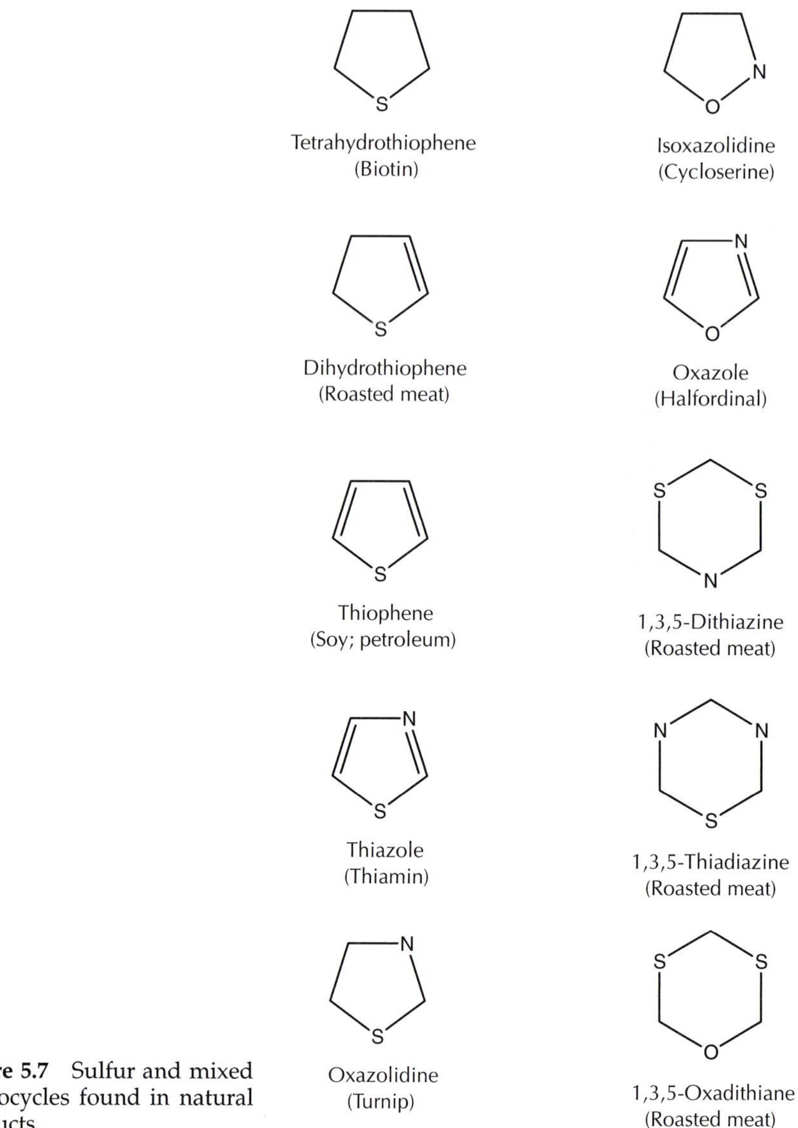

Figure 5.7 Sulfur and mixed heterocycles found in natural products.

structures, and the compounds they compose, are shown in Table 5.3. Rings containing from 3 to 17 carbon atoms are depicted, many of them found in plant oils and fragrances. The permethrins, with a C_3 cycloalkane ring, are an emerging class of natural insecticides. Microbial catabolism of cycloaliphatic-ring compounds and the metabolic logic underlying these reactions are discussed in more detail in chapter 8.

Organic Functional Groups: What Is Known with Respect to Biodegradation and Microbial Biocatalysis?

The main point of this chapter has been to highlight the impressive diversity of biologically relevant chemical structures and to begin to provide a framework for categorizing their microbial metabolism. The structural fragments represented in this chapter are all around us, in the air we breathe, the food we eat, and the flowers we admire. We know that the macroscopic

Table 5.3 Naturally produced cycloalkane ring compounds

Name of function	Structure	Approx. no. in nature	Example	Function or source	Reference
Cyclopropyl		At least several	Permethrins	Natural insecticides	
Cyclobutyl		At least several	Caryophyllene	Oil of clove	32
Cyclopentyl	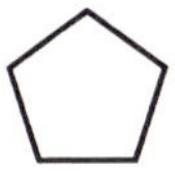	Many	Testosterone	Many steroids	
Cyclohexyl		Many	Testosterone	Many steroids	
Cycloheptanyl		At least several	Thujaplicine	Fungicide of red cedar	32
Cyclononyl	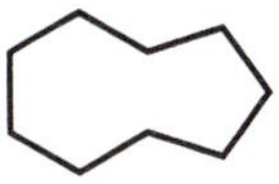	At least one	Carophyllene	Oil of clove	32
Cycloundecanyl	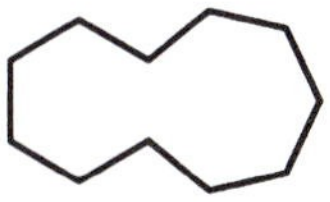	At least one	Humulene	Oil of clove	32
Cyclopentadecanyl	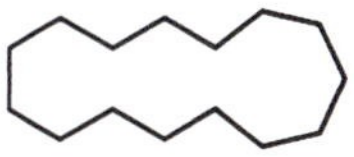	At least one	Muscone	Musk deer	14
Cyclohexadecanyl		At least one	Ambrettolide	Ambrette seed	32
Cycloheptadecanyl	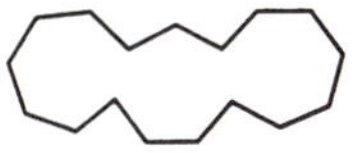	At least one	Civetone	Civet cat	14

materials composed of them are recycled in nature; not just a part, but all of the macroscopic structures. There is no reported crisis due to accumulation of the red cyanin pigment from the rose or the phosphonic acid lipids of the protozoa. These materials become part of the cycles of nature, and this requires that they be biodegraded. Since these functional groups and ring structures are biosynthesized in many environments, their corresponding biodegradation must also be occurring in these environments. It is unlikely that thermal or photocatalytic decomposition would account for more than a small amount of organic-matter turnover.

Given this enormous array of natural-product functional-group diversity, how many functional groups have been studied experimentally with respect to biodegradation? The developers of the University of Minnesota Biocatalysis/Biodegradation Database (UM-BBD) asked that question several years ago and have continually asked it as new compounds are chosen to be depicted in the database. The priority in adding new compounds has been to choose those with functional groups not previously represented in the UM-BBD.

As of 31 December 1999, all of the organic functional groups, and their metabolism by microorganisms, shown in Fig. 5.8, were covered in the UM-BBD. This collection of 47 functional groups represents approximately half of those observed in the biosphere, based on the compilation of Table 5.2, which shows 93 different functional groups. More strikingly, the UM-BBD covers only a very small subset of the cycloaliphatic-ring structures (Table 5.3) and heterocyclic-ring structures (Fig. 5.5 to 5.7) observed in nature.

The UM-BBD does not contain all the biodegradation pathways described in the scientific literature. However, the UM-BBD was developed with the intent of covering as many different types of functional groups as

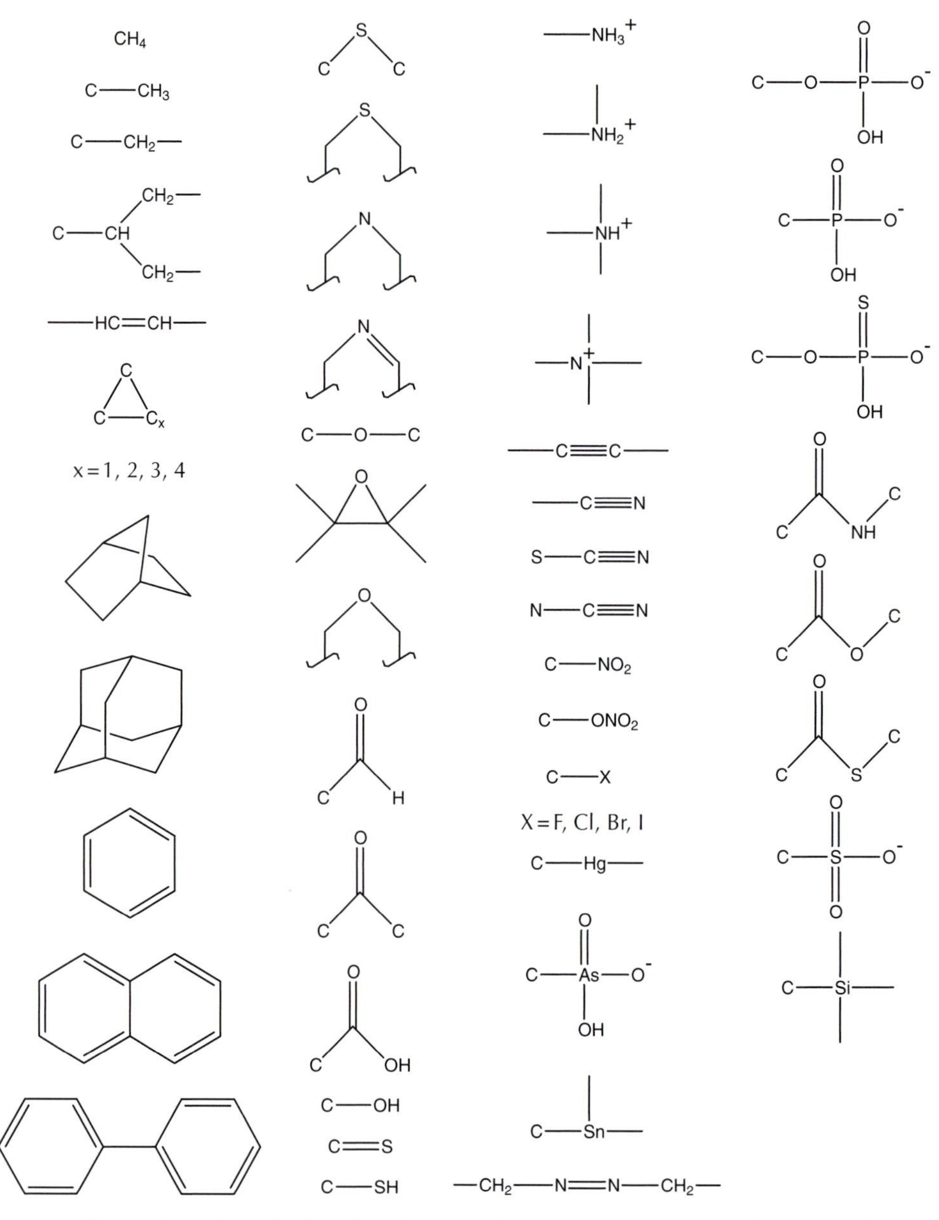

Figure 5.8 Organic functional groups known to undergo transformation by microbes and represented in the UM-BBD.

possible. In this context, the absence of this metabolism is more reflective of a lack of scientific information than the incompleteness of the database. Since compounds containing these functional groups are almost surely biodegradable, it is necessary to continually look for new microbial biocatalysis. This will be one of the goals of functional genomics studies, to identify the many unknown genes emerging in genome-sequencing projects involving soil prokaryotes. This will be discussed further in chapter 11 on genomics.

Summary

The diversity of organic compounds is effectively infinite, and over 10 million compounds are currently described in the chemical literature. Understanding the microbial metabolism of such a broad range of compounds necessitates an efficient categorization of microbial reactions. This chapter discusses an organization based on organic functional group analysis. This approach is common in the organic chemistry literature, where the number of functional groups has been thought to be much greater than that found in biological systems. Here, it was demonstrated that a surprising number of organic functional groups are also found in natural products. There is also an enormous array of organic ring structures biosynthesized by microorganisms and plants. Given that these compounds have likely been present in the environment for millions of years, microbes almost surely have evolved enzymes for their catabolism. There is important work to do in informatics, categorizing the known reactions via a functional-group approach, and experimental science, uncovering new microbial metabolism for the biosynthesis and catabolism of previously unstudied functional groups.

References

1. **Alexander, M.** 1994. *Biodegradation and Bioremediation,* p. 159–176. Academic Press, San Diego, Calif.

2. **Anderson, P. M., Y. C. Sung, and J. A. Fuchs.** 1990. The cyanase operon and cyanate metabolism. *FEMS Microbiol. Rev.* **7:**247–252.

*3. **Blumer, M.** 1976. Polycyclic aromatic compounds in nature. *Sci. Am.* **234:**35–45.

4. **Claiborne, A., J. I. Yeh, T. C. Mallett, J. Luba, E. J. I. Crane, V. Charrier, and D. Parsonage.** 1999. Protein-sulfenic acids: diverse roles for an unlikely player in enzyme catalysis and redox regulation. *Biochemistry* **38:**15407–15416.

5. **Dill, K., and E. L. McGowen.** 1994. The biochemistry of arsenic, bismuth and antimony, p. 695–713. *In* S. Patai (ed.), *The Chemistry of Organic Arsenic, Antimony and Bismuth.* John Wiley and Sons, Chichester, United Kingdom.

6. **Dobbek, H., L. Gremer, O. Meyer, and R. Huber.** 1999. Crystal structure and mechanism of CO dehydrogenase, a molybdo iron-sulfur flavoprotein containing *S*-selanylcysteine. *Proc. Natl. Acad. Sci. USA* **96:**8884–8889.

7. **Fehr, T., J. Kallen, L. Oberer, J. J. Sanglier, and W. Schilling.** 1999. Sanglifehrins A, B, C and D, novel cyclophilin-binding compounds isolated from *Streptomyces* sp. A92-308110. II. Structure, elucidation, stereochemistry and physicochemical properties. *J. Antibiot.* (Tokyo) **52:**474–479.

8. **Ferris, J. P.** 1983. Biological formation and metabolic transformations of compounds containing the cyano group, p. 325–340. *In* S. Patai and Z. Rappaport (ed.), *The Chemistry of Triple-Bonded Functional Groups.* John Wiley and Sons, Chichester, United Kingdom.

9. **Fleming, R. W., and M. Alexander.** 1972. Dimethylselenide and dimethyltelluride formation by a strain of *Penicillium. Appl. Microbiol.* **24:**424–429.

10. **Formica, J. V., and M. A. Apple.** 1976. Production, isolation, and properties of azetomycins. *Antimicrob. Agents Chemother.* **9:**214–221.

11. **Franklin, T. J., and G. A. Snow.** 1975. *Biochemistry of Antimicrobial Action.* Halsted Press, London, United Kingdom.

12. **Gribble, G. W.** 1992. Naturally occurring organohalogen compounds—a survey. *J. Natl. Prod.* **55:**1353–1395.

13. **Gschwend, P. M., J. K. MacFarlane, and K. A. Newman.** 1985. Volatile halogenated organic compounds related to seawater from temperate marine microalgae. *Science* **227:**1033–1035.

*14. **Harbourne, J.** 1988. *Ecological Biochemistry,* 3rd ed. Academic Press, New York, N.Y.

15. **Harper, D. B.** 1994. Biosynthesis of halogenated methanes. *Biochem. Soc. Trans.* **22:**1007–1011.

16. **Hori, T., M. Horiguchi, and A. Hayashi.** 1984. *Biochemistry of Natural C-P Compounds.* Maruzen Ltd., Kyoto, Japan.

17. **Horton, P. A., R. E. Longley, M. Kelly-Borges, O. J. McConnell, and L. M. Ballas.** 1994. New cytotoxic peroxylactones from the marine sponge, *Plakinastrella onkodes. J. Nat. Prod.* **57:**1374–1381.

18. **Kameyama, T., A. Takahashi, H. Matsumoto, S. Kurasawa, M. Hamada, Y. Okami, M. Ishizuka, and T. Takeuchi.** 1988. Thrazarine, a new antitumor antibiotic. I. Taxonomy, fermentation, isolation and biological properties. *J. Antibiot.* (Tokyo) **41:**1561–1567.

19. **Kato, Y., K. Nakamura, H. Sakiyama, S. G. Mayhem, and Y. Asano.** 2000. Novel heme-containing lyase, phenylacetaldoximine dehydratase from *Bacillus* sp. strain OxB-1: purification, characterization, and molecular cloning of the gene. *Biochemistry* **39:**800–809.

20. **Khalil-Rizvi, S., S. I. Toth, D. van der Helm, H. Vidavsky, and M. L. Gross.** 1997. Structures and characteristics of novel siderophores from plant deleterious *Pseudomonas fluorescens* A225 and *Pseudomonas putida* ATCC 39167. *Biochemistry* **36:**4163–4171.

21. **Kurobane, I., L. Dale, and L. C. Vining.** 1987. Characterization of new viridomycins and requirements for production in cultures of *Streptomyces griseus. J. Antibiot.* (Tokyo) **40:**1131–1139.

22. **Kyung, K. H., and H. P. Fleming.** 1997. Antimicrobial activity of sulfur compounds derived from cabbage. *J. Food Prot.* **60:**67–71.

23. **Lander, S. P.** 1982. Naturally occurring allenes, p. 681–703. *In* S. R. Lander (ed.), *The Chemistry of Allenes.* Academic Press, New York, N.Y.

*24. **Laskin, A. I., and H. A. Lechevalier.** 1977. *Handbook of Microbiology,* vol. 9, part A. *Antibiotics.* CRC Press, Boca Raton, Fla.

25. **Lee, T. M., M. M. Siegel, G. O. Morton, J. J. Goodman, R. T. Testa, and D. B. Borders.** 1995. Sufinemycin, a new antihelmintic antibiotic: fermentation, isolation and structure determination. *J. Antibiot.* (Tokyo) **48:**282–285.

26. **Lewis, B. L., and H. P. Meyer.** 1993. Biogeochemistry of methylgermanium species in natural waters, p. 79–99. *In* H. Sigel and A. Sigel (ed.), *Metal Ions in Biological Systems,* vol. 29. Marcel Dekker, New York, N.Y.

27. **Lim, H., K. Kubota, A. Kobayashi, T. Seki, and T. Ariga.** 1999. Inhibitory effect of sulfur-containing compounds in *Scorodocarpus borneenis* Becc. on the aggregation of rabbit platelets. *Biosci. Biotechnol. Biochem.* **63:**298–301.

28. **McCay, D. S., E. K. J. Gibson, K. K. Thomas-Keprta, H. Vali, C. S. Romanek, S. J. Clemett, X. D. F. Chillier, C. R. Maechling, and R. N. Zare.** 1996. Search for past life on Mars: possible relic biogenic activity in martian meteorite ALH84001. *Science* **273:**924–930.

29. **Miyadera, T.** 1975. Biological formation and reactions of hydrazo, azo and azoxy group, p. 495–539. *In* S. Patai (ed.), *The Chemistry of Hydrazo, Azo and Azoxy Groups*. John Wiley and Sons, Chichester, United Kingdom.

30. **Nishino, S. F., and J. C. Spain.** 1993. Degradation of nitrobenzene by a *Pseudomonas pseudoalcaligenes*. *Appl. Environ. Microbiol.* **59:**2520–2525.

31. **Nojiri, M., H. Nakayama, M. Odaka, M. Yohda, K. Takio, and I. Endo.** 2000. Cobalt-substituted Fe-type nitrile hydratase of *Rhodococcus* sp. N-771. *FEBS Lett.* **465:**173–177.

32. **Noller, C. R.** 1957. *Chemistry of Organic Compounds*. Saunders, Philadelphia, Pa.

33. **Oldfield, C., O. Pogrebinsky, J. Simmonds, E. S. Olson, and C. F. Kulpa.** 1997. Elucidation of the metabolic pathway for dibenzothiophene desulphurization by *Rhodococcus* sp. strain IGTS8 (ATCC 53968). *Microbiology* **143:**2961–2973.

34. **Reid, K. A., J. T. Hamilton, R. D. Bowden, D. O'Hagan, L. Dasaradhi, M. R. Amin, and D. B. Harper.** 1995. Biosynthesis of fluorinated secondary metabolites by *Streptomyces cattleya*. *Microbiology* **141:**1385–1393.

35. **Rocha, E. R., and C. J. Smith.** 1999. Role of the alkyl hydroperoxide reductase (*ahpCF*) gene in oxidative stress defense of the obligate anaerobe *Bacteroides fragilis*. *J. Bacteriol.* **181:**5701–5710.

36. **Rosner, B. M., F. A. Rainey, R. M. Kroppenstedt, and B. Schink.** 1997. Acetylene degradation by new isolates of aerobic bacteria and comparison of acetylene hydratase enzymes. *FEMS Microbiol. Lett.* **148:**175–180.

37. **Schmidt, U., and F. Huber.** 1976. Methylation of organolead and lead(II) compounds to $(CH_3)_4Pb$ by microorganisms. *Nature* **259:**157–158.

38. **Schwartz, D., J. Recktenwald, S. Pelzer, and W. Wohlleben.** 1998. Isolation and characterization of the PEP-phosphomutase and the phosphonopyruvate decarboxylase genes from the phosphinothricin tripeptide producer *Streptomyces viridochromogenes* Tu494. *FEMS Microbiol. Lett.* **163:**149–157.

39. **Soda, K., and N. Esaki.** 1987. Biochemistry of physiologically active selenium compounds, p. 349–365. *In* S. Patai (ed.), *The Chemistry of Organic Selenium and Tellurium Compounds*. John Wiley and Sons, Chicester, United Kingdom.

40. **Stadtman, T. C.** 1990. Selenium biochemistry. *Annu. Rev. Biochem.* **59:**111–127.

41. **Stierle, A.** 1993. Taxol and taxane production by *Taxomyces andreanae,* an endophytic fungus of Pacific yew. *Science* **260:**214–216.

42. **Takai, H., M. Yoshida, T. Iida, I. Matsubara, and K. Shirahata.** 1976. Natural product hydrazide. *J. Antibiot.* (Tokyo) **29:** 1253–1257.

43. **Vahter, M., and E. Marafante.** 1993. Metabolism of alkyl arsenic and antimony compounds, p. 161–184. *In* H. Sigel and A. Sigel (ed.), *Metal Ions in Biological Systems,* vol. 29. Marcel Dekker Inc., New York, N.Y.

44. **Vannelli, T., M. Messmer, A. Studer, S. Vuilleumer, and T. Leisinger.** 1999. A corrinoid-dependent catabolic pathway for growth of a *Methylobacterium* strain with chloromethane. *Proc. Natl. Acad. Sci. USA* **96:**4615–4620.

45. **Walton, K., M. M. Coombs, R. Walker, and C. Ioannides.** 1997. Bioactivation of mushroom hydrazines to mutagenic products by mammalian and fungal enzymes. *Mutat. Res.* **381:**131–139.

46. **Weber, A. A.** 1993. Direct inhibition of platelet function by organic nitrates via nitric oxide formation. *Eur. J. Pharmacol.* **247:**29–37.

47. **Xue, Y., L. Zhao, H.-W. Liu, and D. H. Sherman.** 1998. A gene cluster for macrolide antibiotic biosynthesis in *Streptomyces venezuelae:* architecture of metabolic diversity. *Proc. Natl. Acad. Sci. USA* **95:**12111–12116.

48. **Yonezawa, Y., M. Fukui, T. Yoshida, A. Ochi, T. Tanaka, Y. Noguti, T. Kowata, Y. Sato, A. S. Masunaga, and Y. Urushigawa.** 1994. Degradation of tri-*n*-butyltin in Ise Bay sediment. *Chemosphere* **29:**1349–1356.

6

Physiological Processes: Enzymes, Emulsification, Uptake, and Chemotaxis

> "Chance," "discrimination," "memory," "learning," "instinct," "judgment," and "adaption," are words we normally identify with higher neural processes. Yet, in some sense, a bacterium can be said to have each of those properties.
>
> — *Daniel Koshland*

Biochemically, biodegradation of plant and animal material requires enzymatic transformation of proteins, nucleic acids, lipids, simple sugars, and carbohydrate polymers such as cellulose. Such transformations now seem mundane to some people. But the power of microorganisms to transform many of the disparate chemical structures made in nature and by organic chemists continues to astound everyone. It also adds to our astonishment that bacterial enzymes catalyze reactions for which no precedent exists in the organochemical literature.

The chemistry has arisen in biological systems because bacteria and their ancestors have been in the world for 3.6 billion years, an enormous time to evolve new enzymatic activities. As discussed in chapter 4, there are approximately 5×10^{30} prokaryotes (62), offering a tremendous number of individual biocatalytic entities, occupying virtually every conceivable niche on Earth. Each bacterium contains thousands of enzymes. In some soil bacteria, perhaps as many as several hundred enzymes may be involved in catabolizing proteins, nucleic acids, lipids, carbohydrates, and an array of seemingly exotic organic compounds made as secondary metabolites or via human chemical syntheses.

In this chapter, we will consider mainly single-enzyme-catalyzed biodegradation reactions and other physiological processes that microbes use to compete successfully for scarce nutritional resources in soil and water. Many sources of organic carbon in nature are hydrocarbons, for example, in petroleum deposits, ocean seepages, and oil spills. To metabolize hydrocarbons, prokaryotes may need to move toward them through soil or water, to emulsify them to increase their bioavailability, to transport them into the cell, and to protect their own membranes from being extracted via a hydrophobic phase. These physiological processes will be considered in this chapter. Other relevant physiological aspects of biodegradation will be covered elsewhere, for example, enzyme evolution in chapter 7 and metabolic pathways in chapter 8.

General Physiological Responses to Environmental Chemicals

> Physiology is that part of biochemistry which we do not yet understand.
> — *Bo Malmstrom*

All living things are exposed to a sea of chemicals for message transmission, for food, fortuitously, and through chemical warfare perpetrated by a competing organism. The responses to those chemicals are varied, even among very closely related organisms. For example, 4-methylpyrrole-2-carboxylate is a trail hormone for the Attini tribe of ants while other ants completely ignore it (21). It has been postulated that the compound is produced from the catabolism of tryptophan by bacteria in the ants' guts.

There is the saying that one person's meat is another's poison, and this is true for bacteria as well. For example, some bacteria catabolize 2-fluoroacetate as their sole source of carbon and energy while other bacteria are killed by very low concentrations of the compound (33). The key enzyme needed is fluoroacetate halohydrolase (44) for catabolism of 2-fluoroacetate to glycolate, a growth-sustaining metabolite. Bacteria lacking this metabolism incorporate 2-fluoroacetate into 2-fluorocitrate via action of the tricarboxylic acid cycle enzymes. The latter metabolite is a potent inhibitor of the tricarboxylic acid cycle and thus interferes with energy metabolism (58). This constitutes an important economic problem in Australia, where some plants eaten by livestock contain fluoroacetate at levels of up to 5 g per kg (dry weight), which constitutes a lethal dose for the animals. One countermeasure to this has been the cloning and expression of fluoroacetate halohydrolase in bacteria which inhabit the guts of the animals to detoxify the poison during digestion (17).

There is a tendency for people to seek "meaningful" explanations for all observations, meaning that explanations proposing a fortuitous occurrence are often rejected. In fact, it is likely that a lot of reactions that fall under the general heading of biodegradation are fortuitous. This does not mean that such reactions do not contribute to the overall catabolic recycling of organic matter. Rather, the reactions may have no particular benefit to the organism carrying them out, or useful, neutral, and even harmful reactions may be carried out by the same enzyme system under different conditions.

There are numerous examples of this with insects or fungi that biosynthesize broad-specificity enzymes, such as cytochrome P450 monooxygenases, for detoxifying biological toxins. For example, plant-pathogenic fungi are sometimes warded off with toxic chemicals manufactured in the leaves of the plant being attacked. In some cases, the fungi protect themselves by detoxifying the chemical via an oxygenation reaction which renders the plant chemical nontoxic. The fungus *Nectria* is combated by the pea plant with a toxin called pisatin. An oxidative demethylation reaction renders pisatin nontoxic to the fungus, thus allowing it to feed on the plant (46). This also provides an example in which an enzyme produced for protection can also mediate unexpected events. Fungal cytochrome P450 monooxygenases, as discussed below, oxidize an enormous array of compounds, some of which are unlikely to prove toxic, and thus these reactions may well fall into the fortuitous category. If a catalyst with the ability to broadly protect an organism against many plant toxins is produced, it will likely catalyze many other reactions fortuitously.

The fortuitous metabolism described above has sometimes been called cometabolism. We believe that this term has been detrimental to the field because it presupposes a biological meaning for a reaction when often there are inadequate data to make that judgment. Moreover, the term cometabolism has been used with several different definitions over the past 30 years, adding to the confusion. Several people have advocated discontinuing use of the term (10, 15, 28), and we will adhere to that viewpoint in this book.

Enzymes

Proteins are the preeminent biological catalysts in the modern world. RNA catalysis is known but is much more limited in scope (7). For example, some mRNA molecules cleave themselves hydrolytically, but since only one reaction occurs, it is technically not catalysis. RNA molecules participate with proteins in the ribosome to catalyze amide bond formation in protein synthesis. It has been proposed that catalysts may have been RNA based in early evolutionary history, during which time RNA catalysis may have been more diverse. If so, only the remnants of those biocatalysts persist into the modern world, where protein-based systems are predominant.

Enzymes are extremely versatile catalysts that meet the needs of a modern world filled with an enormous diversity of organic structures. This property derives largely from the different amino acid chains, particularly those that contain oxygen, nitrogen, or sulfur atoms which act in coordination or as nucleophiles in catalysis. Additionally, approximately one-third of enzymes coordinate a metal atom that plays an essential role in catalysis or structural maintenance. On top of that, enzymes bind a range of organic or organometallic cofactors that contribute to the diversity of catalytic reactions accessible to biological systems.

Most of the microbial enzymes of interest in biodegradation, and microbial biotechnology in general, fall under the broad heading of catabolic enzymes. This is true not only for cases in which the reaction product is of interest but also when the enzyme itself is the commodity. For example, the biggest group of commercial enzymes is the microbial proteases, amylases, and lipases (see, for example, Table 4.4). These enzymes serve in nature to hydrolyze biopolymers, thus generating the monomeric units available to support cell growth. In industry, the enzymes act on the same substrates but in environments fairly different from their native soil. For example, the protease subtilisin from *Bacillus subtilis* is an important constituent of laundry detergents, where it acts to remove proteinaceous material from clothing. α-Amylases are used to hydrolyze starch for such products as high-fructose corn syrup, a major food additive.

Catabolic enzymes are so useful in large part because many have been found. In fact, catabolic enzymes may be the major group of enzymes catalyzing unique reactions found on the Earth. We can think of enzymes as generally falling into three functional categories involved in (i) intermediary metabolism, (ii) regulation and response to the environment, and (iii) catabolism. Why would we expect catabolic enzymes to be more numerous than biosynthetic enzymes, particularly in light of the ability of plants to biosynthesize hundreds of thousands of novel chemical structures? The answer is that for all the enzymes involved in the biosynthesis of natural products, there are a roughly equivalent number of catabolic

Enzyme Classification

Since 1897, when Büchner prepared the first cell-free enzyme extract from pressed yeast cells (5), our understanding of microbial biocatalysis has largely derived from the reductionist practice of purifying and characterizing enzymes. The human need to classify is great, and the field of enzymology has not escaped the compulsion to bring order to a century of discoveries. The Enzyme Commission (EC) was established for the purpose of categorizing enzymes in classes based on the type of reaction they catalyze (see A. Bairoch, Enzyme Nomenclature Database [http://www.expasy.ch/enzyme/]).

Currently, there are six major divisions of enzymes, as shown in Table 6.1. However, the fractional distribution of the six enzyme classes contained in the University of Minnesota Biocatalysis/Biodegradation Database (UM-BBD) is not equal: oxidoreductases are predominant. This is consistent with the logic of catabolism as discussed in chapter 8. The catabolism of organic compounds containing C-H bonds to yield carbon dioxide requires enzymes to catalyze oxidation reactions, thus providing reducing equivalents for oxidative phosphorylation to make ATP. Electrons may be removed from the substrate directly by dehydrogenases. Another major class of oxidoreductases in aerobic organisms is oxygenases. Oxygenases catalyze the insertion of oxygen atoms derived from atmospheric dioxygen (O_2) directly into organic substrates. Subsequent reactions are typically oxidation reactions in which electrons are used to generate ATP.

Enzyme-catalyzed oxidation-reduction reactions are also important in the assimilation of cell material because the oxidation state of carbon atoms must be adjusted to produce the carbohydrates, lipids, amino acids, and nucleic acids needed for the cell. For example, methanotrophs oxidize methane, CH_4, as their sole source of carbon and energy. Methane is oxidized by eight electrons to yield carbon dioxide and ATP. Some of the methane molecules are oxidized by only four electrons to the oxidation level of formaldehyde, which then enters biosynthetic pathways to make carbohydrates, lipids, amino acids, and nucleic acids.

Table 6.1 EC major divisions and distribution of enzymes in the UM-BBD as of 1 September 1999

EC no.	Enzyme class	No. of class in UM-BBD
1	Oxidoreductase	260
2	Transferase	30
3	Hydrolase	60
4	Lyase	50
5	Isomerase	14
6	Ligase	10

The largest EC class of enzymes after oxidoreductases is hydrolases. This reflects the fact that there are a reasonable number of organic functional groups which are susceptible to hydrolysis. Examples of such groups are amides, esters, phosphate esters, monohaloalkanes, and epoxides. The hydrolysis of such functional groups is often coupled with subsequent oxidation-reduction reactions.

Lyases catalyze the formation or loss of multiple bonds. For example, fatty acid oxidation proceeds via double-bond formation by an oxidation reaction, followed by addition of water across the double bond by a lyase enzyme, followed by additional oxidation reactions. Another addition reaction is that catalyzed by nitrile hydratase, a prominent industrial enzyme. In the relevant reaction, acrylonitrile, a high-volume commodity chemical (see Table 9.1), is hydrated with 1 mol of water to generate acrylamide, an important precursor for acrylamide polymers.

Transferases are reasonably well represented in biosynthesis but are not very prominent in catabolism. The major group of transferases in the UM-BBD is glutathione *S*-transferases. These enzymes participate in the catabolism of a broad range of compounds. The common feature in most of the reactions is the transfer of the thiol group of the tripeptide glutathione to some acceptor group. The acceptor group varies from halocarbons to ethers and epoxides.

Isomerases and ligases in catabolism catalyze mainly reactions which set up subsequent oxidation reactions. Mandelate racemase is used as an example in the discussion of enzyme and pathway evolution in chapter 7.

enzymes which take the natural products apart. Additionally, there are the known catabolic enzymes which catabolize industrial chemicals. Taken in total, one can reasonably argue that the catabolic enzyme set is bigger than the biosynthetic enzyme set.

Enzyme Substrate Specificity

Studies of intermediary metabolism fostered the paradigm that each enzyme catalyzes one, and only one, specific reaction. By this view, each enzyme transforms one substrate or a unique set of cosubstrates. This is often true with enzymes which function in core metabolic pathways, although more exceptions are being uncovered in which enzymes are multifunctional. Nonetheless, intermediary-metabolism enzymes often need to be highly substrate specific to control pathway fluxes in the presence of a cytoplasmic sea of similar chemical structures. For example, isomers of simple sugars and sugar phosphates may be earmarked for different pathways in cells. Enzymes must be highly selective to discriminate among the similar structures.

The logic changes when one looks at enzymes and metabolic pathways involved in the catabolism of aromatic-ring compounds, for example. In

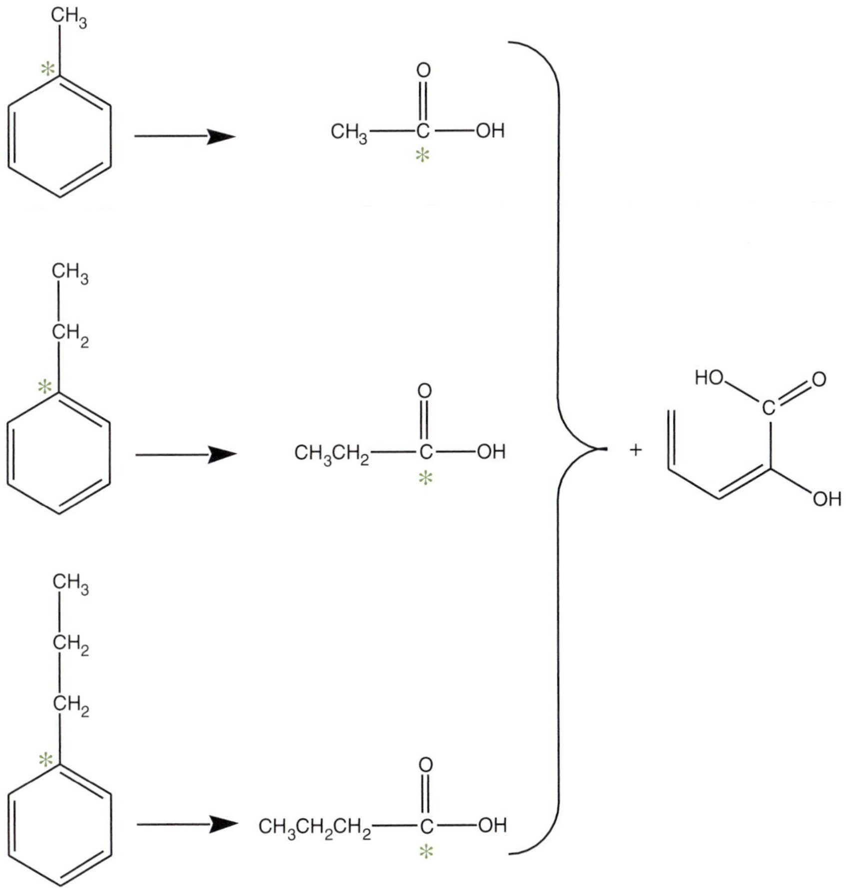

Figure 6.1 Metabolism of alkylbenzenes showing the commonality in processing reactions with release of correspondingly larger organic acids with larger alkyl side chains.

that case, an organism which oxidizes methylbenzene (toluene) to carbon dioxide may also oxidize ethyl and propylbenzene by a parallel pathway using the same enzymes. The enzymes catalyzing sequential reactions must each be able to accommodate a mixed set of alkyl side chains attached to a benzene ring. Moreover, the central metabolism of the organism must be able to utilize a variable set of organic-acid fragments (Fig. 6.1). This metabolic strategy likely arose because these and similar chemical structures are often found together in nature, for example, in petroleum.

How broad can the substrate specificities of enzymes be? In Table 6.2, the known substrate range of one enzyme, naphthalene dioxygenase from *Pseudomonas* sp. strain NCIB9816, is shown. Not only will this enzyme act on at least 72 substrates, but somewhat different reactions are catalyzed: substrate dioxygenation, monooxygenation, or desaturation. These are all oxidation reactions, and they require the generation of an activated oxygen species, which can then take different substrates down one or more of these three reaction pathways.

Naphthalene dioxygenase is not unique. It is prototypical of a large class of aromatic hydrocarbon dioxygenases. The class of enzymes thought to be homologous and mechanistically similar to naphthalene dioxygenase comprises 37 of the 442 enzymes listed on the UM-BBD (http://umbbd.ahc.umn.edu/cgi-bin/page.cgi?ptype=allenzymes) as of 7 November 1999. In this context, the recently solved structure of naphthalene dioxygenase (32) can be used as a guide to probe the structures, functions, and substrate specificities of this large enzyme class. Figure 6.2 highlights the active site of naphthalene dioxygenase. Naphthalene dioxygenase is a trimer of dimers ($\alpha_3\beta_3$ subunit structure). Each β-subunit contains the redox active groups, an Fe_2S_2 center and a mononuclear-iron center, which function in catalysis. The active site is composed of two β-subunits, of which one contributes the Fe_2S_2 center and the other contributes the mononuclear-iron center. The amino acids lining the active site are largely hydrophobic residues, consistent with the observation that the enzyme reacts with hydrophobic substrates. One exception is an essential aspartate residue which bridges the Fe_2S_2 and mononuclear-iron centers and may participate in electron transfer from the former to the latter. Sequence comparisons suggest that a large group of catabolic oxygenases are structurally and mechanistically related to naphthalene dioxygenase.

Another distinct class of microbial oxygenases is composed of those with a binuclear iron cluster and includes methane monooxygenase, xylene monooxygenase, and alkane monooxygenase. Methane monooxygenase is estimated to react with over 100 different substrates (43).

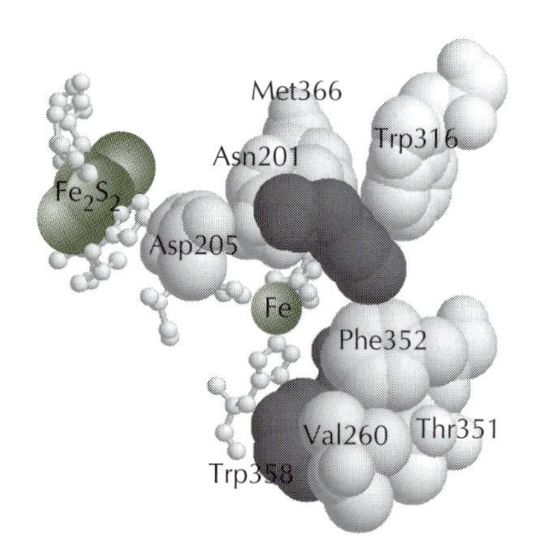

Figure 6.2 Space-filling model of the active site of naphthalene dioxygenase from *P. putida*. The redox active groups, the iron-sulfur cluster and the mononuclear-iron center, are shown in green. (Courtesy of R. E. Parales and D. T. Gibson.)

Uptake: Getting Substrates to the Enzymes

There is a major barrier between bacterial enzymes in vivo and their substrates. That barrier is the cell wall. Cytoplasmic membranes and other cell wall and membrane components play fundamental roles in maintaining a living cell. It is essential that important molecules be retained intracellularly. In his classic review paper entitled "Why nature chose phosphates," Frank Westheimer postulated that the prevalence of phosphorylated metabolic intermediates inside a cell evolved for the purpose of rendering molecules highly charged so that they could not leak through a lipid bilayer (61). This prevents loss of the precious energy locked in the structures of

Table 6.2 Reactions catalyzed by naphthalene 1,2-dioxygenase[a]

Dioxygenation reactions

Substrates are aromatic hydrocarbons

1. Naphthalene → (+)-*cis*-(1*R*,2*S*)-dihydroxy-1,2-dihydronaphthalene
2. Indene → *cis*-(1*R*,2*S*)-indandiol
3. 1,2-Dihydronaphthalene → (−)-*cis*-(1*R*,2*S*)-dihydroxy-1,2,3,4-tetrahydronaphthalene
4. Benzocyclohept-1-ene → (−)-*cis*-(1*R*,2*S*)-dihydroxybenzocycloheptane
5. Biphenyl → (+)-(2*R*,3*S*)-dihydro-2,3-dihydroxybiphenyl
6. Anthracene → (+)-*cis*-(1*R*,2*S*)-dihydroxy-1,2-dihydroanthracene
7. 9,10-Dihydroanthracene → (+)-*cis*-(1*R*,2*S*)-dihydroxy-1,2,9,10-tetrahydroanthracene
8. Phenanthrene → (+)-*cis*-(3*S*,4*R*)-dihydroxy-3,4-dihydrophenanthrene
9. Phenanthrene → (+)-*cis*-(1*S*,2*R*)-dihydroxy-1,2-dihydrophenanthrene
10. Acenaphthylene → *cis*-acenaphthene-1,2-diol
11. 9,10-Dihydrophenanthrene → *cis*-(3*S*,4*R*)-dihydroxy-3,4,9,10-tetrahydrophenanthrene
12. Fluorene → *cis*-(3*S*,4*R*)-dihydroxy-3,4-dihydrofluorene

Substrates are substituted aromatic compounds

13. Styrene → (*R*)-1-phenyl-1,2-ethanediol
14. (1*R*)-Indenol → *cis*-1,2,3-indantriol
15. 2-Methoxynaphthalene → (+)-(1*R*,2*S*)-dihydroxy-7-methoxy-1,2-dihydronaphthalene
16. 2-Methoxynaphthalene → (+)-(1*R*,2*S*)-dihydroxy-6-methoxy-1,2-dihydronaphthalene
17. 2-Naphthoic acid → (+)-*cis*-1,2-dihydroxy-1,2-dihydronaphthalene-2-carboxylic acid
18. 2,6-Dimethylnaphthalene → (+)-*cis*-1,2-dihydro-(1*R*,2*S*)-dihydroxy-3,7-dimethylnaphthalene
19. 2,3-Dimethylnaphthalene → (+)-*cis*-1,2-dihydro-(1*R*,2*S*)-dihydroxy-6,7-dimethylnaphthalene
20. 1-Carbomethoxynaphthalene → *cis*-3,4-dihydro-(2*S*)-hydroxynaphthalene-8,(1*R*)-lactone
21. 1-Methylnaphthalene → *cis*-1,2-dihydro-(1*R*,2*S*)-dihydroxy-8-methylnaphthalene
22. 1-Methoxynaphthalene → *cis*-1,2-dihydro-(1*R*,2*S*)-dihydroxy-5-methoxynaphthalene
23. 2-Nitronaphthalene → *cis*-1,2-dihydro-(1*R*,2*S*)-dihydroxy-7-nitronaphthalene

Substrates are heterocyclic aromatic compounds

24. Indole → indigo
25. Dibenzo-1,4-dioxin → *cis*-1,2-dihydroxy-1,2-dihydrodibenzo[1,4]dioxan
26. Dibenzothiophene → (+)-*cis*-(1*R*,2*S*)-dihydroxy-1,2-dihydrodibenzothiophene
27. Dibenzofuran → *cis*-1,2-dihydroxy-1,2-dihydrodibenzofuran
28. Dibenzofuran → *cis*-3,4-dihydroxy-3,4-dihydrodibenzofuran

Monooxygenation reactions

29. Benzocyclobutene → benzocyclobutene-1-ol
30. Indan → (+)-(1*S*)-indanol
31. Indene → (+)-(1*S*)-indenol
32. 1-Indanone → (*R*)-3-hydroxy-1-indanone
33. 1-Indanone → (*R*)-2-hydroxy-1-indanone
34. 2-Indanone → (*S*)-2-hydroxy-1-indanone

(continued next page)

Table 6.2 *Continued*

35. (1*R*)-Indanol → *cis*-(1*R*,3*S*)-indandiol
36. (1*S*)-Indanol → *trans*-(1*S*, 3*R*)-indan-1,3-diol
37. Fluorene → 9-fluorenol
38. Acenaphthene → 1-acenaphthenol
39. Acenaphthen-1-ol → *cis*-acenaphthene-1,2-diol
40. Acenaphthen-1-ol → *trans*-acenaphthen-1,2-diol
41. 9,10-Dihydrophenanthrene → (+)-9*S*-hydroxy-9,10-dihydrophenanthrene
42. Toluene → benzyl alcohol
43. Ethylbenzene → (*S*)-1-phenethyl alcohol
44. (*S*)-1-Phenethyl alcohol → acetophenone
45. Acetophenone → 2-hydroxy-acetophenone
46. 1,2,4-Trimethylbenzene → 2,4-dimethylbenzoic alcohol
47. 1,2,4-Trimethylbenzene → 3,4-dimethylbenzoic alcohol
48. 1,2,4-Trimethylbenzene → 2,5-dimethylbenzoic alcohol
49. 1,2,4-Trimethylbenzene → 1,2-bis(hydroxymethyl)-4-methylbenzene
50. 1,2,4-Trimethylbenzene → 1,4(3)-bis(hydroxymethyl)-3(4)-methylbenzene
51. 3-Methylbenzothiophene → 3-hydroxymethylbenzothiophene
Desaturation reactions
52. Indan → indene
53. 1,2-Dihydronaphthalene → naphthalene
54. Indoline → indole
55. Phenetole → ethenyloxybenzene
56. Ethylbenzene → styrene
57. (1*R*)-Indanol → (1*S*)-indenol
58. (1*S*)-Indanol → (1*R*)-indenol
O- and N-dealkylation reactions
59. Anisole → phenol
60. Phenetole → phenol
61. *N*-Methylindole → indole
62. *N*-Methylaniline → aniline
63. *N,N*-Dimethylaniline → aniline
Sulfoxidation reactions
64. Methylphenyl sulfide → methylphenyl (*S*)-sulfoxide
65. Ethylphenyl sulfide → ethylphenyl (*S*)-sulfoxide
66. Methyl *p*-tolyl sulfide → methyl *p*-tolyl (*S*)-sulfoxide
67. *p*-Methoxyphenylmethyl sulfide → *p*-methoxyphenylmethyl (*S*)-sulfoxide
68. Methyl *p*-nitrophenyl sulfide → methyl *p*-nitrophenyl (*S*)-sulfoxide
69. 2-Methylbenzo-1,3-dithiole → *cis*-(1*R*, 2*S*)-2-methylbenzodithiole sulfoxide
70. 2-Methylbenzo-1,3-dithiole → *trans*-(1*R*, 2*R*)-2-Methylbenzodithiol sulfoxide
71. 3-Methylbenzothiophene → 3-methylbenzothiophene sulfoxide
72. Dibenzothiophene → dibenzothiophene sulfoxide

[a]From reference 57.

metabolic intermediates. Cell walls and membranes also provide a barrier to protect the cell against potentially toxic agents. For example, a major outer porin in the *Escherichia coli* cell wall is just small enough to prevent the permeation of toxic bile salts found in the animal intestinal environment, a major ecological niche of *E. coli.*

But it is essential for obtaining metabolic energy, vitamins, and essential metals that a membrane be selectively permeable; this requires the participation of active transport. It is well known that sugars, peptides, and nucleotides are transported through the cell membrane by specific transport proteins found in those membranes. It is thus sometimes assumed that all trafficking of substrates into cells is controlled by specific transport proteins. Is this the case?

To answer this question, we need to consider the rates at which selected organic compounds will move through a lipid bilayer. Such measurements have been made and are shown in Fig. 6.3. It is somewhat surprising to consider that glycerol and urea are highly water soluble and yet will transfer across a lipid bilayer at a moderate rate even in the absence of any active transport process (59). Thus, when more hydrophobic compounds are considered, going in a series from indole to long-chain alcohols to alkanes and aromatic hydrocarbons, the rates of spontaneous transfer will increase to the point that transfer will not impose any barrier to the metabolism of these compounds. Observations support this idea. First, hydrocarbon degradation operons have generally not been found to contain genes encoding transport functions. Second, the transfer of hydrocarbon-degrading genes from *Pseudomonas* sp. into *E. coli* strains allows efficient metabolism of compounds like toluene and trichloroethylene without any apparent need for an uptake system (68).

There are examples of specific transport for aromatic-ring compounds, for those compounds containing a charged group that would mitigate against passive transport across a membrane. Recently, genes involved in the facilitated uptake of commonly occurring aromatic acids, 4-hydroxybenzoate (23) and protocatechuate (11), have been identified. One of the genes, *pcaK*, in *Pseudomonas putida* is proposed to play roles in both transport and chemotaxis toward 4-hydroxybenzoate. Chemotaxis will be discussed below.

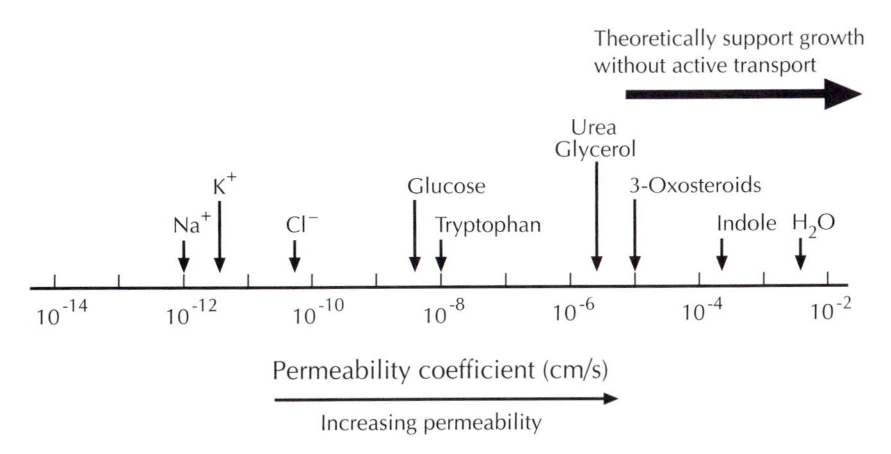

Figure 6.3 Permeation rates across a lipid membrane bilayer by different compounds. Those compounds further to the right transfer across a membrane correspondingly faster.

Emulsification: Overcoming Poor Availability of Substrate

As discussed previously, the metabolism of hydrophobic substrates such as hydrocarbons is not limited by their permeation through a cell membrane. However, their metabolism in natural environments can be limited by adsorption to solid matrices, thus reducing the amount that can get to the cell. This common limiting factor in microbial metabolism, known as bioavailability, is of premier importance in the applied biodegradation of such compounds as polycyclic aromatic hydrocarbons or polychlorinated biphenyls (PCBs). In such classes of compounds, as the water solubility increases, so typically does the bioavailability of the compound.

The physical explanation for the lack of availability is that soils or sediments are complex matrices with many small crevices containing hydrophobic surfaces. Hydrophobic compounds, such as PCBs, slowly diffuse into the crevices. The diffusion back out is an equally slow process, and this can make the step overall rate determining in biodegradation. For example, in a field test to treat PCBs in the Hudson River, at most 60% of the PCBs determined to be present by exhaustive chemical extraction of the sediments were biodegraded (22). The failure to biodegrade the remainder was attributed to their lack of bioavailability.

While PCBs are industrially synthesized, petroleum hydrocarbons occur naturally, are hydrophobic, and also pose bioavailability problems. Are microbes helpless to extract these energy-rich materials from the hydrophobic crevices in soil or particulate material in water? In fact, they are not, and one solution to this problem appears to be the production of biosurfactants.

Surfactants are compounds which are amphipathic; that is, they contain hydrophilic and hydrophobic chemical groups linked together in the same molecule. People use surfactants as soaps and detergents and as emulsifying agents in food. Their commercial applications are based on the ability to solubilize hydrophobic materials from hands or clothes or to suspend hydrophobic compounds in an aqueous medium. A familiar example of the latter is the use of microbial cell wall material, known as xanthan gum, in salad dressings.

Microbes are known to produce surfactants and thus render hydrophobic compounds more bioavailable. Biosurfactants can be thought of as decreasing the surface tension of water or as stabilizing emulsions between water and an organic liquid. With respect to the latter, consider a two-phase system of a water-immiscible alkane and water. When static, the alkane is only available to bacteria suspended in water when they reside at the alkane-water interface. However, the formation of an emulsion leads to the dispersion of many small alkane droplets in the water, greatly increasing the surface area of the alkane-water interface. Bacteria can be seen by microscopy to adhere to the surfaces of the alkane droplets (Fig. 6.4).

The role of biosurfactants in microbial metabolism has been investigated primarily with petroleum or with purified alkanes. Like synthetic surfactants, microbial surfactants contain hydrophilic and hydrophobic parts, but many different compounds serve that function. Many are glycolipids, acylpolyols, or lipopeptides. Table 6.3 shows a representative set of microorganisms, both prokaryotes and yeasts, which are known to produce biosurfactants. Many of the organisms in Table 6.3 were studied for their biodegradation of hydrophobic compounds, but some likely use biosurfactants for other functions as well. Biosurfactants have been implicated

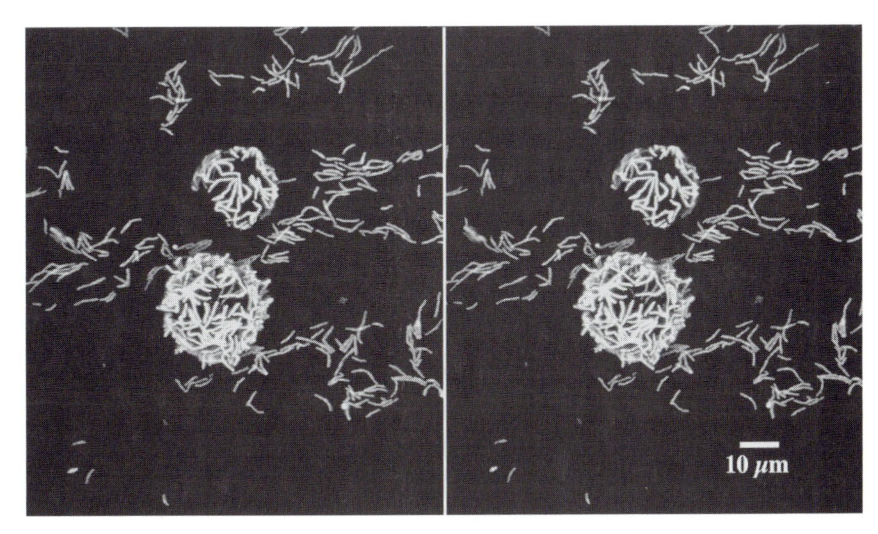

Figure 6.4 Stereo view of bacterial cells adhering to an oil droplet. (From reference 63 with permission.)

in plant pathogenesis, for example, in the penetration of plant cell walls by *Pseudomonas syringae* (29). In another example, an extracellular lipopeptide biosurfactant was implicated in the swarming motility of *Serratia liquefaciens* (42).

Organic-Solvent Resistance

In their natural environment, or in an industrial fermentor, bacteria may encounter organic solvents at concentrations above their solubilities in water. This has generally been considered to be a lethal stress. For example, an often-used experimental protocol for rendering *E. coli* membranes permeable to charged compounds is to treat the cells with toluene. This makes the *E. coli* cell into a nonviable porous enzyme bag. Moreover, to obtain bacteria capable of growth on toluene, it is typically found necessary to conduct enrichment cultures with low concentrations of toluene in the growth medium or to supply toluene in the vapor phase (see Fig. 3.1). In another example, chloroform treatment has been used as a method of soil sterilization (4). These and other observations led to the idea that two-phase systems composed of an organic layer and an aqueous layer are completely inhospitable to microbial life. Under those conditions, microbial death presumably results from being drawn to the water–organic-solvent interface and having the cell membrane lipids extracted into the organic phase. When cells cannot maintain proton and other chemical gradients, they quickly die.

The inability to survive solvent stress is a practical problem in fermentation systems designed to generate hydrophobic products. This occurs when the solvent(s) is a fermentation end product, such as in the acetone-butanol fermentation. It also occurs when the organic compound is a substrate to be transformed and it is practically important to add the compound in large quantities. Solvent stress may also pose a practical problem when a solvent phase is needed to continually extract a toxic product, for example, a reactive epoxide. In that case, the substrate may be a relatively insoluble alkene which is added to the fermentor as both the substrate and

Table 6.3 Examples of biosurfactant-producing microbes[a]

Biosurfactant class	Microbial species or strain	Reference
Glycoconjugates		
Glycoconjugate	*Rhodococcus* sp. strain Q15	63
Glycoprotein	*Acinetobacter radioresistens* KA53	3
Glycoprotein	*Acinetobacter radioresistens* KA53	51
Glycoproteolipid	*Corynebacterium hydrocarboclastus*	66
Glycolipid	Gram-positive actinomycete	12
Glycolipid	*Alcanivorax borkumensis*	1
Glycolipid	*Candida bombicola* ATCC 22214	25
Glycolipid	*Rhodococcus aurantiacus*	56
Glycolipid	*Rhodococcus* sp. strain H13A	13
Glycolipid	*Torulopsis apicola*	26
Trehalose lipid	Gram-positive, cocci-shaped	65
Trehalose mycolate	*Rhodococcus erythropolis*	39
Trehalose-mono, dicoryno-	*Nocardia corynebacteroides*	36
Sophorolipid	*Torulopsis bombicola*	8
Lipopolysaccharides		
Lipopolysaccharide	*Pseudomonas marginalis*	6
Pentasaccharide lipid	*Nocardia corynebacteroides*	36
Rhamnolipid	*Pseudomonas aeruginosa* VPJ-80	40
Rhamnolipid	*Pseudomonas* sp. strain ATCC 15524	24
Rhamnolipid	*Pseudomonas aeruginosa* UG2	60
Rhamnolipid	*Pseudomonas aeruginosa* ATCC 9027	67
Rhamnolipid	*Pseudomonas aeruginosa*	20
Rubiwettin	*Serratia rubidaea*	48
Fatty acids		
Fatty acid	*Corynebacterium lepus*	9
Fatty acid and neutral lipid	*Nocardia erythropolis*	45
Lipodepsipeptides		
Cyclic lipodepsipeptide	*Pseudomonas fluorescens* DR54	53
Cyclic lipodepsipeptide	*Pseudomonas syringae*	29
Cyclic lipopeptide	*Bacillus subtilis* OKB105	37
Serrawettin	*Serratia marcescens*	48
Viscosin	*Pseudomonas fluorescens*	52
Lipopeptides		
Lipopeptide	*Bacillus subtilis*	16
Lipopeptide	*Serratia liquefaciens*	42
Lipopeptide	*Bacillus licheniformis* BAS50	64
Lipopeptide	*Arthrobacter* sp. strain MIS38	49
Lipopeptide	*Bacillus subtilis*	50
Lipopeptide	*Bacillus licheniformis* JF2	31
Lipopeptide	*Bacillus licheniformis* 86	27
Surfactin	*Bacillus subtilis*	2
Others		
Phosphatidylethanolamine	*Rhodococcus erythropolis*	38
Protein-carbohydrate complex	*Pseudomonas fluorescens*	55

[a]Data from reference 14.

the extracting solvent for the product. The practical problem is that a bacterium with the desired enzyme activity to make a valuable industrial product may be killed by the organic solvent.

Several years ago, a bacterium was isolated which grew on toluene as its carbon source when the toluene was in sufficient quantities to form an insoluble organic phase in the growth medium (30). This was an unexpected observation at the time. More recently, the molecular basis of survival under such extreme conditions has begun to be elucidated. Solvent-tolerant *Pseudomonas* spp. were found to be able to protect their membranes from disruption by transforming *cis*- to *trans*-fatty acids via an enzyme present in the membrane or the periplasm (54). The *trans*-fatty acids form a more stable configuration of the lipids and thus maintain membrane integrity better in the presence of solvent. The enzyme is expressed constitutively because the cells must adapt immediately to an environment imposing a solvent stress. Moreover, the *cis*-fatty acids are shielded from the enzyme under normal conditions. When the cell is exposed to solvent, the membrane becomes more permeable, thus allowing contact between the fatty acids and the enzyme, and isomerization is thus catalyzed.

An additional solvent survival mechanism has been found, and it is linked to the resistance of certain bacteria to antibiotics via efflux pumps (34, 35). Similarly, solvents are thought to be pumped outside the cell, thus maintaining a sublethal concentration in the membranes. Such membrane pumps have rather broad specificities in some cases.

Survival in solvent is likely to be a very important phenotype in the expansion of fermentation processes to include specialty and commodity organic-chemical production. Either naturally solvent-tolerant organisms will be used or solvent tolerance genes will be cloned and expressed in important fermentation organisms which would otherwise be solvent sensitive. Some of the genes involved in the biochemical reactions described above have been cloned and shown to be transferable to other host cells.

Chemotaxis: Getting to the Substrates

Bacteria move by gliding on solid surfaces or by flagellum-dependent motion in free water or on soil particles covered with a sphere of hydration (41). The flagellum superstructure consists of a protein filament, sometimes referred to as the flagellum, and a hook structure which anchors the flagellum to the basal motor. The basal motor rotates, which in turn rotates the flagellum. The action of a moving flagellum is comparable to a seagoing ship's propeller.

Bacterial motion in one direction occurs when the flagellum, or multiple flagella, rotate in one uniform direction, either clockwise or counterclockwise. How do bacteria change their direction of motion? For a relatively simple example, consider a pseudomonad, which has flagella at only one end of the cell. In that case, the direction of movement is altered by reversing the direction of the basal rotor, which throws the flagellar bundle into disarray. The orientation of the new direction is random. Thus, bacterial flagellar movement is considered to be a random walk.

Movement is not uncontrolled, however, as it possible for the bacterium to swim in the same direction for relatively long periods when it is moving toward a positive stimulus and to throw its motion into disarray frequently when it is moving away from the stimulus. This physiological control of movement is known as taxis. The positive stimuli for attracting bacteria

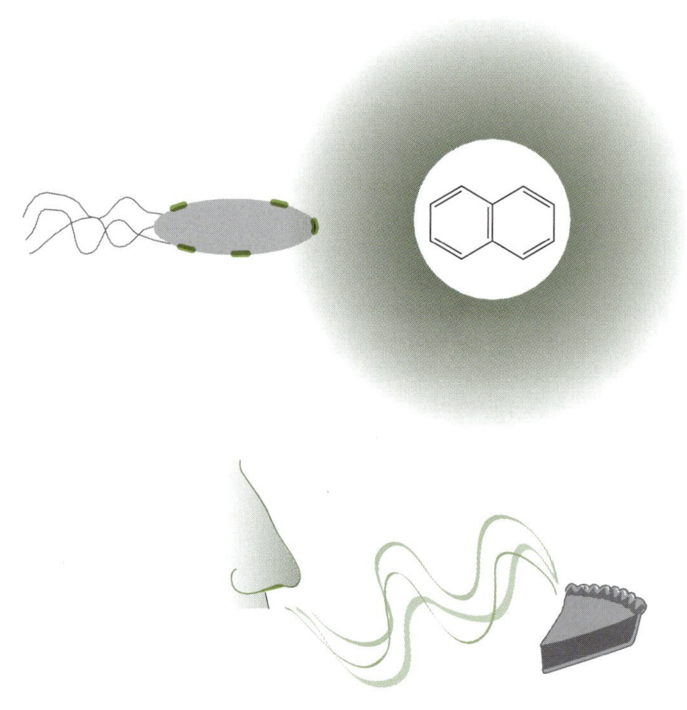

Figure 6.5 Bacterial sensing and chemotaxis (top) are comparable to the human sense of smell (bottom).

include light, the Earth's magnetic field, oxygen gradients, or organic compounds. The last is important in the context of a bacterium's physiological response to chemicals in the environment, and it is known as chemotaxis.

The bacterial chemotactic response is probably the most primitive of the senses, akin to the human sense of smell (Fig. 6.5). The sense of smell has been considered to be the most primitive of our five senses because the nasal neurons connect directly to the limbic system at the base of the brain. In this context, the sensations of smell are least processed cerebrally and are most connected to our emotional, instinctual responses. One of the things smell does for us is lead us to food, and in that sense there is a direct correlation to chemotaxis in bacteria.

Chemotaxis has been best studied in *E. coli*. *E. coli* and other enteric bacteria respond to glucose and amino acids in a positive chemotactic sense. *E. coli* cells have been observed to have a net motion away from some chemicals, typically those it is unable to metabolize and which may prove toxic to it. The sensing of chemical compounds starts with binding at the cell membrane to a methyl-accepting chemotaxis protein (MCP). The extracellular sensing is transmitted through the MCP, which spans the membrane, to its cytoplasmic domain. The protein confirmation induced by chemical binding allows the slow methylation of glutamate residues in the protein. During this period, the cell is "locked" into a chemosensing configuration, and the flagella will continue to rotate in the same direction, which causes the cell to continue to swim in one direction. When the chemical signal abates, or long-term adaption to the signal sets in, the rotation of the flagella switches and the cells move in a new direction. Extensive knowledge of the biochemical basis of bacterial chemotaxis has been accumulated, and the present treatment is but a very brief summary of what is known.

In terms of environmental responses to chemicals, it has only recently emerged that compounds other than intermediary metabolites can serve as positive chemotactic signals. For example, when *P. putida* G7 was grown on naphthalene, or its metabolite salicylate, it was chemotactically attracted to the polycyclic aromatic hydrocarbon naphthalene or biphenyl (18). It was not attracted to other structurally analogous aromatic hydrocarbons, indicating a specificity in the response. Moreover, when the cells were cured of their naphthalene catabolic plasmid, the cells no longer showed a chemotactic response to naphthalene. Additionally, Grimm and Harwood (19) have identified a gene, *nahY,* which is genetically implicated in the chemotactic response to naphthalene. The *nahY* gene was sequenced, and it was proposed, based on homology to other chemotactic receptors, to encode a membrane-spanning receptor protein.

At present, the finding of chemotactic responses to environmental chemicals suggests that this function has a physiological role to facilitate metabolism in nature. Additional chemotactic receptor genes will no doubt be sequenced in the many ongoing genome-sequencing projects involving soil prokaryotes (see chapter 11 on genomics). This should facilitate the discovery of additional chemotactic responses to a diverse array of chemicals.

Summary

The transformation of organic compounds by microbes, in natural or engineered environments, occurs via a coordinated set of physiological responses. Microbes sense, transport, and enzymatically process the organic compounds they have evolved to handle. Microbial "behavior" revolves around these responses to the chemicals in their environment. A knowledge of this behavior can have practical significance. For example, organic-solvent resistance genes may be employed in fermentors to produce a high yield of organic products. Most industrially relevant microbial transformations use intact organisms, and the total response of the organism affects the desired outcome. Recent efforts in whole-genome sequencing will move us toward a more holistic view of microbial responses to chemicals in their environment.

References

1. **Abraham, W. R., H. Meyer, and M. Yakomov.** 1998. Novel glycine containing glucolipids from the alkane using bacterium *Alcanivorax borkumensis. Biochim. Biophys. Acta* **1393:**57–62.

2. **Arima, K., A. Kakinuma, and G. Tamura.** 1968. Surfactin, a crystalline peptidelipid surfactant produced by *Bacillus subtilis:* isolation, characterization and its inhibition of fibrin clot formation. *Biochem. Biophys. Res. Commun.* **31:**488–494.

3. **Barkay, T., S. Navon-Venezia, E. Z. Ron, and E. Rosenberg.** 1999. Enhancement of solubilization and biodegradation of polyaromatic hydrocarbons by the bioemulsifier alasan. *Appl. Environ. Microbiol.* **65:**2697–2702.

4. **Baumann, A., W. Schimmack, H. Steindl, and K. Bunzl.** 1996. Association of fallout radiocesium with soil constituents: effect of sterilization of forest soils by fumigation with chloroform. *Radiat. Environ. Biophys.* **35:**229–233.

5. **Büchner, E.** 1897. Alcoholische Gärung ohne Hefezellen. *Ber. Dtsch. Chem. Ges.* **30:**117–124.

6. **Burd, G., and O. P. Ward.** 1996. Physicochemical properties of PM-factor, a surface-active agent produced by *Pseudomonas marginalis*. *Can. J. Microbiol.* **42:**243–251.

7. **Cech, T. R.** 1993. The efficiency and versatility of catalytic RNA: implications for an RNA world. *Gene* **135:**33–36.

8. **Cooper, D. G., and D. A. Paddock.** 1984. Production of a biosurfactant from *Torulopsis bombicola*. *Appl. Environ. Microbiol.* **47:**173–176.

9. **Cooper, D. G., J. E. Zajic, and D. F. Gerson.** 1979. Production of surface-active lipids by *Corynebacterium lepus*. *Appl. Environ. Microbiol.* **37:**4–10.

10. **Dagley, S.** 1979. Summary of the conference, p. 534–542. *In* A. W. Bourquin and P. H. Pritchard (ed.), *Proceedings of the Workshop on Microbial Degradation of Pollutants in Marine Environments.* United States Environmental Protection Agency, Washington, D.C.

11. **D'Argenio, D. A., A. Segura, W. M. Coco, P. V. Bunz, and L. N. Ornston.** 1999. The physiological contribution of *Acinetobacter* PcaK, a transport system that acts upon protocatechuate, can be masked by the overlapping specificity of VanK. *J. Bacteriol.* **181:**3505–3515.

12. **Esch, S. W., M. D. Morton, T. D. Williams, and C. S. Buller.** 1999. A novel trisaccharide glycolipid biosurfactant containing trehalose bears ester-linked hexanoate, succinate, and acyloxyacyl moieties: NMR and MS characterization of the underivatized structure. *Carbohydr. Res.* **319:**112–123.

13. **Finnerty, W. R., and M. E. Singer.** 1984. A microbial biosurfactant—physiology, biochemistry, and applications. *Dev. Ind. Microbiol.* **25:**31–46.

*14. **Georgiou, G., S. C. Lin, and M. M. Sharma.** 1992. Surface-active compounds from microorganisms. *Bio/Technology* **10:**60–65.

15. **Gibson, D. T.** 1993. Biodegradation, biotransformation, and the Belmont. *J. Ind. Microbiol.* **12:**1–12.

16. **Grau, A., J. C. Gomez Fernandez, F. Peypoux, and A. Ortiz.** 1999. A study on the interaction of surfactin with phospholipid vesicles. *Biochim. Biophys. Acta* **141:**307–319.

17. **Gregg, K., B. Hamdorf, K. Henderson, J. Kopecny, and C. Wong.** 1998. Genetically modified ruminal bacteria protect sheep from fluoroacetate poisoning. *Appl. Environ. Microbiol.* **64:**3496–3498.

18. **Grimm, A. C., and C. S. Harwood.** 1997. Chemotaxis of *Pseudomonas* spp. to the polyaromatic hydrocarbon naphthalene. *Appl. Environ. Microbiol.* **63:**4111–4115.

19. **Grimm, A. C., and C. S. Harwood.** 1999. NahY, a catabolic plasmid-encoded receptor required for chemotaxis of *Pseudomonas putida* to the aromatic hydrocarbon naphthalene. *J. Bacteriol.* **181:**3310–3316.

20. **Guerra-Santos, L. H., O. Kappell, and A. Fiechter.** 1986. Dependence of *Pseudomonas aeruginosa* continuous culture biosurfactant production on nutritional and environmental factors. *Appl. Microbiol. Biotechnol.* **24:**443–448.

21. **Harbourne, J.** 1988. *Ecological Biochemistry,* 3rd ed. Academic Press, New York, N.Y.

22. **Harkness, M. R., J. B. McDermott, D. A. Abramowicz, J. J. Salvo, W. P. Flanagan, M. L. Stephens, F. J. Mondello, R. J. May, J. H. Lobos, K. M. Caroll, M. J. Brennan, A. A. Bracco, K. M. Fish, G. L. Warner, P. R. Wilson, D. K. Dietrich, D. T. Lin, C. B. Morgan, and W. L. Gately.** 1993. *In situ* stimulation of aerobic PCB biodegradation in Hudson River sediments. *Science* **259:**503–507.

23. **Harwood, C. S., N. N. Nichols, M. K. Kim, J. L. Ditty, and R. E. Parales.** 1994. Identification of the *pcaRKF* gene cluster from *Pseudomonas putida:* involvement in chemotaxis, biodegradation, and transport of 4-hydroxybenzoate. *J. Bacteriol.* **176:**6479–6488.

24. **Herman, D. C., Y. Zhang, and R. M. Miller.** 1997. Rhamnolipid (biosurfactant) effects on cell aggregation and biodegradation of residual hexadecane under saturated flow conditions. *Appl. Environ. Microbiol.* **63:**3622–3627.

25. **Hommel, R., and C. Ratledge.** 1990. Evidence for two fatty alcohol oxidases in the biosurfactant-producing yeast *Candida (Torulopsis) bombicola. FEMS Microbiol. Lett.* **58:**183–186.

26. **Hommel, R., O. Stuwer, W. Stuber, D. Haferburg, and H.-P. Kleber.** 1987. Production of water-soluble surface-active exolipids by *Torulopsis apicola. Appl. Microbiol. Biotechnol.* **26:**199–205.

27. **Horowitz, S., J. N. Gilbert, and W. M. Griffin.** 1990. Isolation and characterization of a surfactant produced by *Bacillus licheniformis* 86. *J. Ind. Microbiol.* **6:**243–248.

28. **Hulbert, M. H., and S. Krawiec.** 1977. Cometabolism: a critique. *J. Theor. Biol.* **69:**287–292.

29. **Hutchison, M. L., and D. C. Gross.** 1997. Lipopeptide phytotoxins produced by *Pseudomonas syringae* pv. syringae: comparison of the biosurfactant and ion channel-forming activities of syringopeptin and syringomycin. *Mol. Plant-Microbe Interact.* **10:**347–354.

30. **Inoue, A., and K. Horikoshi.** 1989. A *Pseudomonas* that thrives in high concentration of toluene. *Nature* **338:**264–266.

31. **Jenneman, G. E., M. J. McInerney, R. M. Knapp, J. B. Clark, J. M. Ferro, D. E. Revus, and D. E. Menzie.** 1983. A halotolerant, biosurfactant-producing *Bacillus* species potentially useful for enhanced oil recovery. *Dev. Ind. Microbiol.* **24:**485–492.

*32. **Kauppi, B., K. Lee, E. Carradano, R. E. Parales, D. T. Gibson, H. Eklund, and S. Ramaswamy.** 1998. Structure of an aromatic-ring-hydroxylating dioxygenase—naphthalene 1,2-dioxygenase. *Structure* **6:**571–586.

33. **Kelly, D. P.** 1968. Fluoroacetate toxicity in *Thiobacillus neapolitanus* and its relevance to the problem of obligate chemoautotrophy. *Arch. Mikrobiol.* **6:**59–76.

34. **Kieboom, J., J. J. Dennis, J. A. M. de Bont, and G. J. Zylstra.** 1998. Identification and molecular characterization of an efflux pump involved in *Pseudomonas putida* S12 solvent tolerance. *J. Biol. Chem.* **273:**85–91.

35. **Kieboom, J., J. J. Dennis, G. J. Zylstra, and J. A. M. de Bont.** 1998. Active efflux of organic solvents by *Pseudomonas putida* S12 is induced by solvents. *J. Bacteriol.* **180:**6769–6772.

36. **Kim, J. S., M. Powalla, S. Lang, F. Wagner, H. Lunsdorf, and V. Wray.** 1990. Microbial glycolipid production under nitrogen limitation and resting cell conditions. *J. Biotechnol.* **13:**257–266.

37. **Kowall, M., J. Vater, B. Kluge, T. Stein, P. Franke, and D. Ziessow.** 1998. Separation and characterization of surfactin isoforms produced by *Bacillus subtilis* OKB 105. *J. Colloid Interface Sci.* **204:**1–8.

38. **Kretschmer, A., H. Bock, and F. Wagner.** 1982. Chemical and physical characterization of interfacial-active lipids from *Rhodococcus erythropolis* grown on *n*-alkanes. *Appl. Environ. Microbiol.* **44:**864–870.

39. **Lang, S., and J. C. Philp.** 1998. Surface-active lipids in rhodococci. *Antonie Leeuwenhoek* **74:**59–70.

40. **Lee, Y., S. Y. Lee, and J. W. Yang.** 1999. Production of rhamnolipid biosurfactant by fed-batch culture of *Pseudomonas aeruginosa* using glucose as a sole carbon source. *Biosci. Biotechnol. Biochem.* **63:**946–947.

*41. **Lengler, J. W., and P. W. Postma.** 1999. Global regulatory networks and signal transduction pathways, p. 491–523. *In* J. W. Lengler, G. Drews, and H. G. Schlegel (ed.), *Biology of the Prokaryotes.* Blackwell Science, Stuttgart, Germany.

42. **Lindum, P. W., U. Anthoni, C. Christophersen, L. Eberl, S. Molin, and M. Givskov.** 1998. *N*-Acyl-L-homoserine lactone autoinducers control production of an extracellular lipopeptide biosurfactant required for swarming motility of *Serratia liquefaciens* MG1. *J. Bacteriol.* **180:**6384–6388.

*43. **Lipscomb, J. D.** 1994. Biochemistry of the soluble methane monooxygenase. *Annu. Rev. Microbiol.* **48:**371–399.

44. **Liu, J. Q., T. Kurihara, S. Ichiyama, M. Miyagi, S. Tsunasawa, H. Kawasaki, K. Soda, and N. Esaki.** 1998. Reaction mechanism of fluoroacetate dehalogenase from *Moraxella* sp. B. *J. Biol. Chem.* **273:**30897–30902.

45. **Macdonald, C. R., D. G. Cooper, and J. E. Zajic.** 1981. Surface-active lipids from *Nocardia erythropolis* grown on hydrocarbons. *Appl. Environ. Microbiol.* **41:**117–123.

46. **Maloney, A. P., and H. D. Van Etten.** 1994. A gene from the fungal plant pathogen *Nectria haematococca* that encodes the phytoalexin-detoxifying enzyme pisatin demethylase defines a new cytochrome P450 family. *Mol. Gen. Genet.* **243:**506–514.

47. **Matsuyama, T., M. Fujita, and I. Yano.** 1985. Wetting agent produced by *Serratia marcescens. FEMS Microbiol. Lett.* **28:**125–129.

48. **Matsuyama, T., K. Kaneda, I. Ishizuka, T. Toida, and I. Yano.** 1990. Surface-active novel glycolipid and linked 3-hydroxy fatty acids produced by *Serratia rubidaea. J. Bacteriol.* **172:**3015–3022.

49. **Morikawa, M., H. Daido, T. Takao, S. Murata, Y. Shimonishi, and T. Imanaka.** 1993. A new lipopeptide biosurfactant produced by *Arthrobacter* sp. strain MIS38. *J. Bacteriol.* **175:**6459–6466.

50. **Nakano, M. M., N. Corbell, J. Besson, and P. Zuber.** 1992. Isolation and characterization of *sfp:* a gene that functions in the production of the lipopeptide biosurfactant, surfactin, in *Bacillus subtilis. Mol. Gen. Genet.* **232:**313–321.

51. **Navon-Venezia, S., Z. Zosim, A. Gottlieb, R. Legmann, S. Carmell, E. Z. Ron, and E. Rosenberg.** 1995. Alasan, a new bioemulsifier from *Acinetobacter radioresistens. Appl. Environ. Microbiol.* **61:**3240–3244.

52. **Neu, T. R., T. Hartner, and K. Poralla.** 1990. Surface active properties of viscosin: a peptidolipid antibiotic. *Appl. Microbiol. Biotechnol.* **32:**518–520.

53. **Nielsen, T. H., C. Christophersen, U. Anthoni, and J. Sorensen.** 1999. Viscosinamide, a new cyclic depsipeptide with surfactant and antifungal properties produced by *Pseudomonas fluorescens* DR54. *J. Appl. Microbiol.* **87:**80–90.

*54. **Pedrotta, V., and B. Witholt.** 1999. Isolation and characterization of the *cis-trans*-unsaturated fatty acid isomerase of *Pseudomonas oleovorans* GPo12. *J. Bacteriol.* **181:**3256–3261.

55. **Persson, A., E. Osterberg, and M. Dostalek.** 1988. Biosurfactant production by *Pseudomonas fluorescens* 378: growth and product characteristics. *Appl. Microbiol. Biotechnol.* **29:**1–4.

56. **Ramsay, B., J. McCarthy, L. Guerra-Santos, O. Kappeli, A. Felchter, and A. Margaritis.** 1988. Biosurfactant production and diauxic growth of *Rhodococcus aurantiacus* when using *n*-alkanes as the carbon source. *Can. J. Microbiol.* **34:**1209–1212.

57. **Resnick, S. M., K. Lee, amd D. T. Gibson.** 1996. Diverse reactions caused by naphthalenedioxygenase from *Pseudomonas* sp. strain NCIB9816. *J. Ind. Microbiol.* **17:**438–457.

58. **Rokita, S. E., P. A. Srere, and C. T. Walsh.** 1982. 3-Fluoro-3-deoxycitrate: a probe for mechanistic study of citrate-utilizing enzymes. *Biochemistry* **21:**3765–3774.

59. **Stryer, L.** 1988. *Biochemistry*, 3rd ed. W. H. Freeman and Company, New York, N.Y.

60. **Van Dyke, M. I., P. Couture, M. Brauer, H. Lee, and J. T. Trevors.** 1993. *Pseudomonas aeruginosa* UG2 rhamnolipid biosurfactants: structural characterization and their use in removing hydrophobic compounds from soil. *Can. J. Microbiol.* **39:**1071–1078.

61. **Westheimer, F.** 1987. Why nature chose phosphates. *Science* **235:**1173–1178.

62. **Whitman, W. B., D. C. Coleman, and W. J. Wiebe.** 1998. Prokaryotes: the unseen majority. *Proc. Natl. Acad. Sci. USA* **95:**6578–6583.

63. **Whyte, L. G., S. J. Slagman, F. Pietrantonio, L. Bourbonnier, S. F. Koval, J. R. Lawrence, W. E. Inniss, and C. W. Greer.** 1999. Physiological adaptations involved in alkane assimilation at a low temperature by *Rhodococcus* sp. strain Q15. *Appl. Environ. Microbiol.* **65:**2961–2968.

64. **Yakimov, M. M., H. L. Fredrickson, and K. N. Timmis.** 1996. Effect of heterogeneity of hydrophobic moieties on surface activity of lichenysin A, a lipopeptide biosurfactant from *Bacillus licheniformis* BAS50. *Biotechnol. Appl. Biochem.* **23:**13–18.

65. **Yakimov, M. M., L. Giuliano, V. Bruni, S. Scarfi, and P. N. Golyshin.** 1999. Characterization of antarctic hydrocarbon-degrading bacteria capable of producing bioemulsifiers. *New Microbiol.* **22:**249–256.

66. **Zajic, J. E., H. Guignard, and D. F. Gerson.** 1977. Properties and biodegradation of a bioemulsifier from *Corynebacterium hydrocarboclastus*. *Biotechnol. Bioeng.* **19:**1303–1320.

67. **Zhang, Y., and R. M. Miller.** 1992. Enhanced octadecane dispersion and biodegradation by a *Pseudomonas* rhamnolipid surfactant (biosurfactant). *Appl. Environ. Microbiol.* **58:**3276–3282.

68. **Zylstra, G. J., L. P. Wackett, and D. T. Gibson.** 1989. Degradation of trichloroethylene by *Pseudomonas putida* F1 toluene dioxygenase cloned in *Escherichia coli*. *Appl. Environ. Microbiol.* **55:**3162–3166.

7

Evolution of Catabolic Enzymes and Pathways

Nothing in biology makes sense, except in the light of evolution.

— *Thesodosius Dobzhansky*

To rationalize the structures revealed by the scalpel, sixteenth century anatomists had to invoke God's will. To rationalize the structures revealed by chromatography, twentieth century molecular biologists invoke natural selection.

— *François Jacob*

Chemistry and biology both contribute to studies of microbial biocatalysis. In a very general sense, chemistry offers answers to the "how" questions and biology seeks to answer the "why" questions. The how of chemistry deals with structures and mechanisms. The core of biology is evolution, as wonderfully expressed by Dobzhansky, because it offers an explanation of why living systems arose on Earth and why they have changed over time. When biologists seek to understand why a particular enzyme or pathway came to be, the investigation is best defined as a study in molecular evolution.

Molecular evolution has begun to offer more detailed answers about the relationships of living things to each other. Our knowledge about biological relationships during billions of years of evolution has traditionally been based on fossil records, but fossils are not very useful in the study of single-celled organisms, the predominant biomass and source of genetic diversity on planet Earth. Thus, one must use the sequences of macromolecules—DNA, RNA, and protein—to deduce the evolutionary changes that occurred in the genetic material and its major products. This approach has come to be known as molecular evolution and has had an impact on the study of prokaryotic evolution.

In turn, studies of prokaryotes have contributed a great deal to the field of molecular evolution. In the context of evolution, single-celled prokaryotes may be thought of as metabolic machines that have their enzymatic engines directed toward acquiring carbon, nitrogen, and phosphorus from their environments and assembling those atoms into a new cell. This makes for a relatively direct association between enzymatic activities and bacterial reproductive success, which is sometimes obscured in multicellular organisms by many overlying functions. As François Jacob stated simply, "The goal of a bacterium is to become two bacteria."

The simplicity of studying prokaryotes compared to animals is best illustrated with the following examples. When viewed at the molecular level, a protein involved in a mating display—for example, the antlers on a moose—may help an animal attract a mate and thus enhance its reproductive success. The functioning of this protein would then represent a positive selective force. In this example, multiple genes are probably involved in fashioning the antlers, and the value of large antlers in attracting a mate might be counterbalanced by unwieldiness in movement. Thus, the connection between metabolic performance and reproductive success is indirect. A further disadvantage in studying the selective advantage of the antler protein would be the relatively long generation time of the moose. By contrast, the presence or absence of a catabolic enzyme can bestow on a bacterium the ability to divide or not to divide, and this can be easily measured on the time scale of hours. Thus, the use of bacterial systems and well-described metabolic systems offers the best chance for a meaningful synthesis of the presence of one enzyme and the evolutionary advantage it offers to the organism.

In this chapter, the focus will be on prokaryotic evolution in the context of biodegradation and microbial biocatalysis. With some overlap between topics, we will discuss (i) the major evolutionary families of microbial catabolic enzymes, (ii) current ideas as to how genes are recruited and acquire new functions, and (iii) how new metabolic pathways arise and are disseminated among different prokaryotes. The discussion will start with a brief history of molecular-evolution studies. Then, it will be developed using primarily one example: how soil bacteria have evolved to use the herbicide atrazine as their sole source of nitrogen.

History

> . . . species have changed, and are still slowly changing, by the preservation and accumulation of successive slight favourable variations. I am convinced that Natural Selection has been the main but not exclusive means of modification.
>
> — *Charles Darwin*

> I am convinced that Random Drift acting on Neutral Mutants has been the main but not exclusive means of molecular evolution.
>
> — *Motoo Kimura*

In the latter half of the nineteenth century, Charles Darwin fashioned a comprehensive case for the evolution of new species via natural selection without any knowledge of genes and proteins (9). After Darwin, the science of genetics flourished to explore the basis of phenotype, but genetics also proceeded for decades in ignorance of the chemical structure of the gene. It was only in 1953 that the double-helical structure of DNA was elucidated by James Watson and Francis Crick (55). The structure rapidly provided insight into DNA replication, the chemical basis of mutation, and the central dogma: DNA encodes mRNA, which in turn encodes protein. By the 1960s, it was well appreciated that the genotype was represented by a specific string of DNA nucleotide bases and that the phenotype was determined largely by specific sequences of amino acids in a polypeptide chain.

Since most biologically produced polypeptides are enzymes, a phenotype often results from a specific biochemical reaction or a set of reactions acting in concert. Thus, DNA sequences define the genotype directly, and protein sequences give important information regarding the molecular basis of phenotype. In the 1960s, proteins were being sequenced, and an era of DNA sequencing was about to begin. The historical development of DNA sequencing will be discussed in greater detail in chapter 11, which discusses genomics.

In 1965, Emile Zuckerkandl and Linus Pauling advanced the idea that protein sequences could be considered to be molecular fossils (59). They analyzed the amino acid sequences of hemoglobin molecules from several species and observed that they differed to an extent that was generally consistent with the proposed evolutionary time scale separating those species (Fig. 7.1). This concept became known as the molecular evolutionary clock.

The ticking of the clock became the subject of major lines of investigation, since time translates into evolutionary distance between species. One major question was whether different genes evolve at different rates. Resoundingly, the answer was yes (16). Perhaps not surprisingly, the clock differed markedly with different gene products, as shown in Fig. 7.1. In

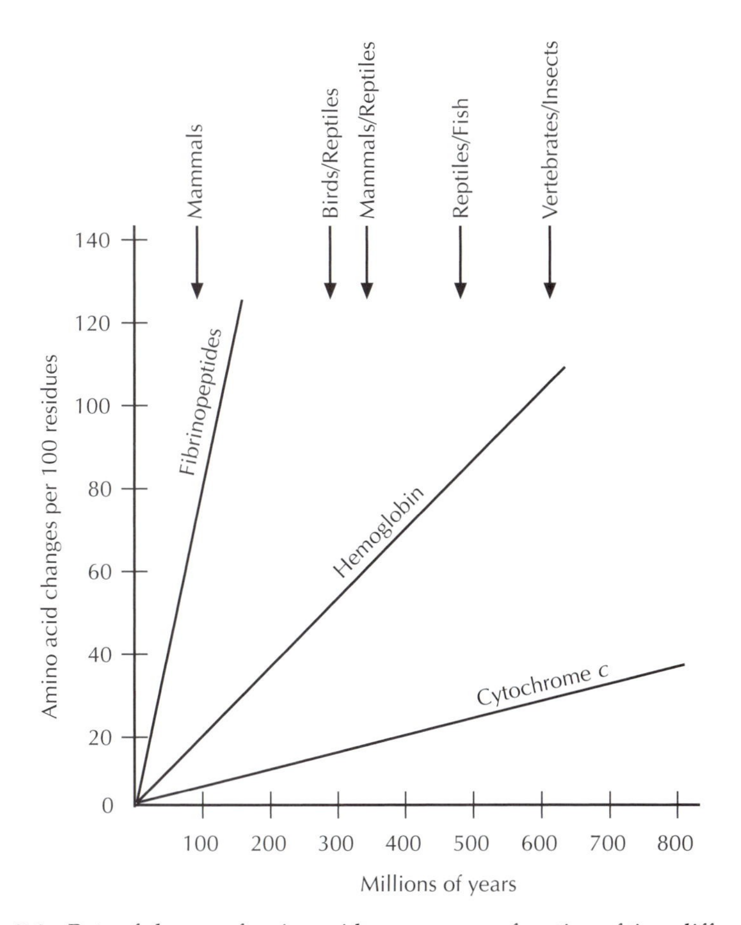

Figure 7.1 Rate of change of amino acid sequence as a function of time differs with different proteins. Adapted from reference 16.

fact, extrapolating the rate of change of fibrinopeptides for more than 100 million years gives on the order of 100% change, a very high rate of evolution. In general, gene products which are more biochemically complex and fundamental to the core functions of a cell have been observed to change at a lower rate.

Ultimately, this discovery led Carl Woese and his colleagues to pioneer the use of 16S rRNA to track the phylogenetic relationships among prokaryotes (57). The concept was based on 16S rRNA being (i) relatively easy to isolate and sequence, (ii) part of the essential protein-synthesizing apparatus of all microbes, and (iii) observed to undergo only very minor sequence changes even among very divergent prokaryotic species. The very slow divergence of 16S rRNA was essential for using one molecule as a phylogenetic marker for all prokaryotes because the prokaryotic evolutionary clock spans billions of years. With most proteins, the sequence difference over such a long evolutionary period would be so great as to show virtually unrecognizable sequence relatedness (Fig. 7.1). In a pairwise comparison of protein sequences, the presence of less than 20% identical amino acids is generally considered low enough to be in what is sometimes known as the twilight zone (18). That is, proteins in which the aligned sequences share less than 20% sequence identity have a high probability of being evolutionarily unrelated. Another way to state this is that a significant number of proteins chosen at random from databases will share 20% sequence identity and yet will be evolutionarily unrelated as shown by other criteria. At this low level of relatedness, other parameters must be used to establish relatedness, such as sharing a common three-dimensional structure or having key mechanistically relevant amino acids conserved.

Major Protein Families in Microbial Biocatalysis

With the current focus on functional genomics, there have been numerous efforts to classify proteins into structurally related families. Since protein structure is thought to be largely conserved during evolution, this classification also demarcates classes of evolutionarily related proteins. The term most commonly used for such a grouping is a protein superfamily. Members of a protein superfamily are said to be homologous, or to derive from a common ancestor.

There is currently a debate as to how many protein families might exist or, stated in a different way, how many different fundamental protein structures, or folds, exist. There is some disagreement about the precise number, but many people think that there might be on the order of 1,000 distinct folds. By this concept, if the many millions of proteins found in all the bacteria, plants, and animals were crystallized and had their structures solved, they would each fall into one of the 1,000 or so classes. This is not to imply that all members of a family are structurally identical but that the general three-dimensional arrangements of structural elements, such as α-helices and β-sheets, are conserved. The variation among family members thus exists mostly in the loops connecting the elements and in the overall sizes of the proteins.

Some microbial protein superfamilies important in biodegradation and biocatalysis are shown in Table 7.1. The major headings in the table generally follow the major Enzyme Commission (EC) headings: (i) oxidoreductases, (ii) transferases, (iii) hydrolases, (iv) lyases, (v) isomerases, and (vi) ligases; except that lyases and isomerases are clustered together. How-

Table 7.1 Microbial biodegradative-enzyme families based on signature sequences[a]

Oxidoreductases

B-class cytochrome P450 monooxygenases
- Cytochrome P450-cam, camphor 5-monooxygenase; *Pseudomonas putida* G786
- Cytochrome P450-BM-1, fatty acid monooxygenase; *Bacillus megaterium*
- Cytochrome P450-soy; *Streptomyces griseus*
- Cytochrome P450-nor, nitric acid reductase; *Fusarium oxysporum*
- Cytochrome P450-terp, terpene monooxygenase; *Pseudomonas* sp.
- Cytochrome P450-lin, linalool 8-monooxygenase; *Pseudomonas incognito*

Aromatic-ring hydroxylases (flavoprotein monooxygenase)
- 4-Hydroxybenzoate hydroxylase; *Pseudomonas fluorescens*
- 2,4-Dichlorophenol 6-monooxygenase; *Streptomyces purpurascens*
- Pentachlorophenol-4-monooxygenase; *Flavobacterium* sp.
- Salicylate hydroxylase; *P. putida*
- Phenol hydroxylase; *Trichosporon cutaneum*

Fungal lignin peroxidase family
- Lignin peroxidase; *Arthromyces ramosus*
- Lignin peroxidase; *Phanerochaete chrysosporium*
- Lignin peroxidase; *Trametes versicolor*

Ring-hydroxylating dioxygenase alpha subunit
- Naphthalene 1,2-dioxygenase alpha subunit; *Pseudomonas* sp. strain NCIB9816
- Toluene 1,2-dioxygenase alpha subunit; *P. putida* F1
- Biphenyl dioxygenase alpha subunit; *Pseudomonas* sp.
- Toluate 1,2-dioxygenase alpha subunit; *P. putida* mt-2
- Benzoate 1,2-dioxygenase alpha subunit; *Pseudomonas arvilla* C-1

Binuclear-iron monooxygenases
- Methane monooxygenase; *Methylosinus trichosporium* OB3B
- Toluene 2-monooxygenase; *Burkholderia cepacia* G4
- Toluene 4-monooxygenase; *Pseudomonas mendocina* KR1
- Alkene monooxygenase; *Xanthobacter* sp. strain Py2

ortho-ring cleavage dioxygenase family
- Catechol 1,2-dioxygenase; *P. putida* mt-2
- Chlorocatechol 1,2-dioxygenase; *Pseudomonas* sp. strain P51
- Protocatechuate 3,4-dioxygenase; *Pseudomonas aeruginosa*

meta-ring cleavage dioxygenase family
- Catechol 2,3-dioxygenase, *P. putida*
- 2,3-Dihydroxybiphenyl-1,2-dioxygenase; *Burkholderia* sp. strain LB400
- 3,4-Dihydroxyphenylacetate 2,3-dioxygenase; *Arthrobacter globiformis* CM-2

Alcohol dehydrogenase superfamily signature
- 3-Beta-hydroxysteroid dehydrogenase; *Pseudomonas testosteroni*
- 15-Hydroxyprostaglandin dehydrogenase; *Streptomyces* sp.
- 7-Alpha-hydroxysteroid dehydrogenase; *Eubacterium* sp.
- *cis*-1,2-Dihydro-1,2-dihydroxynaphthalene dehydrogenase; *Pseudomonas* sp.
- *cis*-2,3-Dihydrobiphenyl-2,3-diol dehydrogenase; *Pseudomonas* sp. strain KKS102
- Halohydrin hydrogen-halide-lyase; *Corynebacterium* sp.

(continued next page)

Table 7.1 *Continued*

Transferases
Glutathione transferase family
Dichloromethane dehalogenase; *Methylophilus* sp. strain DM11
Tetrachlorohydroquinone reductase; *Flavobacterium* sp.
Arylglycerol beta-aryletherase; *Sphingomonas paucimobilis*
Hydrolases
Amidohydrolase superfamily
Atrazine chlorohydrolase; *Pseudomonas* sp. strain ADP
Hydroxyatrazine *N*-ethylaminohydrolase; *Pseudomonas* sp. strain ADP
N-Isopropylammelide isopropylaminohydrolase; *Pseudomonas* sp. strain ADP
Phosphotriesterase; *Pseudomonas diminuta*
Urease; *Klebsiella aerogenes*
Alpha-beta hydrolase fold
Haloalkane dehalogenase; *Xanthobacter autotrophicus*
Haloacetate dehalogenase H-1; *Pseudomonas* sp. strain YL
Epoxide hydrolase; *Agrobacterium radiobacter* AD1
Carboxylesterase; *P. fluorescens*
Polyhydroxybutyrate depolymerase; *Alcaligenes faecalis*
2-Hydroxymuconic semialdehyde hydrolase; *P. putida*
Lyases and isomerases
Enolase/enoyl-CoA superfamily
Muconate cycloisomerase; *P. putida*
Chloromuconate cycloisomerase; *Alcaligenes eutrophus* JMP134 (pJP4)
Dichloromuconate cycloisomerase; *Xanthobacter flavus* 14p1
Mandelate racemase; *P. putida*
D-Glucarate dehydratase; *P. putida*
4-Chlorobenzoyl–CoA dehalogenase; *Pseudomonas* sp. strain CBS3
Feruloyl-CoA hydratase/lyase; *Pseudomonas* sp.
2-Ketocyclohexanecarboxyl–CoA hydrolase; *Rhodopseudomonas palustris*
Ligases
Acyl-adenylate/thioester-forming enzyme family
4-Chlorobenzoate–CoA ligase; *Pseudomonas* sp. strain CBS3
2-Aminobenzoate–CoA ligase; *Pseudomonas* sp. strain KB 740

[a]From various sources, including the PRINTS Fingerprint Database (http://www.biochem.ucl.ac.uk/bsm/dbbrowser/PRINTS/PRINTS.html).

ever, it should be realized that the protein family classification is based on homology, which makes for common structures but not necessarily common function. For example, this is apparent in the enolase superfamily (Table 7.1), which contains hydrolases, hydratase, lyases, and isomerases. In evolution, common structural and mechanistic features of a protein may be used to do what seem to be rather different overall reactions, and thus proteins in the same family can be seen under different EC headings. This is discussed in greater detail in the section "Gene Transfer in the Evolution of Catabolic Pathways" below and in chapter 12 in the section "Enzyme Plasticity and New Biocatalysts."

The list in Table 7.1 is far from exhaustive. In general, the best-studied members of the protein families are shown. Several families, for example the aromatic hydrocarbon dioxygenases and cytochrome P450 monooxygenases, each contain hundreds of known members. For each specific protein identified under the major headings, only one bacterial source for each protein is shown, although other examples may exist. More information on many of these enzymes can be found in the University of Minnesota Biocatalysis/Biodegradation Database (http://umbbd.ahc.umn.edu).

Principles of Evolution Applied to Microbial Catabolism

Prokaryotes have recycled organic matter on Earth over eons, but in a very short period on an evolutionary time scale, organic chemists have fashioned millions of new organic structures, posing new challenges to microbial metabolism. Microbes have responded to the challenge. The scientific literature suggests that in some cases, industrial chemicals initially evade microbial catabolism, as evidenced by their persistence in the environment, but are later found to be readily biodegraded. Examples include polychlorinated biphenyls (5), tetrachloroethene (35), and atrazine (7); all are compounds that in early studies were poorly, if at all, cleared from the environment. All of these compounds have more recently been well documented to be metabolized by bacteria and are readily cleared in some environments. These are, however, not systematic comparisons. The early and later studies were conducted by different researchers using different methods, and thus, direct comparisons are difficult. One explanation for the data is merely that scientists get more clever in identifying bacteria capable of biodegrading new compounds over time and thus experience greater success with additional experience. It could be that microorganisms collectively capable of biodegrading tens of millions of compounds are always present in the soil, waiting for the right mix to fall on them. Do the observations of "new biodegradation" reflect the evolution of new enzymes to handle new compounds, or might preexisting enzymes have proliferated due to selective pressure posed by the organic pollutant?

Like the fossil record, there are gaps in the records of molecular evolution. For example, there are scores of serine proteases or catabolic Reiske-iron center dioxygenases mentioned in the literature; other enzymes, such as acetylene hydratase or styrene oxide isomerase, have been discovered in at most a few bacteria. This is no doubt attributable to the vastness of the microbial world and the fact that we have sampled such a small part of it and that in a nonhomogeneous manner.

Another problem is the fact that the known protein sequences encoding a certain biochemical function have been found incompletely and irregularly in nature. For example, isofunctional catabolic enzymes from different microbes are often found to have 30 to 70% sequence identity in pairwise comparisons, with changes interspersed throughout the sequence. This reflects a clear ancestral relationship. However, it is also clear that they are separated by millions of years of evolutionary change. It may be that the sequences intermediate between the two have become extinct; however, given that greater than 10^{30} prokaryotes exist on Earth, it is more likely that those sequences are lying in the soil and water waiting to be discovered.

In light of these gaps in molecular information, how are we to infer what changes evolution has wrought on bacterial catabolic enzymes? First, general principles have been deduced for thinking about naturally occurring enzymes which have diverged from a common ancestor. Second, directed-evolution studies have been conducted to learn how microbes can adapt to well-defined conditions which can be imposed in the laboratory.

One can think of changes to an enzyme as falling into several categories. First, there are neutral mutations in the gene which do not change an amino acid in the protein due to the redundancy in the genetic code. Then, there are changes in a DNA base which change a codon into a different one

Where Do New Catabolic Enzymes Come From?

An influential paper from over 50 years ago postulated that catabolic pathways would likely evolve from the last reaction of the pathway out toward the first one (27). That is, the acquisition of a single new reaction would have a selective advantage if its product funneled into an existing catabolic pathway or intermediary pathway. Then, another reaction could be added to feed the substrate into the previous new enzyme, thus expanding the range of growth substrates a little at a time. It had been thought that enzymes in a pathway likely arose from a common ancestral gene via duplication of the primordial gene (27). This would give rise to the often-observed operonic structure of catabolic genes which function to encode enzymes in the same pathway. Moreover, the idea was advanced that an enzyme able to bind to a given compound could readily give rise to an enzyme that would transform a new substrate into that compound, since that enzyme would already have a compatible binding site for that general structure.

Since 1945, there have been many examples in which all enzymes in a microbial catabolic pathway have been sequenced, and in some cases, structures have been determined for a number of the enzymes. In most of those examples, the genes in the pathway, even when clustered in an operon, largely encode proteins in different superfamilies and are thus not derived from each other (Fig. 7.2). Figure 7.2 shows one representative pathway found in a *Pseudomonas* species involved in catabolism of the aromatic acid mandelate. The enzymes in the pathway shown take mandelate to benzoate, a common intermediate in bacterial aromatic-acid metabolism. The reactions catalyzed are, consecutively, isomerization, dehydrogenation, decarboxylation, and dehydrogenation. Each enzyme, in all pairwise sequence comparisons, has been deduced to bear no discernible homology to the others. This is true even for the two dehydrogenases, which come from two distinct dehydrogenase protein families.

If they are not related to each other, where did these enzymes come from? The right side of Fig. 7.2 shows enzymes known to be homologous to the mandelate pathway enzymes. In at least one case, the comparison is somewhat surprising. The first homolog to mandelate racemase determined by X-ray structure comparisons of the enzymes was muconate-lactonizing enzyme (40). While the sequence identity between the two enzymes is relatively low, the backbone structures are nearly superimposable. Moreover, the active sites are fairly conserved, and mechanistic studies show how adaptations at the active site can make an epimerization reaction into a ring cyclization reaction or vice versa, depending on which came first. In comparing other enzymes in the pathway, the mandelate dehydrogenase was found to be homologous to glycolate oxidase. Both enzymes use flavin mononucleotide in redox catalysis, so this makes sense. Benzylformate decarboxylase and benzaldehyde dehydrogenase are homologous to other enzymes catalyzing analogous reactions with different substrates.

The ideas represented in Fig. 7.2 have been developed further. The versatility of enzyme active sites is now seen to allow the development of slightly different constellations of active-site amino acids to fashion new catalysts which carry out what appear to be fundamentally different reactions. For example, a class of enzymes has become known as the vicinal oxygen chelate fold (see the PRINTS database [http://www.biochem.ucl.ac.uk/bsm/dbbrowser/PRINTS/PRINTS.html]), which contains an active site motif known as the 2-His-1-carboxylate facial triad (33). There are enzymes in this family catalyzing distinct reactions which bear no apparent similarity to each other. For example, members of this enzyme family catalyze dioxygen addition to a substrate, a glutathione transferase reaction, and an isomerization reaction. These reactions can be viewed in chapter 12 (Fig. 12.3), where they are discussed in more detail. The key point to be made here is that similar active sites can catalyze very different overall reactions and that it is the key active-site features, encased in the supporting overall structural scaffolding, that provide the raw materials for a random evolutionary process giving rise to new needed chemical reactivities. Put another way, in the fashioning of new catalysts, the choice of precursor enzyme is likely to be guided by the chemical mechanism capable of providing the desired catalytic function (24).

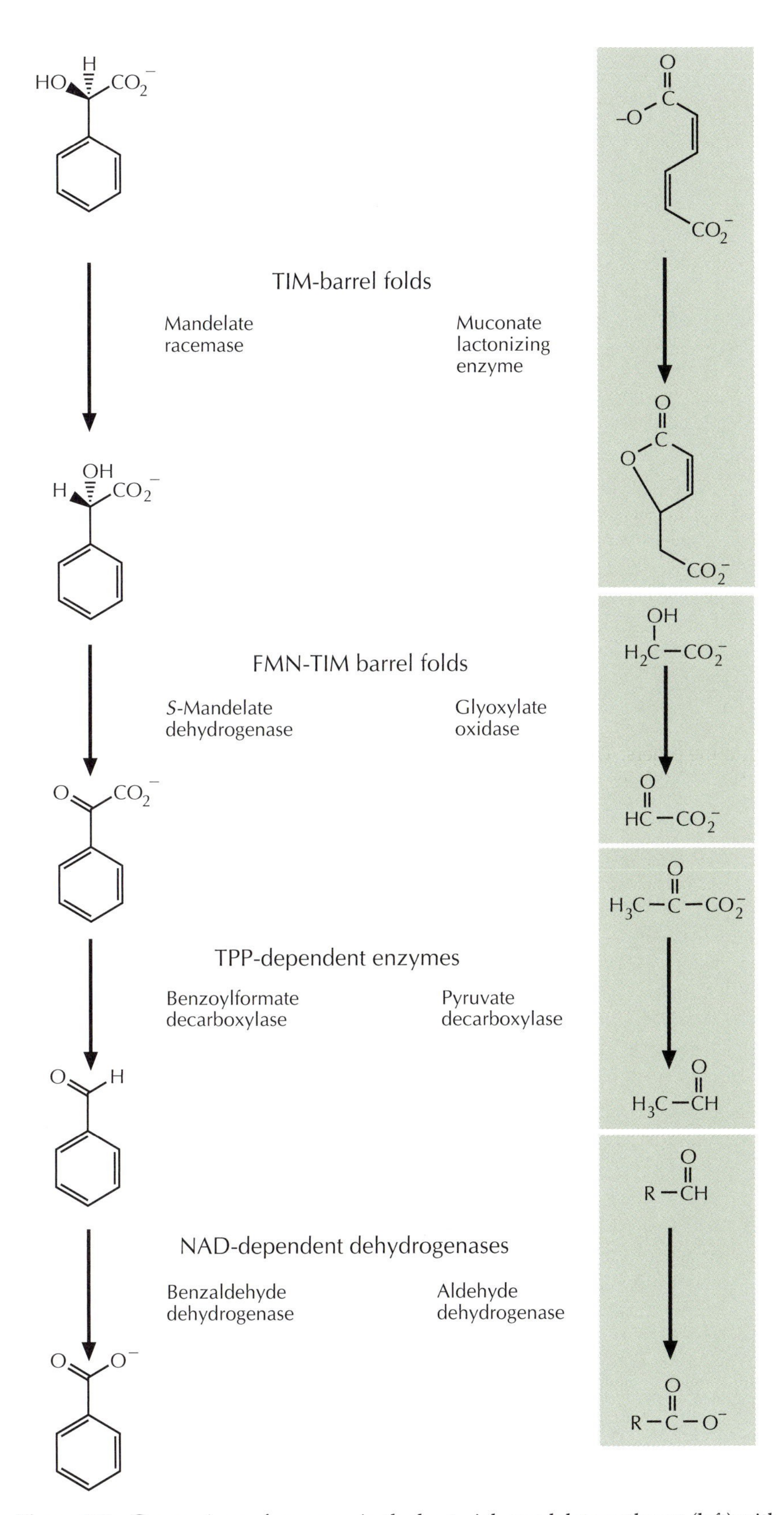

Figure 7.2 Comparison of enzymes in the bacterial mandelate pathway (left) with other evolutionarily related enzymes which catalyze reactions with structurally distinct substrates.

encoding a similar amino acid, for example, a leucine to an isoleucine, in a region of the protein not affecting the folding pathway or the catalytic mechanism. Such a conservative change may have no discernible effect on the protein's function. Then, there might be changes in the protein which have no selective advantage or disadvantage. An example would be the diminution in the upper temperature limit from 70 to 60°C for a protein expressed in a mesophilic soil bacterium. If all other properties were the same, this trait would not be selected against in that organism.

It should be considered that most enzyme mutations, if not deleterious, are neutral. In *Drosophila melanogaster,* 39 out of 40 DNA polymorphisms were observed to yield no difference in the protein sequence of the alcohol dehydrogenase (32). In *Escherichia coli,* in studies involving four genes, most (a total of 24) naturally occurring enzyme mutations yielded no detectable selective advantage (19). In another example, even under strong selection pressure, many laboratory-generated mutations in β-galactosidase were selectively neutral (10). All of these changes fall under the umbrella of Motoo Kimura's neutral theory of evolution. Trivial changes in amino acid sequence can accrue, not due to selective pressure but to random survival of some changes rather than others that become fixed in bacterial populations over time.

Of course, mutations are known to affect enzyme function, with most of the effects being adverse. This has become manifest from decades of mutagenesis induced by X-rays, chemical agents, and, more recently, site-directed mutagenesis. Deleterious mutations may affect the active site in a protein folded essentially correctly or may alter the folding pathway of the enzyme such that the active site is never formed.

What about mutations that do not change the enzyme's folding or catalytic mechanism but alter the substrate specificity? With the enzymes of intermediary metabolism—in which many enzymes transform one substrate, or a small set of cosubstrates, to a fixed product(s)—alterations in substrate specificity may be strongly selected against. In catabolism, or biodegradation, alterations in substrate specificity may be neutral or have a minor positive or negative selective value. As we have seen in the previous chapter, some catabolic enzymes have a very broad substrate specificity; over 70 substrates are known for naphthalene dioxygenase (see Table 6.2), and over 100 are known for toluene dioxygenase (http://umbbd.ahc.umn.edu/tol/tdo.html). In these examples, many soil bacteria expressing catabolic dioxygenases are increasingly being identified as expressing a set of homologous proteins with distinct but overlapping substrate specificities (43). This suggests that broad specificity in the catabolism of certain classes of compounds, such as aromatic hydrocarbons, is selectively advantageous to the organism. Since naturally occurring petroleum contains thousands of aromatic hydrocarbons, it makes sense that it is better to be able to feed on many of the components in the mixture rather than a few.

In the context of evolutionary adaptation to new chemical compounds, another important functional change in a catabolic enzyme is its alteration to catalyze a different chemical reaction. The meaning of "different reaction" is defined here as one occurring with a different functional group (as defined in chapter 5). An example with two homologous enzymes will help define this further. The enzymes enoyl-coenzyme A (CoA) hydratase and 4-chlorobenzoyl–CoA dehalogenase catalyze the addition of water to a double bond and the hydrolytic dehalogenation of a chlorinated aromatic-ring compound, respectively (45). Neither member of the pair will catalyze the other's reaction, so this is not a case of overlapping reactivity. These two enzymes show 56% amino acid sequence identity in pairwise comparison and have very similar three-dimensional structures (58), but they clearly manifest different biological functions.

Gene Transfer in the Evolution of Catabolic Pathways

In the section above, the evolution of new functions de novo was addressed. This is clearly important to expand the range of organic compounds which microbes, collectively, can catabolize. When mutational events have succeeded in generating an effective new catalyst, the genes encoding that catalyst will proliferate if selection is imposed in that environment. More accurately, the successful microorganism will proliferate. But we see more than one species of bacteria with many of the catabolic genes and enzymes which have been studied. While this may partly reflect ancient evolutionary adaptation and maintenance of the gene during speciation, there are clearly other forces at work. A major facet of catabolic-pathway evolution is the mixing and matching of genes among organisms.

Likely, gene acquisition by microbes is more important in new pathway construction than gene duplication and mutation giving rise to the next needed enzyme to broaden a pathway. And, in fact, there is a large and growing body of information on the mechanisms of catabolic-gene transfer and acquisition in bacteria. Catabolic plasmids were discovered over 25 years ago (6, 42). Plasmids are defined as independently replicating genetic elements, often circular, and mostly between 1 and 500 kbp long. Some plasmids contain genes encoding their transfer to other strains, and the host range of transfer can be quite broad; for example, some transfer broadly among gram-positive bacteria. Plasmids often contain genes which help the host organism to survive under specific conditions by conveying resistance to antibiotics or mercury, or they contain genes for the catabolism of nonintermediary organic compounds. In the 1970s, Chakrabarty and others showed that plasmids can contain complete operons for the metabolism of such compounds as camphor and naphthalene. Most catabolic transposons (see below) are conjugative; that is, they move via a conjugationlike process to another cell. While this can confer new pathways on a bacterium en masse, it does not explain the construction of new pathways to meet the demands of a new chemical structure which might appear in the environment. One mechanism for new catabolic-plasmid construction is the occurrence of catabolic transposons (49).

Transposons are mobile genetic elements; they are transposable to other genetic elements within or among organisms. Simple transposons have insertion sequence elements and a transposase gene which codes for an enzyme which mediates transposition. Composite transposons have the same elements as the simple transposons and carry additional genes. Catabolic transposons (Table 7.2) are by definition composite transposons, having both the vehicle for movement and the catabolic genes which undergo passive transfer.

The transfer of catabolic genes can have several effects. Within a cell, it can rearrange genes into clusters and bring genes under the control of new regulatory elements. Thus, transposons could play a role in developing new operonic structures. Transposons can also play a role in constructing new metabolic pathways by bringing new genes into a cell that complement

Table 7.2 Catabolic transposons in bacteria[a]

Transposon	Size (kb)	Catabolic marker (genes)	IS[b]	Location	Organism	Reference
Class I transposons						
Tn*5271*	17	Chlorobenzoate (*chaABC*)	IS*1071*	pBRC60	*Alcaligenes* sp. strain BR60	38
Tn*5280*	8.5	Chlorobenzene (*tcbAB*)	IS*1066* IS*1067*	pP51	*Pseudomonas* sp. strain P51	53
Tn*5542*	12	Benzene (*hedDC1C2BA*)	IS*1489*	pHMT112	*P. putida* ML2	22
		Chlorinated aliphatic acids (*deh112*)	IS*1071*	pUO1	*Moraxella* sp. strain B	31
		Nylon oligomers (*nylABC*)	IS*6100*	pOAD2	*Flavobacterium* sp. strain K172	30
Transposons of unknown class and transposon-like elements						
Tn*4371*	59	Chlorobiphenyl (*bph* and *cbpABCD*)	ND	Chromosome	*Alcaligenes eutrophus* A5	48
	90	Biphenyl and salicylate (*bph* and *sal*)	ND	Chromosome	*P. putida* KF715	39
	26	Aniline (*tdnOTA1A2BR*)	IS*1071*	pTDN1	*P. putida* UCC22	23
		p-Toluenesulfonate and sulfobenzoate (*tsaMBCDR* and *pshAC*)	IS*1071*	pTSA	*Comamonas testosteroni* T2	29
		4-Carboxydiphenyl ether (*pobAB*)	IS*1071*	pPOB	*Pseudomonas pseudoalcaligenes* POB310	11
		Monobromoacetate (*dhlB*)	IS*1247*	Chromosome	*Xanthobacter autotrophicus* GJ10	54
Class II transposons						
Tn*4651*	56	Toluene (*xyl*)	46[c]	pWW0	*P. putida* mt-2	50
Tn*4652*	17	None		Chromosome	*P. putida* PaW85	26
Tn*4653*	70	Toluene (*xyl*)	38[c]	pWW0	*P. putida* mt-2	51
Tn*4655*	38	Naphthalene (*nah*)	38[c]	NA117	*P. putida* G7	52
Tn*4656*	39	Toluene (*xyl*)	38[c]	pWW53	*P. putida* MT53	56

[a]Data from reference 49 with permission of Springer-Verlag.
[b]IS, insertion sequence. ND, not determined.
[c]Size of inverted repeats in base pairs.

genes already present. In the section below, transposons will be discussed in the context of the evolution of the catabolism of the herbicide atrazine. This example will serve to illustrate potential new enzyme evolution and how the enzymes can be brought together in the same cell to constitute a new metabolic pathway.

Case Study: Enzyme Evolution in the Amidohydrolase Protein Superfamily

In biodegradation, the choice of which compounds to study is inevitably skewed toward those of greatest economic and environmental significance. The herbicide atrazine is a major agricultural chemical used in the United States, Europe, and Israel. Moreover, over 1 billion pounds of atrazine have been applied globally; it is considered only moderately biodegradable, and it is sometimes found in ground and surface waters. These issues have spurred research on the microbial catabolism of atrazine.

Atrazine, 2-chloro-4-(ethylamino)-6-(isopropylamino)-1,3,5-triazine, is one member of a class of *s*-triazine pesticides which have been used commercially since 1960. We are not aware of reports of natural-product *s*-triazine ring compounds, although the natural-product antibiotic fervenulin contains an asymmetrical *s*-triazine ring. In this context, it was not surprising to many people that atrazine was considered to be poorly to moderately biodegradable for the first several decades of its use. Moreover, the molecular genetic basis of atrazine catabolism to carbon dioxide, ammonia, and chloride was largely undefined before 1993. Earlier reports indicated that atrazine was catabolized largely via oxidative dealkylation reactions (Fig. 7.3) that removed either or both of the ethyl and isopropyl side chains (7, 21). These metabolites were often observed in soils, and thus, oxidative dealkylation was thought of as a dead-end metabolism. The genes encoding bacterial oxidative dealkylation reactions, *thcBCD*, and the cytochrome P450 monooxygenase enzyme system they encode, have been described by Shao and Behki (47). A similar bacterial system catalyzing enzymatic dealkylation has been described by Nagy and coworkers (37).

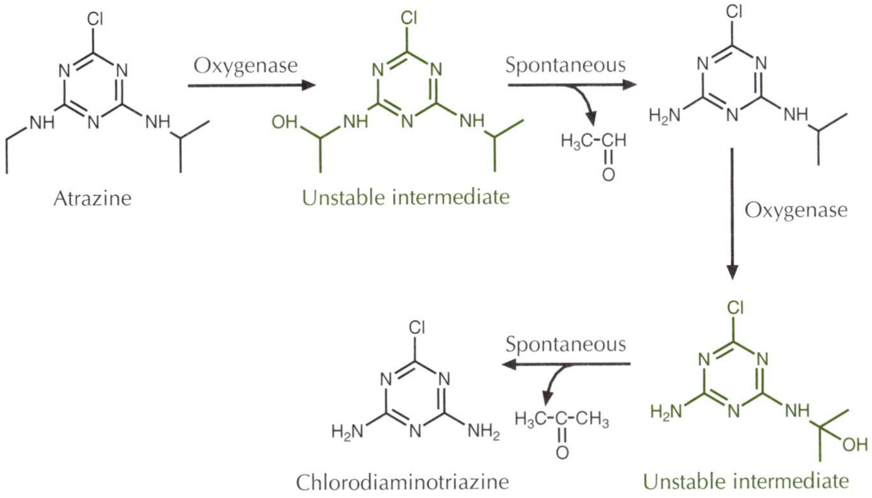

Figure 7.3 Oxygenative route of microbial atrazine catabolism, which generally results in the accumulation of aminotriazine metabolites.

A

AtzA → Atz B → Atz C

Atrazine Hydroxyatrazine *N*-Isopropylammelide Cyanuric acid

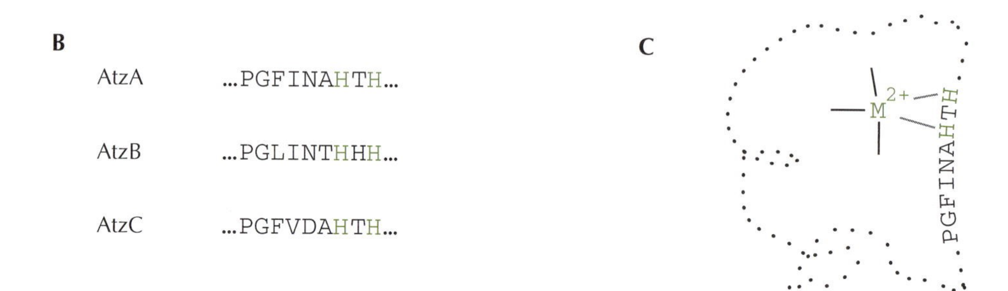

Figure 7.4 Pathway of atrazine catabolism catalyzed by *Pseudomonas* sp. strain ADP and other atrazine-catabolizing bacteria. (A) First three steps in atrazine catabolism. (B) Sequence identity in a short stretch of AtzABC. (C) Hypothetical divalent metal coordination by AtzABC.

Most recently, a number of bacteria were described that rapidly catabolize atrazine with liberation of ammonia and carbon dioxide derived from the triazine ring carbon atoms (34, 41). Some of those strains were shown to produce hydroxyatrazine as an intermediate during atrazine metabolism (Fig. 7.4A). A *Rhizobium* sp. stoichiometrically metabolizes atrazine to hydroxyatrazine (4). These and other studies convincingly countered the prevailing hypothesis that hydroxyatrazine in the environment is derived exclusively via abiotic soil-catalyzed hydrolysis of the carbon-chlorine bond (1, 21).

Subsequently, the molecular genetics and enzymology underlying hydrolytic atrazine catabolism were revealed by using *Pseudomonas* sp. strain ADP (Fig. 7.4A). First, *E. coli* clones containing the *atzA* gene from *Pseudomonas* sp. strain ADP stoichiometrically convert atrazine to hydroxyatrazine (13). An *E. coli* clone containing a somewhat larger DNA fragment from *Pseudomonas* sp. strain ADP transiently produces hydroxyatrazine, which is subsequently converted to *N*-isopropylammelide (14). The *atzB* gene is responsible for this transformation and encodes an enzyme catalyzing a hydrolytic deamidation reaction (3). The *atzC* gene was subsequently cloned and shown to encode another hydrolytic deamidase that transforms *N*-isopropylammelide to cyanuric acid (44).

Elucidation of the first three steps in atrazine catabolism by *Pseudomonas* sp. strain ADP (Fig. 7.4A) revealed the metabolic and evolutionary logic underlying atrazine catabolism. These three consecutive enzyme-catalyzed hydrolysis reactions transform atrazine, a compound few bacteria can catabolize, to cyanuric acid, a compound many soil bacteria can metabolize (7, 8, 20). Common soil bacteria with a cyanuric acid pathway that acquire enzymes isofunctional to AtzA, AtzB, and AtzC will be able to metabolize atrazine to carbon dioxide and liberate ammonia to use as a nitrogen source. But what are the evolutionary origins of atrazine chlorohydrolase (AtzA), hydroxyatrazine *N*-ethylaminohydrolase (AtzB), and *N*-isopropylammelide *N*-isopropylaminohydrolase (AtzC)?

The answers come from sequence comparisons. The protein sequences of AtzA, AtzB, and AtzC were derived from the corresponding gene sequences. Pairwise comparisons of these sequences and those in the GenBank database did not reveal any striking evolutionary relationships; percent sequence identities were only on the order of 20%. It was revealing, however, to compare the sequences to conserved motifs of different hydrolytic-enzyme families (41). Significant sequence identity was observed in comparing AtzA, -B, or -C (a very minor stretch is shown in Fig. 7.4B) with the conserved sequence motif discovered in a large amidohydrolase protein superfamily (25). For the structurally defined superfamily members, the conserved region includes amino acids involved in providing ligands to a functionally important transition metal (Fig. 7.4C). This enzyme class includes bacterial adenine deaminase, adenosine deaminase, and cytosine deaminase. Thus, there is a potential mechanistic connection between metalloenzymes that catalyze purine or pyrimidine hydrolytic deamination reactions and the atrazine enzymes AtzABC that catalyze hydrolytic deamination or dechlorination reactions, also with nitrogen heterocyclic substrates. The extensive sequence divergence of AtzA, -B, and -C indicates that one did not give rise to another recently. Rather, AtzA, -B, and -C all arose from protein members of an ancient amidohydrolase superfamily and have presumably reunited in *Pseudomonas* sp. under the selective pressure imposed by the presence of greater than 1 billion pounds of atrazine in the environment. That all the enzymes in the pathway are in the same enzyme superfamily is not typical. This differs from the more typical example given previously for the (*R*)-mandelate pathway of *Pseudomonas putida* (see Fig. 7.2).

How do genes and enzymes of diverse mechanisms and divergent evolutionary origins come together to constitute new metabolic pathways? And how do these pathways disseminate globally once formed? The atrazine catabolic pathway offers one example of how this occurs. Initially, it was discovered that DNA which is highly homologous to the *Pseudomonas* sp. strain ADP *atzABC* genes is present in six recently isolated atrazine-catabolizing bacteria belonging to disparate genera: *Ralstonia, Alcaligenes, Agrobacterium,* and *Clavibacter*. It was striking that the DNA sequences in the 600-bp PCR-amplified DNA product were 99% or more identical, virtually within the margin of error of DNA-sequencing methods. This is even more striking when one considers the geographical locations of the atrazine-catabolizing strains (now over 1 dozen) for which at least some of the *atz* genes have been sequenced (Fig. 7.5). Highly homologous genes have been shown to be present in soil microorganisms from around the United States, as well as Europe, Israel, Asia, and Australia.

In *Pseudomonas* sp. strain ADP, the *atzABC* genes reside on a 96-kb broad-host-range plasmid that has been denoted pADP-1 (15). Plasmids have been detected in the other recently isolated atrazine-catabolizing bacteria described above (Fig. 7.6). In fact, in-well lysis methods and gel electrophoresis establish that three to six plasmids are present in all the atrazine bacteria. The atrazine catabolism genes *atzA, atzB,* and *atzC* are present on these plasmids. It is particularly illuminating that the atrazine plasmids do

Figure 7.5 Global distribution of bacteria isolated for their abilities to catabolize atrazine and known to process atrazine via an initial dechlorination reaction.

not have identical sizes in different bacteria. Additionally, there are different clusterings of atrazine genes on the plasmids in the different bacteria. Some plasmids have all three *atzABC* genes, and others have only one or two.

The sizes and gene distributions of the plasmids are consistent with observations that IS*1071* elements flank genes on plasmid pADP-1 in *Pseudomonas* sp. strain ADP. In fact, IS*1071* elements flank the *atzA* gene and the *atzA* and *atzB* gene pair. The two atrazine genes are separated by approximately 10 kb of DNA having an unknown function. This arrangement suggests that transposable cassettes can exist in which *atzA*, *atzB*, or *atzAB* together could come out of, or move into, a replicon. There is substantial evidence that this can occur. First, the IS*1071*-like elements in plasmid pADP-1 have greater than 99% sequence identity to those on known catabolic transposons. Second, selective losses of *atzA* and *atzAB* have been observed to occur spontaneously, suggesting the IS*1071* elements are active. Third, we observe that IS*1071* elements are present in the different atrazine-metabolizing bacteria with plasmids, and some of the plasmids contain *atzA* or *atzC* in isolation. Transposons have been linked to the global distribution of other catabolic genes. For example, identical or nearly identical gene sequences have been detected in geographically diverse bacteria ca-

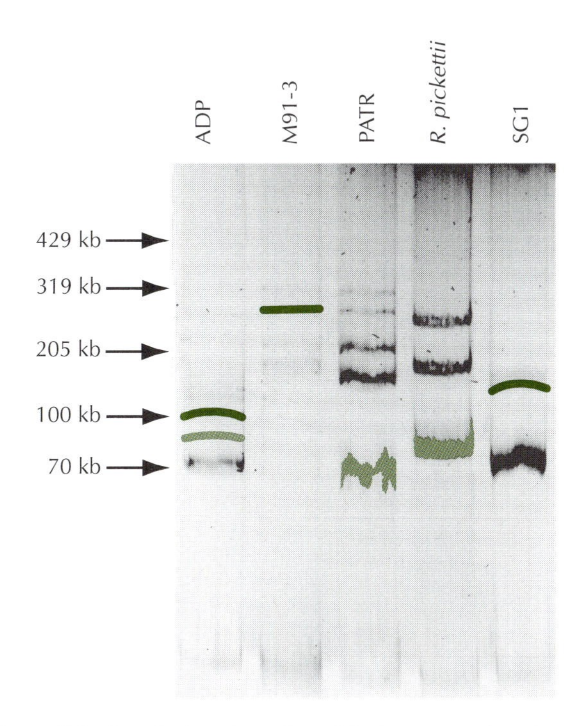

Figure 7.6 In-well lysis gel showing plasmids from atrazine-degrading bacteria containing atrazine catabolism genes (shown in green). From left to right, the DNAs in the wells are derived from the following bacterial strains: *Pseudomonas* sp. strain ADP; *Ralstonia* sp. strain M91-3, *Rhizobium* sp. strain PATR 2, *Ralstonia pickettii*, and *Alcaligenes* sp. strain SG1.

tabolizing phosphotriester pesticides such as parathion (36), 3-chlorobenzoate (17), and naphthalene (46).

The high sequence identity of the atrazine genes suggests gene transfer between strains. The presence of the genes on plasmids suggests that catabolic plasmids are important in global atrazine catabolism. The facts that the plasmids are different sizes and that transposable elements flank the genes suggest that the transposons move and establish different replicons in different bacteria. The same genes have transposed, perhaps carried by the same flanking sequences, into different genera of bacteria to ultimately construct duplicate metabolic pathways.

Summary

Evolution is at the heart of our understanding of biology. Molecular evolution has become an incisive tool for deducing evolutionary relationships. Molecular evolution is a relatively young field; it began with the advent of protein and nucleic acid sequencing. It is particularly powerful when applied to microorganisms, whose high rate of reproduction makes them prime subjects for study. It is especially instructive to study microbial catabolic enzymes to learn how new enzymes evolve and are transferred globally. We now appreciate that microbial genes can evolve quickly, in several as opposed to millions of years. Microbial catabolic genes are also highly mobile, and this further contributes to the rapid adaptation of microbes to a changing environment. This has important implications for our responses to environmental pollution and for the use of microbes as catalysts in fermentors. In this context, the study of evolution is expanding from its historical role of providing fundamental explanations to become a biological tool that offers great promise for using microbes in our efforts to develop a sustainable human society.

References

1. **Armstrong, D. E., and G. Chesters.** 1968. Adsorption catalyzed chemical hydrolysis of atrazine. *Environ. Sci. Technol.* **2:**683–689.

*2. **Babbitt, P. C., and J. A. Gerlt.** 1997. Understanding enzyme superfamilies: chemistry as the fundamental determinant in the evolution of new catalytic activities. *J. Biol. Chem.* **272:**30591–30594.

3. **Boundy-Mills, K. L., M. L. de Souza, R. T. Mandelbaum, L. P. Wackett, and M. J. Sadowsky.** 1997. The *atzB* gene of *Pseudomonas* sp. strain ADP encodes the second enzyme of a novel atrazine degradation pathway. *Appl. Environ. Microbiol.* **63:**916–923.

4. **Bouquard, C., J. Ouazzani, J.-C. Prome, Y. Michel-Briand, and P. Plesiat.** 1997. Dechlorination of atrazine by a *Rhizobium* sp. isolate. *Appl. Environ. Microbiol.* **63:**862–866.

5. **Brown, J. F., D. L. Bedard, M. J. Brennan, J. C. Carnahan, H. Feng, and R. E. Wagner.** 1987. Polychlorinated biphenyl dechlorination in aquatic sediments. *Science* **236:**709–712.

6. **Chakrabarty, A. M., G. Chou, and I. C. Gunsalus.** 1973. Genetic regulation of octane dissimilation plasmid in *Pseudomonas. Proc. Natl. Acad. Sci. USA* **70:**1137–1140.

7. **Cook, A. M.** 1987. Biodegradation of s-triazine xenobiotics. *FEMS Microbiol. Rev.* **46:**93–116.

8. **Cook, A. M., P. Beilstein, H. Grossenbacher, and R. Huetter.** 1985. Ring cleavage and degradative pathway of cyanuric acid in bacteria. *Biochem. J.* **231:**25–30.

9. **Darwin, C.** 1859. *On the Origin of Species by Means of Natural Selection, or The Preservation of Favoured Races in the Struggle for Life.* John Murray, London, England.

10. **Dean, A. M., D. E. Dykhuizen, and D. L. Hartl.** 1986. Fitness as a function of β-galactosidase activity in *E. coli. Genet. Res.* **48:**1–8.

11. **Dehmel, U., K. H. Engesser, K. N. Timmis, and D. F. Dwyer.** 1995. Cloning, nucleotide sequence, and expression of the gene encoding a novel dioxygenase involved in metabolism of carboxydiphenyl ethers in *Pseudomonas pseudoalcaligenes* POB310. *Arch. Microbiol.* **163:**35–41.

12. **de Souza, M. L., D. Newcombe, S. Alvey, D. E. Crowley, A. Hay, M. J. Sadowsky, and L. P. Wackett.** 1998. Molecular basis of a bacterial consortium: interspecies catabolism of atrazine. *Appl. Environ. Microbiol.* **64:**178–184.

13. **de Souza, M. L., M. J. Sadowsky, and L. P. Wackett.** 1996. Atrazine chlorohydrolase from *Pseudomonas* sp. strain ADP: gene sequence, enzyme purification, and protein characterization. *J. Bacteriol.* **178:**4894–4900.

14. **de Souza, M. L., L. P. Wackett, K. L. Boundy-Mills, R. T. Mandelbaum, and M. J. Sadowsky.** 1995. Cloning, characterization, and expression of a gene region from *Pseudomonas* sp. strain ADP involved in the dechlorination of atrazine. *Appl. Environ. Microbiol.* **61:**3373–3378.

15. **de Souza, M. L., L. P. Wackett, and M. J. Sadowsky.** 1998. The genes encoding atrazine catabolism are located on a self-transmissible plasmid in *Pseudomonas* sp. strain ADP. *Appl. Environ. Microbiol.* **64:**2323–2326.

16. **Dickerson, R. E.** 1971. The structures of cytochrome c and the rates of molecular evolution. *J. Mol. Evol.* **1:**26–45.

17. **Di Giola, D., M. Peel, F. Fava, and R. C. Wyndham.** 1998. Structures of homologous composite transposons carrying *cbaABC* genes from Europe and North America. *Appl. Environ. Microbiol.* **64:**1940–1946.

*18. **Doolittle, R. F.** 1987. *Of URFs and ORFs: a Primer on How to Analyze Derived Amino Acid Sequences.* University Science Books, Mill Valley, Calif.

19. **Dykhuizen, D. E., and D. L. Hartl.** 1983. Functional effects of PGI allozymes in *Escherichia coli. Genetics* **105:**1–18.

20. **Eaton, R. W., and J. S. Karns.** 1991. Cloning and comparison of the DNA encoding ammelide aminohydrolase and cyanuric acid amidohydrolase from three *s*-triazine-degrading bacterial strains. *J. Bacteriol.* **173:**1363–1366.

21. **Erickson, L. E., and K. H. Lee.** 1989. Degradation of atrazine and related *s*-triazines. *Crit. Rev. Environ. Control* **19:**1–14.

22. **Fong, P. Y., C. Goh, G. Tan, and H. M. Tan.** 1997. Identification and genetic analysis of Tn*5542,* a transposable element carrying the *bedD* and *bedC1C2BA* genes in *Pseudomonas putida* ML2, p. 53. *In Abstracts of the Sixth International Congress on Pseudomonas: Molecular Biology and Biotechnology,* Madrid, Spain.

23. **Fukumori, F., and C. P. Saint.** 1997. Nucleotide sequence and regulational analysis of genes involved in conversion of aniline to catechol in *Pseudomonas putida* UCC22 (pTDN1). *J. Bacteriol.* **179:**399–408.

24. **Gerlt, J. A., and P. C. Babbitt.** 1998. Mechanistically diverse enzyme superfamilies: the importance of chemistry in the evolution of catalysis. *Curr. Opin. Chem. Biol.* **2:**607–612.

25. **Holm, L., and C. Sander.** 1997. An evolutionary treasure: unification of a broad set of amidohydrolases related to urease. *Proteins Struct. Funct. Genet.* **28:**72–82.

26. **Horak, R., and M. Kivisaar.** 1998. Expression of the transposase gene *tnpA* of Tn*4652* is positively affected by integration host factor. *J. Bacteriol.* **180:**2822–2829.

27. **Horowitz, N. H.** 1945. On the evolution of biochemical synthesis. *Proc. Natl. Acad. Sci. USA* **31:**153–157.

28. **Hudlicky, T., D. Gonzalez, and D. T. Gibson.** 1999. Enzymatic dihydroxylation of aromatics in enantioselective synthesis: expanding asymmetric methodology. *Aldrichim. Acta* **32:**35–62.

29. **Junker, F., and A. M. Cook.** 1997. Conjugative plasmids and the degradation of arylsulfonates in *Comamonas testosteroni. Appl. Environ. Microbiol.* **63:**2403–2410.

30. **Kato, K., K. Ohtsuki, H. Mitsuda, T. Yomo, S. Negoro, and I. Urabe.** 1994. Insertion sequence IS*6100* on plasmid pOAD2, which degrades nylon oligomers. *J. Bacteriol.* **176:**1197–1200.

31. **Kawasaki, H., K. Tsuda, I. Matsushita, and K. Tunomura.** 1992. Lack of homology between two haloacetate dehalogenase genes encoded on a plasmid from *Moraxella* sp. strain B. *J. Gen. Microbiol.* **138:**1317–1323.

32. **Kreitman, M.** 1983. Nucleotide polymorphism at the alcohol dehydrogenase locus of *Drosophila melanogaster. Nature* **304:**412–417.

33. **Lange, S. J., and L. J. Que.** 1998. Oxygen activating nonheme iron enzymes. *Curr. Opin. Chem. Biol.* **2:**159–172.

34. **Mandelbaum, R. T., D. L. Allan, and L. P. Wackett.** 1995. Isolation and characterization of a *Pseudomonas* sp. that mineralizes the *s*-triazine herbicide atrazine. *Appl. Environ. Microbiol.* **61:**1451–1457.

35. **Maymo-Gatell, X., Y. Chien, J. M. Gossett, and S. H. Zinder.** 1997. Isolation of a bacterium that reductively dechlorinates tetrachloroethene to ethene. *Science* **276:**1568–1571.

36. **Mulbry, W. W., P. C. Kearney, J. O. Nelson, and J. S. Karns.** 1987. Physical comparison of parathion hydrolase plasmids from *Pseudomonas diminuta* and *Flavobacterium* sp. *Plasmid* **18:**173–177.

37. **Nagy, I., F. Compernolle, K. Ghys, J. Vanderleyden, and R. de Mot.** 1995. A single cytochrome P-450 system is involved in degradation of the herbicides EPTC (ethyl dipropylthiocarbamate) and atrazine by *Rhodococcus* sp. strain NI86/21. *Appl. Environ. Microbiol.* **61:**2056–2060.

38. **Nakatsu, C., J. Ng, R. Singh, N. Straug, and C. Wyndham.** 1991. Chlorobenzoate catabolic transposon Tn5271 is a composite class I element with flanking class II insertion sequences. *Proc. Natl. Acad. Sci. USA* **88:**8312–8316.

39. **Nishi, A., K. Tominaga, and K. Furukuwa.** 1998. Horizontal transfer of the chromosomal gene clusters coding for biphenyl and salicylate metabolism in

Pseudomonas putida KF715, p. 92. *In Abstracts of the First International Conference of the Federation of Asia-Pacific Microbiology Societies,* Singapore.

*40. **Petsko, G. A., G. L. Kenyon, J. A. Gerlt, D. Ringe, and J. W. Kozarich.** 1993. On the origin of enzymatic species. *Trends Biochem. Sci.* **18:**372–376.

41. **Radosevich, M., S. J. Traina, Y. Hao, and O. H. Tuovinen.** 1995. Degradation and mineralization of atrazine by a soil bacterial isolate. *Appl. Environ. Microbiol.* **61:**297–302.

42. **Rheinwald, J. G., A. M. Chakrabarty, and I. C. Gunsalus.** 1973. A transmissible plasmid controlling camphor oxidation in *Pseudomonas putida. Proc. Natl. Acad. Sci. USA* **70:**885–889.

43. **Romine, M. F., L. C. Stillwell, K. K. Wong, S. J. Thurston, E. C. Sisk, C. Sensen, T. Gaasterland, J. K. Fredrickson, and J. D. Saffer.** 1999. Complete sequence of a 184-kilobase catabolic plasmid from *Sphingomonas aromaticivorans* F199. *J. Bacteriol.* **181:**1585–1602.

44. **Sadowsky, M. J., M. L. de Souza, Z. Tong, and L. P. Wackett.** 1998. AtzC is a member of the amidohydrolase protein superfamily and is homologous to other atrazine-metabolizing enzymes. *J. Bacteriol.* **180:**152–158.

45. **Scholten, J. D., K.-H. Chang, P. C. Babbitt, H. Charest, M. M. Sylvestre, and D. Dunaway-Mariano.** 1991. Novel enzymic hydrolytic dehalogenation of a chlorinated aromatic. *Science* **253:**182–185.

46. **Serdar, C. M., and D. T. Gibson.** 1989. Studies of nucleotide sequence homology between naphthalene-utilizing strains of bacteria. *Biochem. Biophys. Res. Commun.* **164:**772–779.

47. **Shao, Z. Q., and R. Behki.** 1996. Characterization of the expression of the *thcB* gene, coding for a pesticide-degrading cytochrome P-450 in *Rhodococcus* strains. *Appl. Environ. Microbiol.* **62:**403–407.

48. **Springael, D., S. Kreps, and M. Mergeay.** 1993. Identification of a catabolic transposon, Tn*4371,* carrying biphenyl and 4-chlorobiphenyl degradation genes in *Alcaligenes eutrophus* A5. *J. Bacteriol.* **175:**1674–1681.

*49. **Tan, H.-M.** 1999. Bacterial catabolic transposons. *Appl. Microbiol. Biotechnol.* **51:**1–12.

50. **Tsuda, M., and T. Iino.** 1987. Genetic analysis of a transposon carrying toluene degrading genes on a TOL-plasmid pWWO. *Mol. Gen. Genet.* **210:**270–276.

51. **Tsuda, M., and T. Iino.** 1988. Identification and characterization of Tn*4653,* a transposon converting the toluene transposon Tn*4651* on TOL plasmid pWWO. *Mol. Gen. Genet.* **213:**72–77.

52. **Tsuda, M., and T. Iino.** 1990. Naphthalene degrading genes on plasmid NAH7 are on a defective transposon. *Mol. Gen. Genet.* **223:**33–39.

53. **van der Meer, J. R., A. J. Zehnder, and W. M. de Vos.** 1991. Identification of a novel composite transposable element, Tn*5280,* carrying chlorobenzene dioxygenase genes of *Pseudomonas* sp. strain P51. *J. Bacteriol.* **173:**7077–7083.

54. **van der Ploeg, J., M. Willemsen, G. van Hall, and D. B. Janssen.** 1995. Adaptation of *Xanthobacter autotrophicus* GJ10 to bromoacetate due to activation and mobilization of the haloacetate dehalogenase gene by insertion element 1-S*1247*. *J. Bacteriol.* **177:**1348–1356.

55. **Watson, J. D., and F. H. C. Crick.** 1953. A structure of deoxyribosenucleic acid. *Nature* **171:**737–738.

56. **Williams, P. A., S. J. Assinder, P. Marco, K. J. O'Donnell, C. L. Poh, L. E. Shaw, and M. K. Winson.** 1992. Catabolic gene duplications in TOL plasmids, p. 341–352. *In* E. Galli, S. Silver, and B. Witholt (ed.), *Pseudomonas: Molecular Biology and Biotechnology.* American Society for Microbiology, Washington, D.C.

*57. **Woese, C. R., and G. E. Fox.** 1977. Phylogenetic structure of the prokaryotic domain: the primary kingdoms. *Proc. Natl. Acad. Sci. USA* **74:**5088–5090.

58. **Xiang, H., L. Luo, K. L. Taylor, and D. Dunaway-Mariano.** 1999. Interchange of catalytic activity within the 2-enoyl-coenzyme A hydratase/isomerase superfamily based on a common active site template. *Biochemistry* **38:**7638–7652.

*59. **Zuckerkandl, E., and L. Pauling.** 1965. Molecules as documents of evolutionary history. *J. Theor. Biol.* **8:**357–366.

8

Metabolic Logic and Pathway Maps

No knowledge is entirely reducible to words.

—*Seymour Papert*

In this chapter, we make extensive use of visual images of chemical compounds, in a familiar format, to help the reader see the underlying patterns of catabolism. This is the visual art of catabolism. Each soil bacterium chooses from an enormous selection of colors and patterns and makes its own canvas. The quality of the final art form is dependent on the complementarity of the shapes and patterns. Similarly, the robustness of a bacterium is dependent on its patterns of metabolic reactions. Complementarity in metabolism means that a significant amount of the free energy available in a growth substrate is captured by the organism to reproduce itself. The apparent elegance of this is sometimes denoted by the term *metabolic logic.* By logic we mean that the guiding hand of natural selection paints a metabolic pathway that almost optimally captures the available free energy in a given organic compound. We look at visual images of the chemical intermediates in a metabolic pathway and see beauty in what nature has constructed.

Overview

Electronic devices, mechanical motors, and living things are similar in the sense that they all require an input of energy and then ultimately dissipate that energy. Of course, the majority of people are most interested in life-dependent energy dissipation; we seem to be endlessly fascinated with ourselves. At the heart of life's energy dissipation process is metabolism: the enzyme-catalyzed reactions that generate heat, work, and stored chemical energy that is later transformed into heat and more work.

The details of metabolism were more recently elucidated than those of electromagnetic radiation or thermodynamics. Over the last century, intermediary metabolism has been revealed, largely through studies of microorganisms, as discussed in chapter 2. Intermediary metabolism is defined as enzyme-catalyzed reactions common to most living things: bacteria, eukaryotic single-celled life, plants, and animals. More specifically, intermediary metabolism deals with (i) common energy metabolism—for example, the catabolism of glucose by the glycolytic, or Embden-Meyerhof, pathway—and (ii) common biosynthetic pathways, such as those that generate necessary amino acids, lipids, and carbohydrates. This is illustrated, in the form of an electronic circuit diagram of the intermediary metabolism of *Escherichia coli,* in Fig. 8.1.

The circuit diagram allows the compactness of intermediary metabolic networks to be seen. The central linear pathway is depicted at the core of Fig. 8.1. The linear pathway, flowing in the downward direction as drawn, represents the glycolytic pathway of glucose metabolism. The circle represents the tricarboxylic acid cycle, sometimes known as the Krebs cycle after Hans Krebs, who elucidated several of its biochemical reactions. The centrality of the pathway arises from the observation that these reactions occur in many phylogenetically diverse bacteria. Moreover, as Fig. 8.1 makes manifest, many other dots are connected to the dots, or chemical intermediates, in the central, or core, glycolytic and tricarboxylic acid pathways. Stated another way, the compounds in those pathways serve as the building blocks to biosynthesize amino acids, nucleotides, and lipids; furthermore, they serve as funneling points during catabolic metabolism,

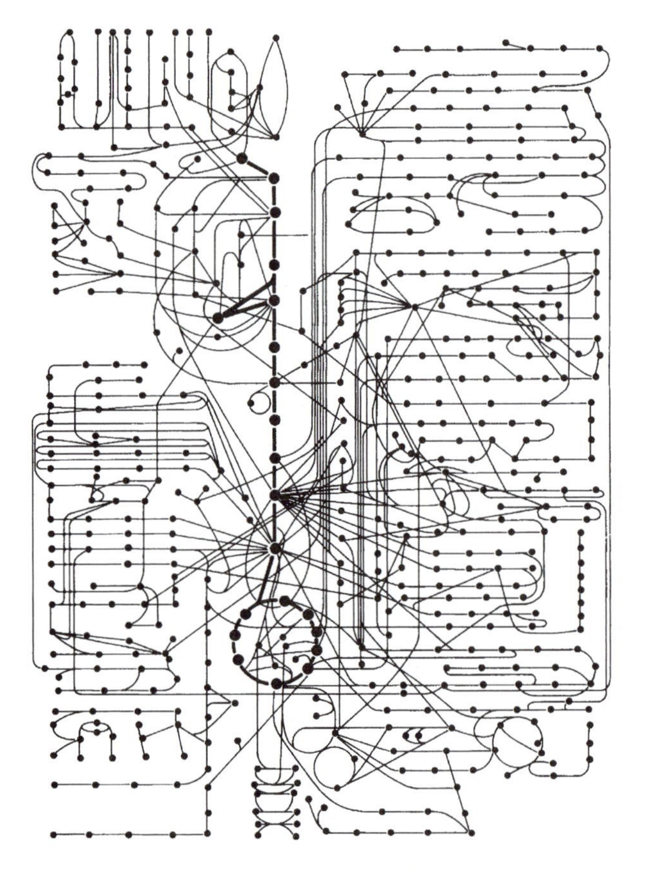

Figure 8.1 Network of intermediary metabolism showing compounds as nodes and interconnecting reactions as lines. Highlighted in boldface is the central linear pathway flowing into the tricarboxylic acid cycle (circle). (From *Proteins, Energy and Metabolism* [Rawn, © 1989], by permission of Prentice-Hall, Inc., Upper Saddle River, N.J.)

through which many compounds are oxidized to carbon dioxide with the capture of ATP.

Many of the reactions of interest in the context of biodegradation and biocatalysis are not part of the core metabolic circuits. Rather, they are circuits that overlie, or feed into, the core metabolism, as shown in Fig. 8.2. This makes sense from an evolutionary context. It would be very improbable for an organism to evolve an entire set of reactions simultaneously to generate a new catabolic pathway. However, new reactions that would connect to an existing core metabolic pathway might develop and thus extend the organism's substrate range incrementally. Even the simultaneous acquisition of several genes which encode enzymes catalyzing sequential reactions brings the greatest benefit if that metabolism is linked to another metabolism, thus maximizing the ability of the cell to capture carbon and energy.

It is also axiomatic that catabolic pathways funnel into a limited set of key intermediates, getting the most metabolic bang from the limited genetic buck. Thus, a new pathway is not required for each compound the organism might metabolize if there is redundancy by generating a common intermediate. This is well illustrated in aromatic-ring metabolism, which often proceeds through catecholic intermediates aerobically and benzoyl-coenzyme A (CoA) anaerobically.

These points will be illustrated in this chapter by using a series of metamaps. These are not single-pathway maps, nor are they like the metabolic-

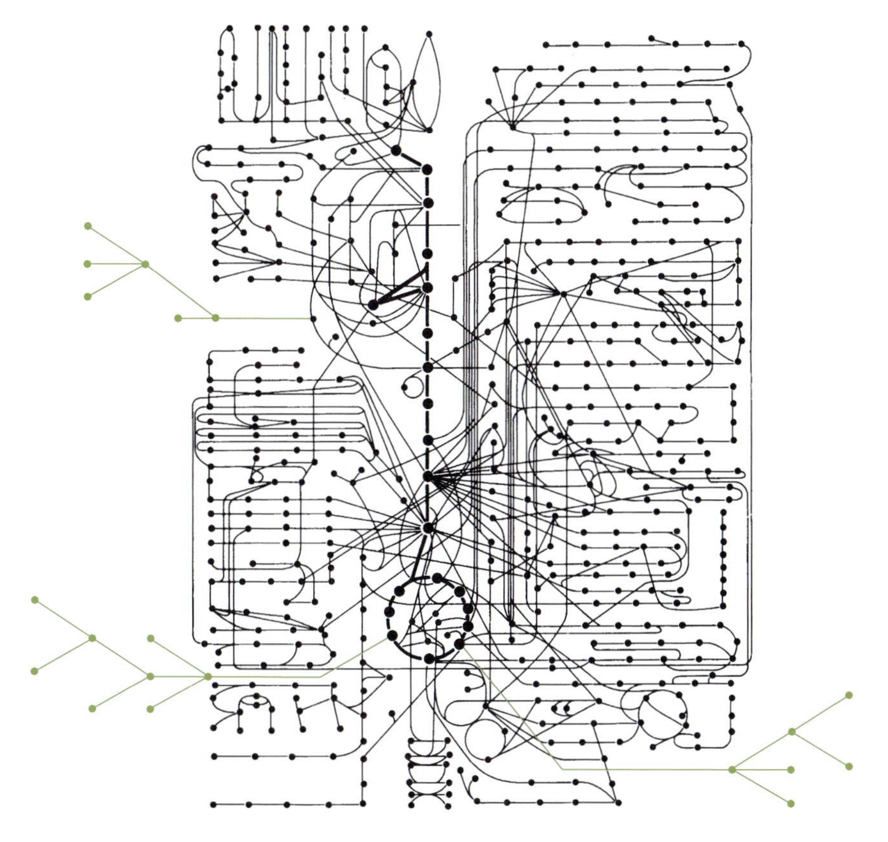

Figure 8.2 Network of intermediary metabolism with novel catabolic reactions (shown in green) which funnel into intermediary metabolites.

pathway wall maps that show all of intermediary metabolism. Rather, the metamaps show clusters of metabolism that are linked by a common metabolic logic and stripped of all interleafing metabolism so the underlying strategy is unveiled.

C_1 Metamap

The first metamap is the simplest (Fig. 8.3). It shows the core metabolism of methanotrophs and methylotrophs and how various pollutants and natural products feed into the core metabolism. Many of the feeder compounds contain a single carbon atom, or they can be larger molecules that split off C_1 intermediates as part of their metabolism. There are a limited number of intermediates to feed into because methane is oxidized to carbon dioxide in discrete two-electron steps. Intermediates generated by one-electron reduction would be radical species, which are not stable. Thus, there are only three intermediates: methanol, formaldehyde, and formate, into which many compounds can be transformed (Fig. 8.3).

In fact, some organisms having a C_1 oxidative pathway as shown in Fig. 8.3 also assimilate more complex organic molecules as a carbon source. In that metabolism, methyl, methylene, and methenyl fragments are shuttled from and into more complex molecules, for example in the synthesis of purines and pyrimidines. This metabolism is for biosynthesis and is not inextricably linked to the carbon assimilation and energy generation of the cell. In many methanotrophic and methylotrophic bacteria, both carbon and

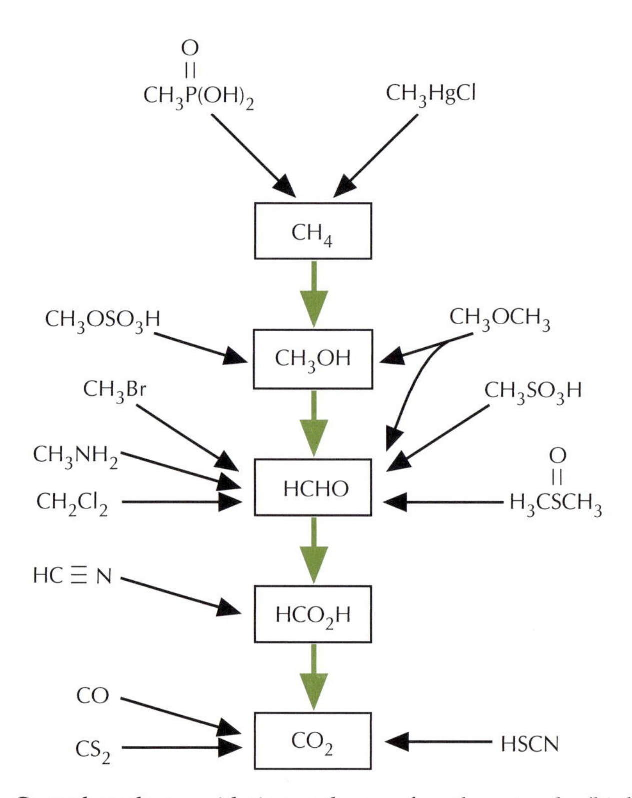

Figure 8.3 Central methane oxidation pathway of methanotrophs (highlighted in green) showing how single-carbon compounds can potentially funnel into the pathway when specific enzymes are present.

energy needs are obligately met by carrying out oxidation reactions on the carbon fragments. Thus, these bacteria are most adept at metabolizing C_1 fragments for biodegradation purposes.

One example of a pathway will illustrate the point made above. Dichloromethane, or methylene chloride, would appear to be an unpalatable growth substrate for a bacterium. First, it is at the oxidation level of formaldehyde, so energy acquisition is limited. Second, if formaldehyde is generated in a high-flux pathway, which would be necessary to meet carbon and energy needs, then the cell must protect itself from the cross-linking reactions of formaldehyde. Remember that formaldehyde is used to "fix" tissues, which it does by cross-linking proteins and other biomolecules. Third, dichloromethane is an excellent solvent for lipids, and the cell must resist the extraction of membrane components which would permeabilize the cell, followed by certain death. Fourth, the organism must assimilate all of its cell carbon from dichloromethane.

Here is how *Methylobacter* sp. strain DM2 handles these biochemical problems. Both the first and the third problems are addressed by having enough enzymatic turnover to derive sufficient carbon and energy and to prevent extraction of the lipids. This would require either a fast enzyme(s) to process dichloromethane or a large amount of the enzyme. In fact, only one enzyme is needed to transform dichloromethane to formaldehyde: dichloromethane dehalogenase. Information on the enzyme, the gene that encodes it, and a postulated reaction mechanism can be found in the University of Minnesota Biocatalysis/Biodegradation Database. Dichloromethane dehalogenase in strain DM2 is a sluggish enzyme, with a catalytic constant (k_{cat}), or turnover number, of approximately 0.6 s^{-1}, but this is counteracted by its presence as 25% of the total soluble protein (9, 18). This processes enough dichloromethane to allow *Methylobacterium* sp. strain DM4, which contains an identical dichloromethane dehalogenase, to multiply with a maximal doubling time of 9.6 h in batch culture. The fact that another bacterium, *Methylophilus* sp. strain DM11, contains an enzyme which has a 5.5-fold-higher k_{cat} (9) and produces the enzyme as a more modest 8% of the total soluble protein (29) is consistent with this. The higher turnover more than offsets the lower expression of the enzyme, allowing *Methylophilus* sp. strain DM11 to grow with a maximal doubling time of 4.3 h in batch culture.

How does the organism protect itself from the reactivity of the intermediate formaldehyde? In fact, the bacteria are already set up to process methane or methanol through formaldehyde. One trick is to have a thiol(s) present in the cytoplasm such that much of the free formaldehyde will be in the equilibrating hydroxymethylthioether form. Significant levels of the thiol-containing compound glutathione have been detected in these organisms. This might also explain the fact that dichloromethane dehalogenase requires glutathione for activity.

Many other C_1 compounds funnel into methanotrophic and methylotrophic bacteria, which are ubiquitous in aerobic environments. These bacteria, belonging to genera such as *Methylobacterium, Methylomonas, Methylosinus,* and *Hyphomicrobium,* are members of the α and γ subdivisions of *Proteobacteria.* More recently, anaerobic bacteria which use C_1 compounds, like dichloromethane, as a sole source of carbon and energy have been identified (21). Much less is known about their biochemistry, and it is not yet possible to deduce if many share a common metabolic route for funneling the carbon in and generating ATP. Comparatively little is known about the taxonomy of the anaerobic dichloromethane utilizers, although present data suggest that they are distinct from the aerobic bacteria described above.

C_2 Metamap

The catabolism of C_1 compounds is relatively simple with respect to the number of metabolic intermediates which can be produced and the number of reactions required to generate those intermediates. Most C_1 compounds are enzymatically transformed into methanol, formaldehyde, and formate, and then oxidation and assimilation reactions can proceed. Can similar metamaps, illustrating the overall metabolic logic, be designed for more complex organic molecules?

To consider the question, it is necessary to think about the complexity of organic compounds. When carbon atoms are added, there is a combinatorial explosion of possibilities, including linear, branched, ring, and aromatic hydrocarbon structures and metabolically distinct situations such as heterocyclic rings. It is probably impossible to write strict metabolic rules for all of those situations at this time. It is possible, however, to show general cases illustrating the ways in which many of those classes of compounds are metabolized; some of these are considered in later sections of this chapter. Moreover, it is possible to demonstrate the key intermediates into which C_2 compounds would funnel metabolically, since C_2 compounds are not complex enough to include branching, cyclic rings, and aromaticity.

To consider C_2 compounds, it is necessary to look at the rule for C_1 compounds requiring that oxidation reactions proceed in discrete two-electron steps. If the same logic is applied to two linked carbon atoms, they can each cycle between two-electron oxidation steps in going from ethane, with both carbon atoms fully reduced, to oxalate, with both carbon atoms largely oxidized (Fig. 8.4). The pairwise combinations of the different oxidation states at each carbon make nine relatively commonly occurring metabolic intermediates, from ethanol to oxalate (Fig. 8.4, box). This includes compounds which form important nodes in metabolism, acetate and glyoxylate.

Ethane, ethene, and ethyne may also be formed; this is known to occur principally during reductive dehalogenation of haloalkanes and haloalkenes. However, these compounds are not considered intermediary metabolites but may be metabolized to one or more of the nine intermediary compounds shown in the box in Fig. 8.4. For example, ethene and ethyne are typically catabolized by bacteria, with functionalization of the carbon atoms leading to the loss of the carbon-carbon multiple bond. Ethene is oxidized to an epoxide, which can be hydrated to ethylene glycol (11). The known mechanism for alkyne metabolism, acting on acetylene, would generate acetaldehyde (26). The other examples of C_2 compounds of interest are largely those in which an alkane or alkene is functionalized and the functional group is removed during catabolism to generate one of the nine common intermediates shown in Fig. 8.4. Examples of high-volume organic compounds whose biodegradative metabolisms are of great interest are 1,2-dichloroethane and trichloroethylene. The former is completely dechlorinated when it reaches the metabolic intermediate glycolate (14). Catabolism of trichlorethylene is more complicated, since it involves oxygenation reactions which generate intermediates that spontaneously form glyoxylate, formate, and carbon monoxide (7).

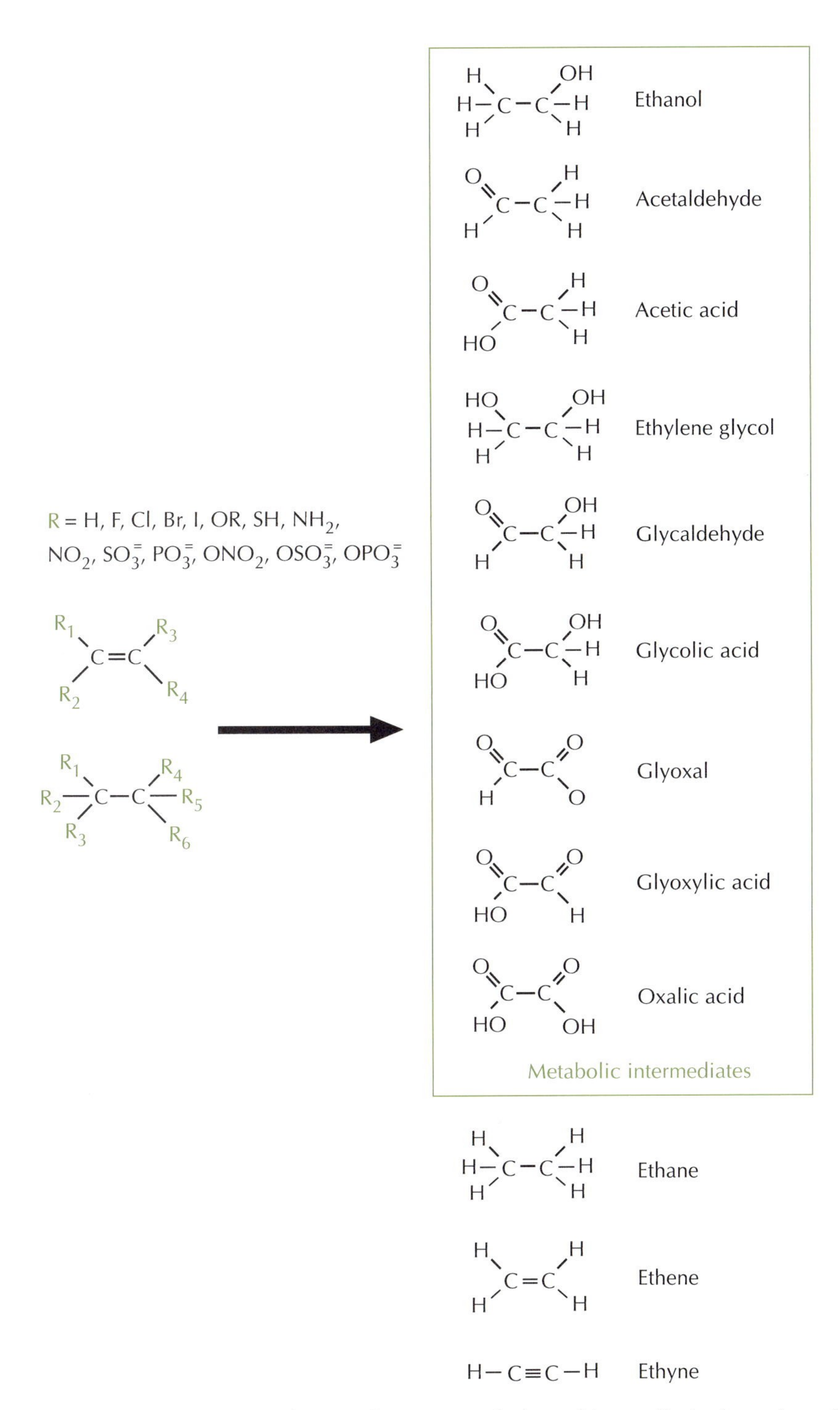

Figure 8.4 Metabolisms of many C_2 compounds funnel into a limited number of intermediary compounds (box) or the C_2 hydrocarbons.

Cycloalkane Metamap

With as few as three carbon atoms, a carbon chain can bend back onto itself and cyclize, given energy input to form the requisite carbon-carbon bonds. When all the carbon atoms are saturated, these compounds are known as cycloalkanes. Cycloalkanes form an enormous class of molecules in the biological world, a fact that is underappreciated. These compounds derive from biosynthesis and from natural diagenesis. The basic carbon skeletons, each representing a large class of different molecules, are shown in the upper left of Fig. 8.5. For example, cyclic terpenes, largely of plant origin, are produced in thousands of different chemical structures (see Table 5.1). Steroid compounds are ubiquitous in biological systems: they are found in bacteria, fungi, plants, and animals. Fossil fuel hydrocarbons, in various stages of maturity, are generated by natural diagenesis, or slow abiotic "burning" of organic matter under the pressures and temperatures occurring in sediments. One set of structures that arise from diagenesis is fused-ring compounds, some of which are aromatic while others are saturated. The saturated rings must be functionalized and cleaved by mechanisms different from the dioxygen addition across aromatic double bonds that has been discussed previously.

Given their common carbon backbone structure, one would anticipate some common mechanisms for functionalization and cleavage of cycloalkane rings (36). All of the compounds discussed above share some common biochemical mechanisms of aerobic microbial attack, and this is illustrated in the metamap with one of the simpler examples, cyclohexane (Fig. 8.5). Cyclohexane is initially functionalized by a monooxygenase-catalyzed reaction to make cyclohexanol. Dehydrogenation yields a ketone. The next step is the defining reaction for this class of compounds, a Baeyer-Villiger

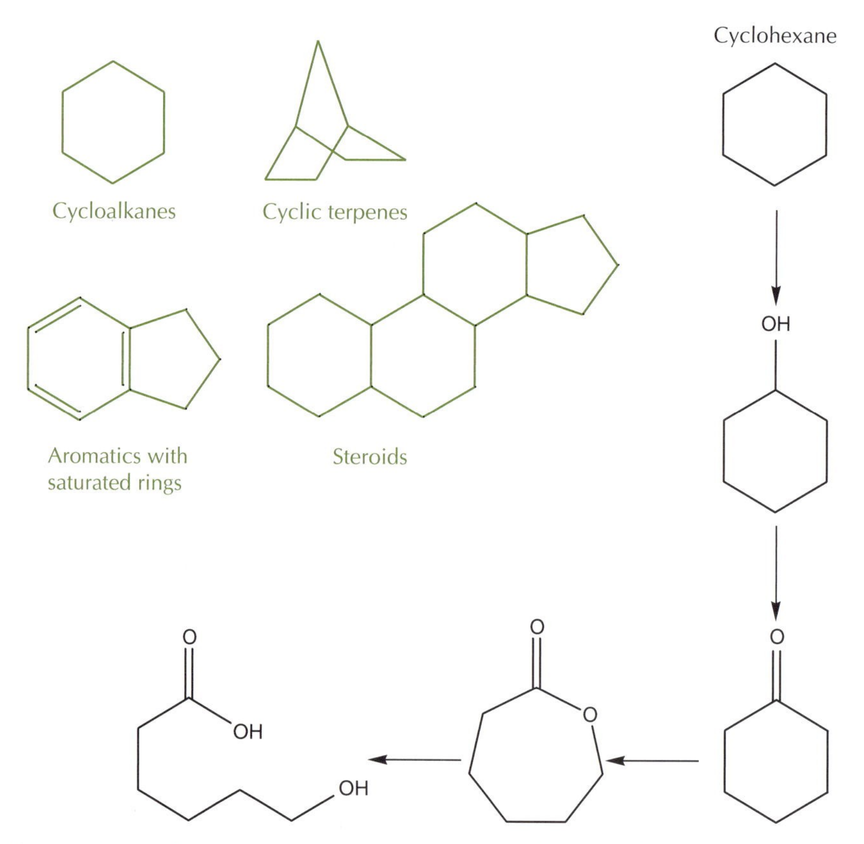

Figure 8.5 Cycloaliphatic compounds are metabolized aerobically via hydroxylation and Baeyer-Villiger-type oxygen insertion reactions.

monooxygenation reaction in which the oxygen atom is inserted between the carbon bearing the carbonyl oxygen and the adjacent carbon. The enzyme cyclohexanone monooxygenase is perhaps the best studied of this very widespread class of enzymes (20, 35). Originally obtained from an *Acinetobacter* strain capable of growing on cyclohexanol and cyclohexanone, the enzyme is now more routinely produced from recombinant *E. coli* strains. The enzyme is a flavoprotein monooxygenase with a very broad substrate specificity. Mechanistic studies suggest the generation of a flavin hydroperoxide intermediate, which gives rise to the species capable of inserting a single atom of oxygen into the cycloalkane ring. After the generation of the lactone, the metabolic strategy is clear: hydrolysis of the ring ester and then further metabolism of the open-chain carboxylic acid. These two reactions, shown in Fig. 8.5, are common to the aerobic catabolism of tens of thousands of natural products. Thus, the Baeyer-Villiger monooxygenases have attracted the attention of synthetic chemists as a useful tool for the production of chiral lactones and sulfoxides (38).

BTEX Metamap: Aerobic Metabolism

The same diagenic reactions which produce cycloalkanes also give rise to benzenoid aromatic-ring structures in petroleum. These compounds are found together in natural environments. Moreover, industry obtains aromatic hydrocarbons largely from the refining of petroleum. In fuels and solvents, these compounds are often used as mixtures. In soil remediation efforts, they are often referred to as BTEX compounds, with the acronym representing the four main aromatic components in a low-boiling-point petroleum mixture: benzene, toluene, ethylbenzene, and xylene (Fig. 8.6). Xylene itself is a mixture of compounds, composed of *ortho-*, *meta-*, and *para-* isomers. Only the *meta*-isomer is shown in Fig. 8.6.

The benzenoid aromatic ring presents both an obstacle and a reward to a microbe. The obstacle is the resonance stability of the aromatic-ring structure, rendering these compounds relatively unreactive. A benzene ring, with three double bonds, is much less reactive than the comparable linear structure, 1,3,5-hexatriene. While the latter will add water and other nucleophiles readily, the benzene ring will not react under similar conditions. This has implications for metabolic strategies directed against an aromatic-ring compound, which must overcome a higher activation energy than occurs during the hydration of nonaromatic double bonds, for example. The reward of benzene metabolism is obtained from the substantial energy available in oxidizing the compound to 6 mol of carbon dioxide.

The fact that the initial benzenoid structure is unreactive dictates that a price of reduced pyridine nucleotides, NADH or NADPH, must be paid to activate the aromatic ring for further metabolism. Since 1 mol each of NADH or NADPH can be oxidized via electron transport to generate 3 mol of ATP, the cost to the cell is substantial. The cost differs, depending upon whether the initiating reactions, under aerobic conditions, are catalyzed by a dioxygenase or a monooxygenase. This is discussed in more detail below.

The microbial catabolism of BTEX compounds has been relatively well studied, particularly under aerobic conditions (8) (Fig. 8.6). A defining characteristic of the reactions is that BTEX must be activated to overcome the resonance stability and subsequently split open the aromatic ring. To do this, microbes must use the potentially most reactive molecule available: molecular oxygen (O_2), which is referred to here as dioxygen. The reaction

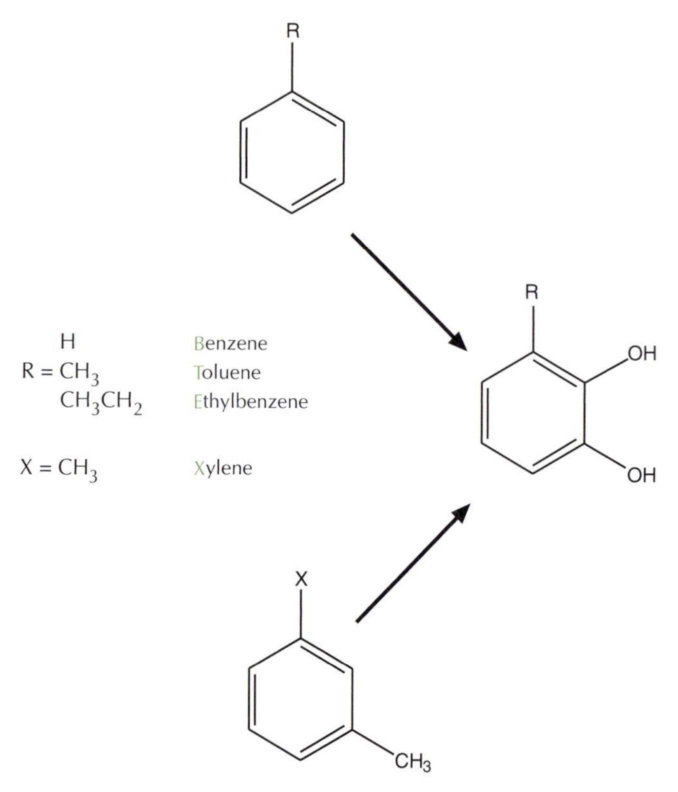

Figure 8.6 Aerobic metabolism of BTEX compounds, and other aromatic hydrocarbons, typically proceeds through catechol (1,2-dihydroxybenzene) intermediates.

of benzene and dioxygen is highly exergonic, or thermodynamically favorable, under standard conditions. This is why benzene combusts explosively. Its oxidation releases heat, which fuels more burning in an explosive manner. But the process requires something to start the reaction by overcoming the resonance stability of the benzene ring; the activation energy requirement is rather steep. There is another feature of the reaction requiring energy input for the reaction to proceed: the activation of dioxygen. Dioxygen in the atmosphere exists as a triplet molecule, a diradical with its two unpaired electrons possessing the same orbital-spin orientation. The ramification of this is that oxygen insertion into a carbon-hydrogen bond requires that one of the orbital spins be inverted. This requires significant input of energy. It also explains why organic matter does not spontaneously burst into flame and why our bodies can exist in a sea of gaseous dioxygen.

The lowering of activation energy barriers is the hallmark of enzyme catalysis, and oxygenase enzymes that attack BTEX compounds use transition metals, typically iron, for dioxygen activation. Binding of dioxygen to an enzyme-coordinated iron atom causes some of the oxygen's electron density to overlap with the metal's *d*-orbitals. *d*-Orbitals typically require small energy input to move electrons between energy states, and this facilitates the effective spin conversion of dioxygen to make it more reactive with organic substrates.

BTEX compounds are oxidized by dioxygenases or monooxygenases in different organisms; both are characterized as iron-containing, oxygen-activating enzymes. In some cases, the metabolic strategy proves to be similar. For example, toluene is known to be oxidized by dioxygenases to yield *cis*-1,2-dihydroxydihydro-3-methyl-3,5-cyclohexadiene. A dehydrogenase oxidizes that intermediate to 3-methylcatechol, which is a common substrate for ring cleavage dioxygenase enzymes. Note that via this strategy, the NADH equivalent expended in the dioxygenase reaction is recouped in the subsequent dehydrogenase-catalyzed step. Thus, the organism oxidizes a benzene to a catechol with no net gain or expenditure of energy, other than the liberation of heat.

A parallel pathway is known in which a monooxygenase oxidizes toluene to *o*-cresol (2-hydroxytoluene) and then takes *o*-cresol to 3-methylcatechol (25). Thereafter, the metabolism of catechol via dioxygenolytic ring cleavage and subsequent reactions is the same. However, the monooxygenase-dependent pathways differ from the dioxygenase pathways in that the former expend 2 mol of NADH or NADPH to go from an unactivated benzenoid ring to a catechol. Thus, more overall energy is captured via the dioxygenase-dependent pathways. This might explain the observation that dioxygenase-dependent pathways are found more frequently in aerobic benzene-ring–catabolizing bacteria.

BTEX Metamap: Anaerobic Metabolism

Shortly after the elucidation of the dioxygenase-catalyzed metabolism of BTEX compounds, the anaerobic metabolism of these compounds was thought by some to be nonexistent. This was based partly on observation (BTEX compounds are more persistent in anaerobic soils) and partly on thermodynamic grounds. What could possibly activate the resonance-stabilized aromatic ring in the absence of diatomic oxygen?

It is now well established that BTEX compounds are readily metabolized anaerobically (2, 6, 19) and that the metabolism is not due to the presence of small amounts of gaseous oxygen and bacteria containing oxygenases with high affinity for dioxygen. Moreover, the various anaerobic pathways for BTEX compounds have been determined to have fundamental metabolic commonality, much like the aerobic pathways (Fig. 8.7). In this case, the common intermediate is benzoyl-CoA (12, 13). Benzoyl-CoA, in turn, is subjected to aromatic-ring reduction and subsequent ring opening with capture of carbon and energy by the cell.

PAH Metamap

PAHs, or polycyclic aromatic hydrocarbons, are ubiquitous products of burning organic matter. The organic matter may be living, as in forests ignited by lightning strikes, or dead, such as the sediments giving rise to liquid petroleum and coal. PAHs are also naturally produced by diagenesis, or slow burning, in sediments. Recently, PAHs have been detected in interstellar space (32) and on a Martian meteorite, where the PAHs were shown to be of Martian origin (22). Industry uses fused-ring aromatic compounds, but typically not hydrocarbons (compounds containing only carbon and hydrogen), in applications as pure substances. One exception to this is naphthalene, which has been marketed as mothballs and other insect and animal repellents.

Bacteria have undoubtedly been exposed to PAHs for millions of years. If a bacterium acquires the ability to metabolize a PAH, there is substantial benefit with respect to obtaining metabolic energy and carbon atoms to

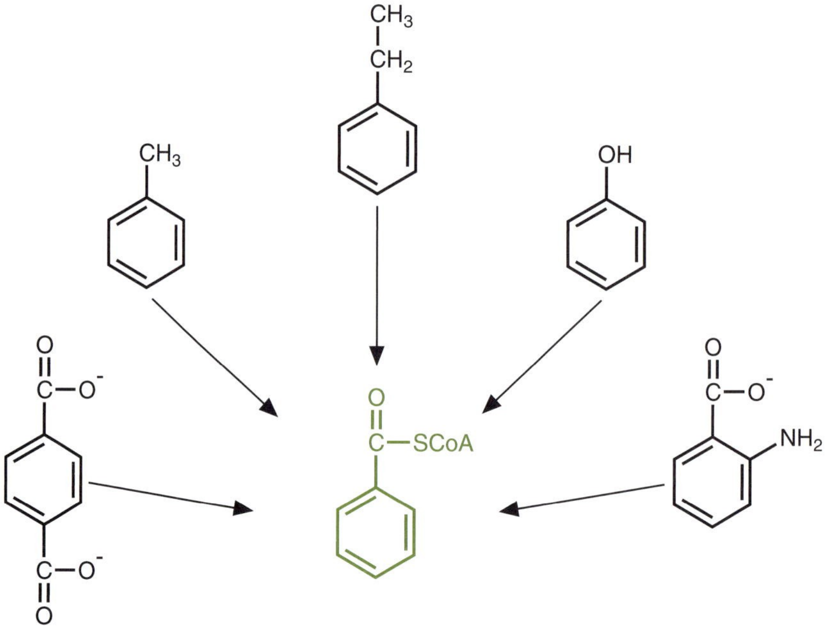

Figure 8.7 Anaerobic metabolism of many substituted aromatic compounds proceeds through the intermediacy of benzoyl-CoA.

fashion biomass. The fundamental strategy to metabolize PAHs is derived from that for single-ring benzenoid aromatic hydrocarbons. We have discussed the reactivity, or lack of reactivity, of the ring carbon atoms and the means by which molecular oxygen, freely available in the atmosphere, can be used for the job.

Using a linearly fused PAH as an example (Fig. 8.8), the aromatic nucleus can undergo dioxygenation by the action of a dioxygenase. In the case of naphthalene, the enzyme naphthalene dioxygenase is the member of the aromatic hydrocarbon dioxygenases for which an X-ray structure is available (16). Following dioxygenation to produce a dihydrodiol (not shown in Fig. 8.8), with saturated carbon atoms at C-1 and C-2, this intermediate undergoes enzymatic dehydrogenation to yield a naphthalene catechol. This is reminiscent of the case with single-ring BTEX compounds shown in Fig. 8.6. The catecholic ring is aromatic but is now activated by the presence of the two hydroxyl groups. This facilitates an enzymatic ring cleavage reaction. Subsequent metabolism to pare away a three-carbon unit leaves an *ortho*-hydroxyaromatic acid, also known as a salicylate, as the product shown in Fig. 8.7. The salicylate can then be oxidatively decarboxylated to yield a second catechol. The process can be repeated to metabolically deconstruct the next aromatic ring.

PAHs are not only composed of carbon atoms tied up in aromatic rings. In petroleum, for example, there are many examples of nonaromatic five- and six-member rings fused to aromatic rings. A relatively simple example of this is the tricyclic-ring compound acenaphthene shown at the bottom of Fig. 8.7. The bacteria implicated in acenaphthene metabolism favor hydroxylation of the nonaromatic ring first. The benzylic carbon atoms are much more chemically reactive. A second hydroxylation and subsequent dehydrogenation yield a diketone. A key subsequent reaction is oxygen

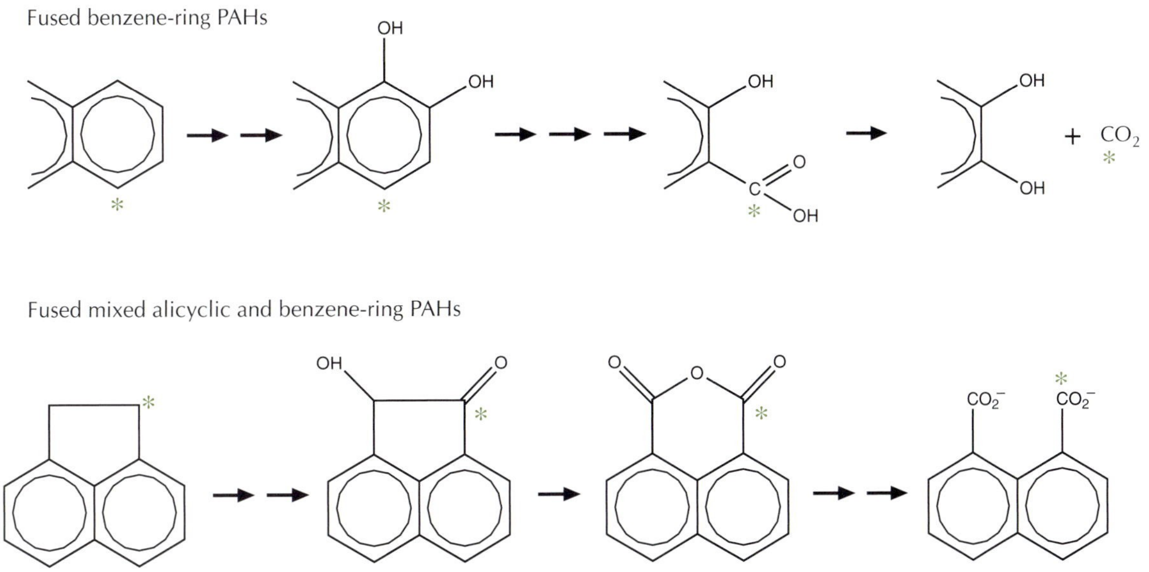

Fused benzene-ring PAHs

Fused mixed alicyclic and benzene-ring PAHs

Figure 8.8 Aerobic catabolism of fused-ring benzenoid compounds (top) and fused alicyclic and benzenoid ring compounds (bottom).

insertion, reminiscent of the oxygen insertion into single-ring cycloalkanes, as shown in Fig. 8.5. The result is a naphthalene dicarboxylic acid, which can undergo dioxygenation with decarboxylation to make a pathway similar to that for naphthalene metabolism.

Fewer bacteria are thought to metabolize PAHs than BTEX compounds. This might be true for several reasons. First, BTEX-type single-ring aromatic compounds are probably more abundant in nature. Second, it requires more genes to completely handle PAHs, because isofunctional reactions in ring functionalization and cleavage sometimes require different enzymes to handle the different-sized substrates. Third, not all PAHs can be metabolized by the same suite of enzymes. This is due to differences in fused-ring architecture, which can be linear or angular, and the degree and positions of alkyl substituents on the rings. The large number of different structures found in nature makes it necessary for many nuances of metabolism to evolve to handle them all. Fourth, PAHs are more poorly water soluble than single-ring BTEX compounds and thus are less bioavailable. PAHs, especially those with four or more fused rings, tend to bind tightly to organic matter in soils and sediments and only slowly become bioavailable. This could limit the diet of a population of PAH-metabolizing bacteria in the environment.

Heterocyclic-Ring Metamap

Heterocyclic-ring compounds contain carbon atoms and one or more heteroatoms, defined here as nitrogen, oxygen, and sulfur. The possible structures which can be made are nearly endless; see Fig. 5.5, 5.6, and 5.7 for examples. There are likely hundreds of thousands of heterocyclic-ring structures in nature, if one considers all the known alkaloids, plant pigments, and porphyrin-like derivatives produced biologically and subsequently generated by diagenic reactions. Some of these structures are not rare in living things: they are found in the intermediary metabolism of all organisms. Our DNA contains the *N*-heterocycles purine and pyrimidine, and nicotinamide is an essential cofactor, either biosynthesized or required in the diet as a vitamin for humans.

It is impossible to represent the various biochemical mechanisms necessary to catabolize all of these compounds simply, and experimental data are available for only a limited number of examples (15). Rather, we will illustrate a metabolic principle which is common to the metabolisms of a significant number of the more commonly occurring heterocyclic compounds (Fig. 8.9). This is a basic mechanism by which the heterocyclic ring can be opened, which is fundamentally the same with nitrogen, oxygen, and sulfur heterocyclic rings. The metabolic strategy is to hydroxylate the ring at a carbon atom adjacent to the heteroatom. Thereafter, two different reaction pathways can ensue, with the same outcome of opening the ring to generate a carboxylic acid. In pathway A, dehydrogenation generates a ring ester, amide, or thioester, depending on the identity of the heteroatom. All of these intermediates can be cleaved by hydrolysis to open the ring (Fig. 8.9). Pathway B in Fig. 8.9 shows an isomerization of the 2-hydroxy-heteroatom compound, with the ring opening to produce an open-ring aldehyde product. The isomerization may be enzyme catalyzed or occur spontaneously, depending upon the reactivity of the compound, which will be determined by ring electron currents.

Triazine-Ring Metamap

We will discuss here a special case of a nitrogen heterocyclic ring in which the three nitrogen atoms are symmetrically displaced around a six-member carbon ring (Fig. 8.10). These *s*-triazine ring compounds form an important class of pesticides which have been used extensively in agriculture. This provides an unusual example in biodegradation in which a ring structure can be completely metabolically deconstructed by hydrolytic reactions (4, 37).

Commercial products have substituents on each of the ring carbon atoms, most typically two alkylamino substituents and a third substituent of

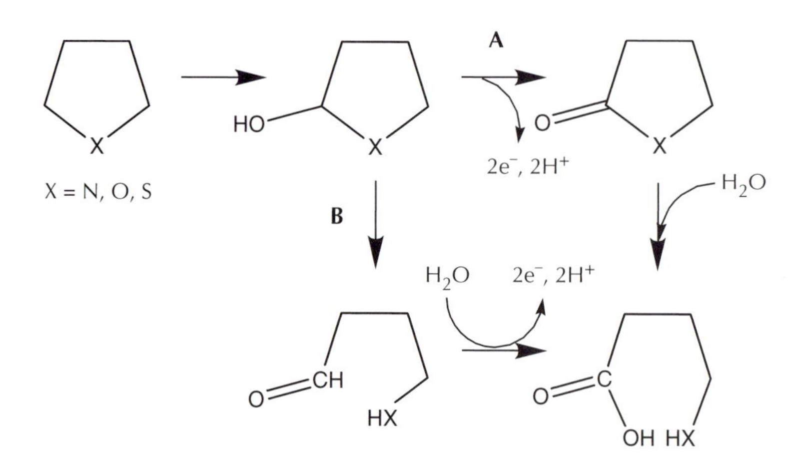

Figure 8.9 Prominent metabolic strategy for opening nitrogen, oxygen, or sulfur heterocycles.

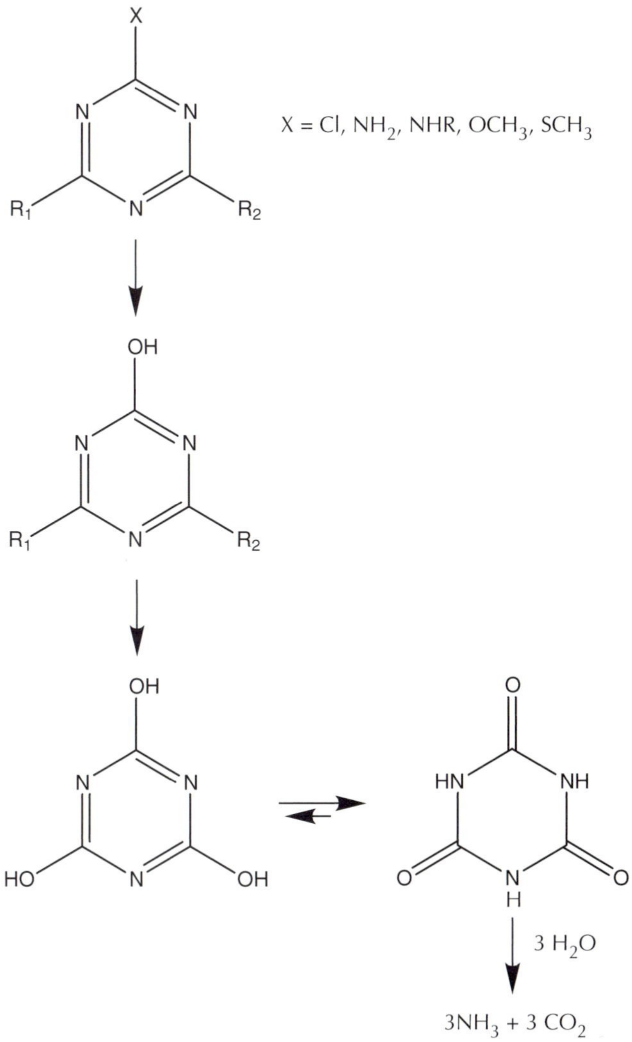
X
N
N
N
R1
R2
X = Cl, NH2, NHR, OCH3, SCH3
OH
N
N
N
R1
R2
OH
N
N
N
HO
OH
O
HN
NH
O
N
H
O
3 H2O
3NH3 + 3 CO2

Figure 8.10 Metabolic strategy for many triazine-ring compounds proceeds through the intermediacy of cyanuric acid (ring structure at lower right).

chlorine, *O*-methyl, or *S*-methyl. The important agricultural products with these substituents include atrazine, simazine, ametryn, and prometryn. It is known that chlorine, amino, and *N*-alkyl substituents can be cleaved from the ring via hydrolytic reactions. In some cases, there are oxidative reactions to remove the alkyl groups first, followed by hydrolytic removal of the remaining free amino group.

In all these cases, the underlying metabolic strategy is to generate trihydroxy-*s*-triazine, or cyanuric acid, which exists under standard conditions in water in one of its tautomeric states as the trione shown in Fig. 8.10. Cyanuric acid is thus a cyclic polyamide. Amide bonds are very prevalent in biological systems, and it is not surprising that this structure is readily susceptible to hydrolysis by soil microorganisms. The potential intermediates in such metabolism, for example, urea, are likewise readily metabolized by many soil bacteria.

Organohalogen Metamap

Organohalides are those compounds containing a carbon bonded to fluorine, chlorine, bromine, and iodine. The halogen column of elements in the periodic table also includes astatine, but there is no literature on organoastatine compounds. There are no known stable isotopes of astatine. As a result, it is estimated that only 50 mg of astatine exists at any given moment on Earth, forming transiently during natural radioactive decay and then quickly undergoing decay in turn (17).

Organohalides have a Jekyll-and-Hyde existence. DDT and other organohalide pesticides have killed malaria-carrying mosquitoes and saved millions of human lives. It is not surprising that immediately after their introduction, organohalide pesticides were considered to be marvelous industrial products. Chlorinated compounds were widely used as solvents and industrial degreasing agents. But decades later, in most of the Western world, organohalides are perceived by the public to be problem pollutants. Even some scientists consider organohalides to be exotic man-made creations which do not resemble natural products. But Gordon Gribble, who studies organohalide natural-product chemistry, reminds us that thousands of organohalides are made in nature without human intervention (10). In fact, all of the properties attributed to organohalides are true. Selected organohalides are wonder compounds, drugs, intermediates in synthesis, natural products, and potent carcinogens, and they often persist in the environment for considerable periods.

Given the natural occurrence of organohalides, it is understandable that some of them are readily metabolized by soil microorganisms. As discussed in chapter 5, 10^{10} lb of chloromethane is biosynthesized annually by soil fungi. Chloromethane is apparently recycled; it is not known to accumulate in the environment. What are the fundamental mechanisms by which organohalides, both naturally occurring and synthetic, are recycled?

There is no single metabolic mechanism that will handle all organohalides. Fundamentally, four different mechanisms are observed in biodehalogenation (Fig. 8.11). First, nucleophilic displacement mechanisms, predominantly formal displacement with water (hydrolysis), often occur with singly chlorinated alkylhalides. This type of mechanism occurs with 1,2-dichloroethane in one well-studied example. Nucleophilic aromatic substitution is also known, for example, that catalyzed by 4-chlorobenzoyl-CoA dehalogenase (28). In this example, the bacterial growth substrate is 4-chlorobenzoate, and the initial generation of a CoA ester of the carboxylate serves to chemically activate the *para*-carbon bearing the chlorine substituent.

In other cases, less activated chlorinated aromatic-ring compounds and alkenes are too unreactive to be processed via a nucleophilic displacement mechanism. In these cases, a reductive dehalogenation is often the preferred mechanism (Fig. 8.11). Reductive dechlorination requires the input of two electrons and two protons to generate a carbon-hydrogen bond and HCl. These reactions are more commonly observed to be catalyzed by anaerobic bacteria. There are at least two reasons for this. One is that anaerobes have greater amounts of low-potential redox cofactors and enzymes which can effect reductive dechlorination chemistry. Another reason is that some anaerobes have evolved the ability to use certain organohalides as their final electron acceptors. By definition, anaerobes do not use dioxygen as the final electron acceptor and need alternative compounds. Both inorganic and or-

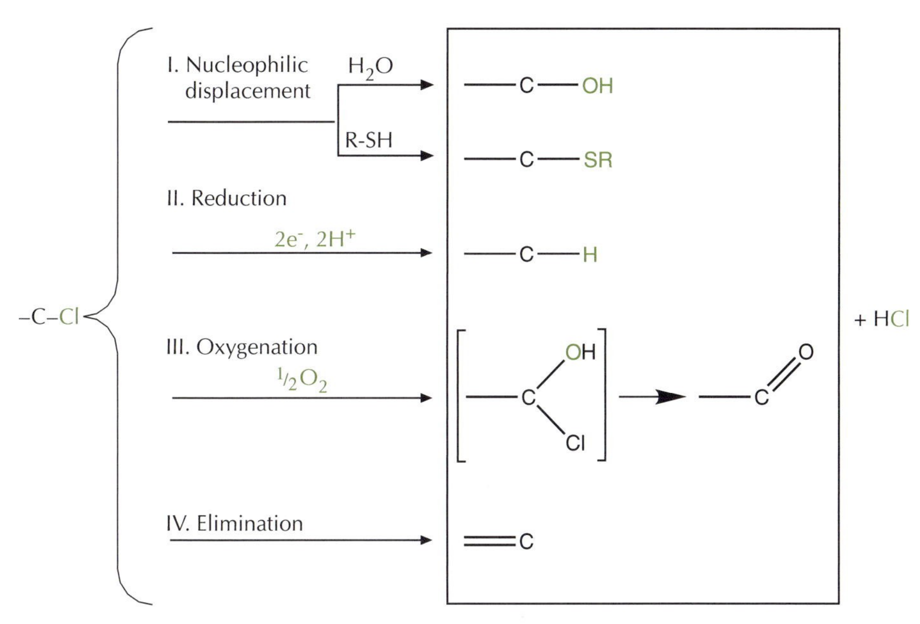

Figure 8.11 Chloroorganic compounds are metabolized via four fundamental mechanisms capable of cleaving the carbon-chlorine bond.

ganic compounds can serve as a sink of electrons generated from oxidative reactions. Organohalides are another class of compounds now known to serve as the final electron acceptor in the metabolism of some bacteria. Organohalides which have been shown to act as final electron acceptors in energy metabolism include 3-chlorobenzoate (5) and tetrachloroethene (24).

As in other types of catabolism, oxygenation has been observed to be a versatile mechanism for dehalogenation. Oxygenation of an alkylhalide generates an unstable *gem*-chloroalcohol which undergoes rapid spontaneous elimination of HCl to yield a carbonyl compound (Fig. 8.11). A similar strategy proves effective with haloalkenes and haloaromatic-ring compounds. Dioxygenation of the double bond bearing a chlorine substituent likewise generates a *gem*-chloroalcohol, which undergoes rapid elimination. An example of this is the 1,2-dioxygenation of 2-chlorobenzoate, which results in the liberation of both HCl and carbon dioxide to leave the readily metabolizable catechol intermediate (30).

The fourth fundamental pathway of dehalogenation is via elimination of a proton and a chloride anion from adjacent carbon atoms to yield an additional bond between the carbon atoms. In microbial catabolism, this typically occurs with substrates containing a carbon-carbon single bond, which undergoes elimination to yield a double bond. A well-studied example of this reaction is the microbial catabolism of γ-hexachlorocyclohexane, or lindane. Lindane is a halogenated organic insecticide that has been used worldwide for agriculture and public health applications. *Sphingomonas* (formerly *Pseudomonas*) *paucimobilis* SS86 was isolated from an experimental field to which lindane had been applied annually over 12 years. *S. paucimobilis* initiates lindane metabolism via two consecutive elimination reactions catalyzed by the LinA dehalogenase (23). The resultant cyclohexadiene undergoes hydrolytic dehalogenation and subsequent aromatization to yield a chlorinated hydroquinone which is known to be metabolized by a number of bacteria. The overall metabolic strategy for lindane can be viewed on the University of Minnesota Biocatalysis/Biodegradation Database (http://umbbd.ahc.umn.edu/ghch/ghch_image_map.html).

Organometallic Metamap

In nature, metalloid compounds from groups 14, 15, and 16 of the periodic table, as well as mercury, are known to undergo alkylation, most notably methylation (Fig. 8.12). There is some evidence of these compounds undergoing biological dealkylation. Some compounds have been produced industrially. For example, butyltin compounds have been used in boat paints to deter the adherence of marine organisms. Organotin compounds are highly toxic to those organisms. Unfortunately, organotin compounds are toxic to most forms of life and have caused pollution problems, particularly in harbors where such paints have been applied. Similarly, tetraethyllead has been used in gasolines to decrease fuel waste due to preignition. It has become clear that the hazards outweigh the benefits for the use of this product. Lead is a neurotoxin, with its greatest effects on children. As a result of this knowledge, many countries have banned the use of alkyllead additives. In contrast to the prominent organotin and -lead compounds, methylmercury is produced naturally by methylation of environmental mercury in anaerobic sediments (33). In the laboratory, it has been shown that sulfate-reducing bacteria will methylate mercury (3). In vitro, methylcobalamin will transfer its alkyl substituent from the cobalt atom of the cofactor to mercury(II) salts. Methylmercury is a potent neurotoxin in humans and has been implicated in deaths and birth deformities. In Minamata Bay, Japan, there occurred a tragic example of mercury pollution followed by biological generation of methylmercury in the anaerobic sediments of the bay. Many people were adversely affected.

With the occurrence, both naturally and industrially, of alkylated metal compounds (Fig. 8.12), it would be anticipated that microbes would respond by metabolizing these compounds. In general, however, there is a lack of data on the detailed mechanisms of bacterially mediated dealkylation, with the exception of organomercury compounds (31, 34). A broad group of bacteria have either broad-spectrum or narrow-spectrum mercury metabolism. Narrow-spectrum mercury metabolism means the ability to detoxify ionic mercury(II) ions but not organomercurials. This is accom-

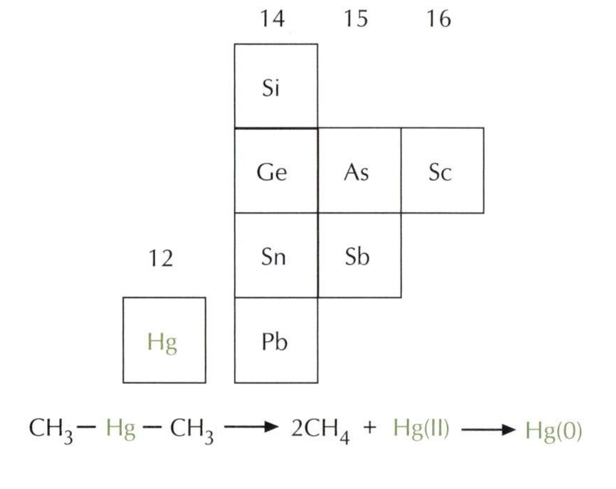

Figure 8.12 The best-studied metabolism of organometallic compounds is that for the catabolism of alkylmercurial compounds. After dealkylation, mercury(II) is reduced to volatile mercury(0), which renders it nontoxic.

plished by the presence of the *mer* operon genes encoding mercury(II) transport through the periplasm and the inner cell membrane and to a cytoplasmic enzyme known as mercuric reductase. Mercuric reductase reduces mercury(II) to mercury(0), which is much less toxic and more volatile, allowing it to diffuse out of the cell and out of the aqueous environment containing the bacterium (Fig. 8.12). Mercuric reductase is a flavoprotein with a redox-active disulfide. At the active site, the coordination environment in which the mercuric ion is reduced contains two cysteines and two tyrosine residues, as evidenced by the high-resolution X-ray structure solved in 1992 (27). Mercuric reductase is highly specific for mercury; other metal ions are not reduced. A proposed reaction mechanism for mercuric reductase is shown on the University of Minnesota Biocatalysis/Biodegradation Database under reaction r0406. Broad-spectrum mercury resistance consists of these same gene products plus an enzyme known as organomercurial lyase. Organomercurial lyase cleaves the carbon-mercury bond of a wide range of compounds: alkyl, alkenyl, and phenyl mercurials have all been shown to be products (1). In the reaction, the carbon-mercury bond is cleaved heterolytically to generate an alkane, alkene, or benzene, respectively. For example, with dimethylmercury, the reaction products are 2 mol of methane and mercury(II) ion (Fig. 8.12). A proposed reaction mechanism for organomercurial lyase is also depicted on the University of Minnesota Biocatalysis/Biodegradation Database (http://umbbd.ahc.umn.edu/core/graphics/r0071_mech.gif).

Summary

It is said that every picture is worth 1,000 words. When we depict a chemical structure that represents thousands of structurally analogous chemicals, the effect is multiplied. When exposed to microbes, structurally analogous compounds often have similar metabolic fates. This makes metabolic logic very powerful. The logic of metabolism allows us to avoid memorizing thousands of chemical structures and their metabolic transformations. With the increasing availability of information in databases on the World Wide Web, a knowledge of metabolic logic will be the key to using this information most effectively.

References

1. **Begley, T. P., A. E. Walts, and C. T. Walsh.** 1986. Mechanistic studies of a protonolytic organomercurial cleaving enzyme: bacterial organomercurial lyase. *Biochemistry* **25:**7192–7200.
2. **Beller, H. R., and A. M. Spormann.** 1997. Anaerobic activation of toluene and *o*-xylene by addition to fumarate in denitrifying strain T. *J. Bacteriol.* **179:**670–676.
3. **Choi, S. C., and R. Bartha.** 1993. Cobalamin-mediated mercury methylation by *Desulfovibrio desulfuricans* LS. *Appl. Environ. Microbiol.* **59:** 290–295.
4. **Cook, A. M.** 1987. Biodegradation of *s*-triazine xenobiotics. *FEMS Microbiol. Rev.* **46:**93–116.
5. **Dolfing, J.** 1990. Reductive dechlorination of 3-chlorobenzoate is coupled to ATP production and growth in an anaerobic bacterium, strain DCB-1. *Arch. Microbiol.* **153:**264–266.
6. **Evans, P. J., D. T. Mang, K. S. Kim, and L. Y. Young.** 1991. Anaerobic degradation of toluene by a denitrifying bacterium. *Appl. Environ. Microbiol.* **57:** 1139–1145.

7. **Fox, B. G., J. G. Borneman, L. P. Wackett, and J. D. Lipscomb.** 1990. Haloalkane oxidation by the soluble methane monooxygenase from *Methylosinus trichosporium* OB3b: mechanistic and environmental implications. *Biochemistry* **29:**6419–6427.

8. **Gibson, D. T., and V. Subramanian.** 1984. Microbial degradation of aromatic hydrocarbons, p. 181–251. *In* D. T. Gibson (ed.), *Microbial Degradation of Organic Compounds.* Marcel Dekker, New York, N.Y.

9. **Gisi, D., L. Willi, H. Traber, T. Leisinger, and S. Vuilleumier.** 1998. Effects of bacterial host and dichloromethane dehalogenase on the competitiveness of methylotrophic bacteria growing with dichloromethane. *Appl. Environ. Microbiol.* **64:**1194–1202.

*10. **Gribble, G. W.** 1992. Naturally occurring organohalogen compounds—a survey. *J. Nat. Prod.* **55:**1353–1395.

11. **Hartmans, S., F. J. Weber, D. P. Somhorst, and J. A. de Bont.** 1991. Alkene monooxygenase from *Mycobacterium:* a multicomponent enzyme. *J. Gen. Microbiol.* **137:**2555–2560.

*12. **Harwood, C. S., G. Burchhardt, H. Herrmann, and G. Fuchs.** 1999. Anaerobic metabolism of aromatic compounds via the benzoyl-CoA pathway. *FEMS Microbiol. Rev.* **22:**439–458.

13. **Heider, J., and G. Fuchs.** 1997. Anaerobic metabolism of aromatic compounds. *Eur. J. Biochem.* **243:**577–596.

14. **Janssen, D. B., A. Scheper, L. Dijkhuizen, and B. Withold.** 1985. Degradation of halogenated aliphatic compounds by *Xanthobacter autotrophicus* GJ10. *Appl. Environ. Microbiol.* **49:**673–677.

*15. **Kaiser, J. P., Y. Feng, and J. M. Bollag.** 1996. Microbial metabolism of pyridine, quinoline, acridine, and their derivatives under aerobic and anaerobic conditions. *Microbiol. Rev.* **60:**483–498.

16. **Kauppi, B., K. Lee, E. Carradano, R. E. Parales, D. T. Gibson, H. Eklund, and S. Ramaswamy.** 1998. Structure of an aromatic-ring-hydroxylating dioxygenase—naphthalene 1,2-dioxygenase. *Structure* **6:**571–586.

*17. **Kirk, K. L.** 1991. *Biochemistry of the Elemental Halogens and Inorganic Halides.* Plenum Press, New York, N.Y.

18. **Kohler-Staub, D., and T. Leisinger.** 1985. Dichloromethane dehalogenase of *Hyphomicrobium* sp. strain DM2. *J. Bacteriol.* **162:**676–681.

19. **Krieger, C. J., H. R. Beller, M. Reinhard, and A. M. Spormann.** 1999. Initial reactions in anaerobic oxidation of *m*-xylene by the denitrifying bacterium *Azoarcus* sp. strain T. *J. Bacteriol.* **181:**6403–6410.

20. **Latham, J., and C. Walsh.** 1986. Bacterial cyclohexanone oxygenase. A versatile flavoprotein oxygen transfer catalyst. *Ann. N. Y. Acad. Sci.* **471:**208–216.

21. **Magli, A., F. A. Rainey, and T. Leisinger.** 1995. Acetogenesis from dichloromethane by a two-component mixed culture comprising a novel bacterium. *Appl. Environ. Microbiol.* **61:**2943–2949.

22. **McCay, D. S., E. K. J. Gibson, K. K. Thomas-Keprta, H. Vali, C. S. Romanek, S. J. Clemett, X. D. F. Chillier, C. R. Maechling, and R. N. Zare.** 1996. Search for past life on Mars: possible relic biogenic activity in Martian meteorite ALH84001. *Science* **273:**924–930.

23. **Nagata, Y., T. Nariya, R. Ohtomo, M. Fukuda, K. Yano, and M. Takagi.** 1993. Cloning and sequencing of a dehalogenase gene encoding an enzyme with hydrolase activity involved in the degradation of *gamma*-hexachlorocyclohexane in *Pseudomonas paucimobilis. J. Bacteriol.* **175:**6403–6410.

24. **Neumann, A., H. Scholz-Muramatsu, and G. Diekert.** 1994. Tetrachloroethene metabolism of *Dehalospirillum multivorans. Arch. Mikrobiol.* **162:**295–301.

25. **Newman, L. M., and L. P. Wackett.** 1995. Purification and characterization of toluene 2-monooxygenase from *Burkholderia cepacia* G4. *Biochemistry* **34:**14066–14076.

26. **Rosner, B. M., F. A. Rainey, R. M. Kroppenstedt, and B. Schink.** 1997. Acetylene degradation by new isolates of aerobic bacteria and comparison of acetylene hydratase enzymes. *FEMS Microbiol. Lett.* **148:**175–180.

27. **Schlering, N., W. Kabsch, M. J. Moore, M. D. Distefano, C. T. Walsh, and E. F. Pai.** 1991. Structure of the detoxification catalyst mercuric ion reductase from *Bacillus* sp. strain RC607. *Nature* **352:**168–172.

28. **Scholten, J. D., K.-H. Chang, P. C. Babbitt, H. Charest, M. M. Sylvestre, and D. Dunaway-Mariano.** 1991. Novel enzymic hydrolytic dehalogenation of a chlorinated aromatic. *Science* **253:**182–185.

29. **Scholtz, R., L. P. Wackett, C. Egli, A. M. Cook, and T. Leisinger.** 1988. Dichloromethane dehalogenase with improved catalytic activity isolated from a fast-growing dichloromethane-utilizing bacterium. *J. Bacteriol.* **170:**5698–5704.

30. **Selifonov, S. A., J. E. Gurst, and L. P. Wackett.** 1995. Regiospecific dioxygenation of *ortho*-trifluoromethylbenzoate by *Pseudomonas aeruginosa* 142: evidence for 1,2-dioxygenation as a mechanism in *ortho*-halogenzoate dehalogenation. *Biophys. Res. Commun.* **213:**759–767.

*31. **Silver, S., and L. T. Phung.** 1996. Bacterial heavy metal resistance; new surprises. *Annu. Rev. Microbiol.* **50:**753–789.

32. **Snow, T. P., V. Le Page, Y. Keheyan, and V. M. Bierbaum.** 1998. The interstellar chemistry of PAH cations. *Nature* **391:**259–260.

33. **Steffan, R., E. T. Korthals, and M. R. Winfrey.** 1988. Effects of acidification on mercury methylation, demethylation, and volatilization in sediments from an acid-susceptible lake. *Appl. Environ. Microbiol.* **54:**2003–2009.

34. **Summers, A. O.** 1988. Biotransformations of mercury compounds. *Basic Life Sci.* **45:**105–109.

35. **Trudgill, P. W.** 1990. Cyclohexanone 1,2-monooxygenase from *Acinetobacter* NCIMB 9871. *Methods Enzymol.* **188:**70–77.

*36. **Trudgill, P. W.** 1990. Microbial metabolism of monoterpenes—recent developments. *Biodegradation* **1:**93–105.

37. **Wackett, J. P., M. J. Sadowsky, M. L. de Souza, and R. T. Mandelbaum.** 1998. Enzymatic hydrolysis of atrazine, p. 82–87. *In* L. Ballantine, J. McFarland, and D. Hackett (ed.), *Triazine Herbicides: Risk Assessment.* Oxford University Press, New York, N.Y.

38. **Willetts, A.** 1997. Structural studies and synthetic applications of Baeyer-Villiger monooxygenases. *Trends Biotechnol.* **15:**55–62.

9

Predicting Microbial Biocatalysis and Biodegradation

Prediction is very difficult, especially about the future.

— *H. L. Mencken*

The conceptual scheme of science is a tool, ultimately, for predicting future experience in light of past experience.

— *Willard Van Orman Quine*

Despite Mencken's pessimism, the words of the American philosopher W. V. O. Quine are perhaps more applicable to the goals of the scientific enterprise. For example, the science of meteorology is sometimes thought of as being equivalent to weather prediction. Historically, this field has had an enormous impact on human cultural development. The importance of shamans and rainmakers was evident in early society, and early religious rituals in agrarian societies were often connected to weather "prediction." Today, the science of meteorology operates somewhat differently, but its importance is still evident. Few would set out to scale Mount Everest or hike through a desert without checking the weather forecasts. They would do this despite the fact that weather prediction is still an imperfect endeavor. Short-term prediction is reasonably good. Long-term prediction, due to the complexity of weather pattern movement, is less reliable. Clearly, however, imperfect information on this subject is preferable to no information.

In general, people might be thought to possess knowledge of a specific subject at the point where they can predict events as yet unobserved. As discussed previously, the biodegradation of most of the 10 million described organic compounds is largely uninvestigated, so the true test of integrity for our knowledge about biodegradation rests on our ability to predict the fate of those 10 million compounds that have not been subjected to biodegradation studies.

This is much more than an esoteric academic exercise. The ability to accurately predict biodegradation has enormous implications. Industries will continue to synthesize new materials faster than they, or regulatory agencies and academic researchers, can study their environmental fates.

Many companies already invest significant resources to predict the biodegradation of their new compounds in order to avoid commercialization of materials that are later found to be dangerous and subsequently need to be withdrawn from commerce. Such events can waste a great deal of capital and engender negative publicity for the company. Thus, improved accuracy of industrial biodegradation predictions will benefit both individual industries and society.

Regulatory agencies, likewise, can do a better job of fairly regulating industry and protecting the public with more accurate models for predicting the environmental fate of organic chemicals. Every year, the U.S. Environmental Protection Agency has to regulate dozens of chemicals for which inadequate microbial metabolism information is available. A classic example of a problem that can arise from environmental metabolism is the microbial reduction of trichloroethene to the human carcinogen vinyl chloride. This metabolism was described in the laboratory and observed in the environment and became an important regulatory concern several years later. This scenario is played out many times, but the time spent in research and administration is substantial. It cannot be done for more than perhaps 0.01% of the organic compounds currently known, so predictive methods become the only recourse to avoid situations where people are exposed to known carcinogens for years before action is taken.

The focus of this chapter will be largely on the prediction of biodegradation pathways. It is recognized that the use of the term "metabolic pathway" is coming under criticism, as pathways do not exist in isolation and occur as many variations on a theme. As such, metabolic pathways as described in biochemistry textbooks are historical constructs. Despite this limitation, the concept of a metabolic pathway is essential for focusing discussion on certain clusters of reactions in isolation from thousands of others which might go on concomitantly in the same cell. As computational methods for modeling metabolism improve, including those described here for predicting biodegradation, the pathway concept will necessarily improve along with them.

Why Is It Necessary To Predict Biodegradation Pathways?

To meet regulatory and other criteria, might it be sufficient to determine merely if a compound is biodegradable? One perspective on that question was given by Stanley Dagley over 20 years ago (3):

> What features of chemical structure determine the biodegradability of a molecule? If he knows the physical conditions and the chemical reagents for a proposed degradation, an organic chemist can make accurate predictions about its course. But microorganisms are not only subject to the laws of chemistry; what they can and cannot do is also determined by their evolutionary history. The conceptual framework, within which many environmental problems must be discussed, can only be constructed from an extensive knowledge of microbial catabolism. All biochemical degradative processes occur by a series of reactions, each of which involves a small free energy change. No step is explosive; and as a consequence, there are usually many steps, liberating the total energy in many packets. It is therefore a matter of common sense that if compound A is to be converted, through the alphabet, to Z which represents CO_2, then the structures of B, C and all the other compounds are also relevant to our considerations of biodegradability.

As Dagley suggested, the issue of biodegradability is best addressed by being able to deduce a pathway(s) for a compound's catabolism and knowing whether that pathway(s) has become widespread by evolutionary natural-selection processes. There are also the important toxicological issues surrounding any accumulating metabolites which might be generated via microbial catabolism in the environment. The microbial reduction of trichloroethylene to the carcinogen vinyl chloride makes matters worse with respect to human health. The hazard would have been predicted if a reductive pathway had been anticipated and toxicological databases had been checked for problems associated with each metabolite.

Biodegradation Prediction Systems

> It is nice to know that the computer understands the problem. But I would like to understand it, too.
>
> — *Eugene Wigner*

Most biodegradation prediction efforts have focused on producing a yes-or-no answer with respect to the compound's biodegradability. This typically consists of surveying experts or trying to deduce rules for computer prediction. In fact, these methods may be combined; experts can be queried for the best information, which can in turn be encoded to make predicting computer software (Fig. 9.1).

In the best efforts, the predictions are then experimentally tested. The tests typically involve incubating the compound of interest with some source of microorganisms, often something like activated sludge. The disappearance of the compound over time is then measured. This would indicate that the compound is transformed, but not necessarily metabolized, to an endpoint like carbon dioxide. An alternative test might measure biological oxygen demand or carbon dioxide evolution.

One approach to prediction has become known as QSAR methods, which stands for quantitative structure activity relationships. In the QSAR method, certain structural features of a molecule are correlated with some outcome, in this case biodegradability (2, 4). For example, conclusions have been reached that the presence of halogen or nitro substituents correlates negatively with biodegradability. The term xenophores has been used for these substituents (1), with the implication that they are exotic, or foreign, to biology and thus nonbiodegradable. However, this is not a very accurate way to explain why some compounds are not biodegraded. It is certainly true that adding four, five, or six chlorine atoms to a benzene ring will slow down its biodegradation, and hence affixing the xenophore label on chlorine is a tempting idea. But adding four, five, or six methyl substituents to a benzene ring would likewise slow its metabolism greatly, and methyl groups are not foreign to biology. The reasons why certain substituents slow down biodegradation are revealed by looking at the possible metabolic pathways, not merely by trying to correlate a substituent with enhancing or inhibiting effects on biodegradation.

Another set of systems have had the goal of predicting metabolic pathways for biodegradation. Two such systems are known as META and BESS. The META metabolic-pathway prediction program is an expert system originally designed to predict pathways for the mammalian detoxification of drugs and environmental toxicants. It was later transformed into a microbial-biodegradation prediction tool by developing DEGR, a knowledge base of microbial biodegradation reactions (6, 7). META is fundamen-

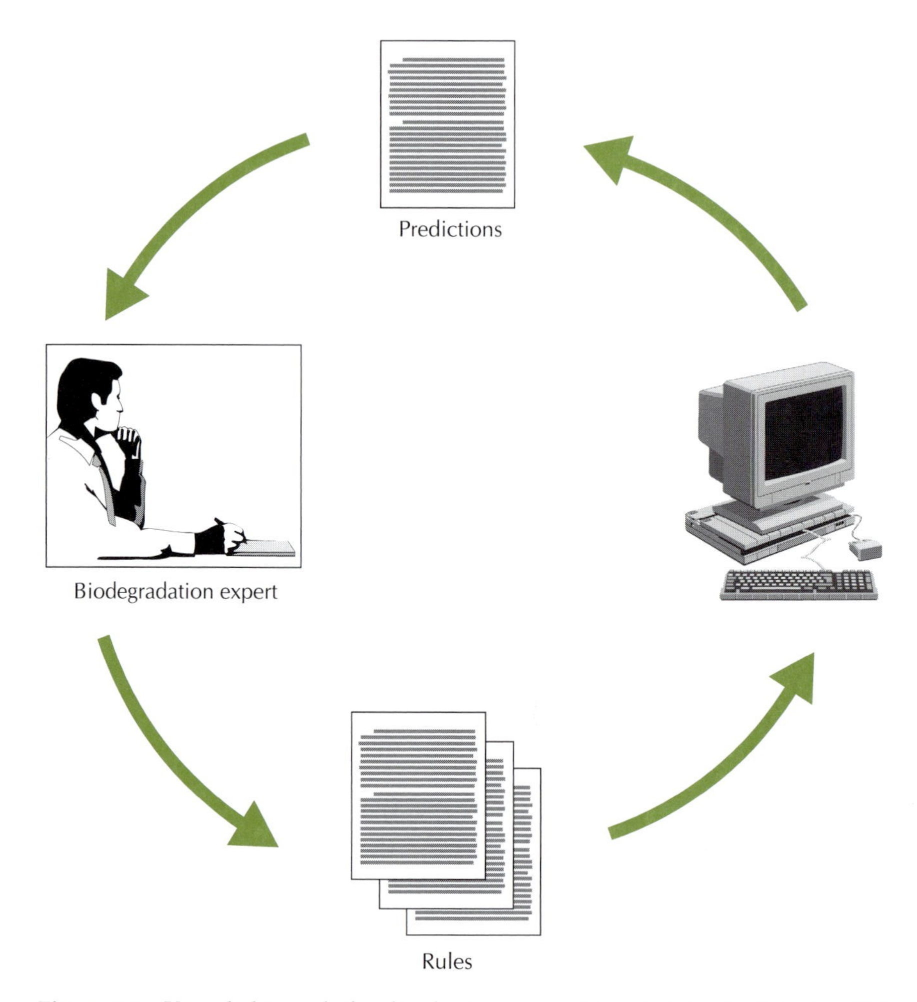

Figure 9.1 Knowledge cycle for developing an expert system to predict microbial catabolism.

tally a computer expert system in which a substructural fragment within a larger chemical entity is identified and then transformed or replaced with another substructural fragment. This is a computational representation of a single enzyme-catalyzed reaction in a metabolic pathway. The rules for the program are thus a series of paired recognition and replacement sets for a large number of beginning structural fragments. The chemical fragment is typically a collection of 2 to 11 atoms, connected by covalent bonds. When a chemical structure is submitted to the META software, its various recognizable fragments are identified and transformed. Each product of a given transformation can be resubmitted for a further round of recognition and transformation. This iterative process can generate a metabolic tree of possible catabolic reactions. Obviously, with a moderate-to-large starting molecule, the catabolic tree can sprout a large number of branches, a problem sometimes referred to as combinatorial explosion. This is addressed by assigning priority numbers to reactions. When two reactions with a given substrate are predicted, the reaction with a lower priority number proceeds first. Both reactions and their priorities come from information obtained from biodegradation experts (Fig. 9.1).

BESS, or Biodegradability Evaluation and Simulation System, is so named because it simulates the transformations of chemical compounds which occur during microbial biodegradation (9). Like META, it starts with expert knowledge. The expert knowledge came from a database compiled at Procter & Gamble which contained information on the biodegradation of compounds of interest to the company. This included compounds in the chemical classes alkanes, alkenes, alicyclic-ring compounds, ethers, alkylsulfonates, and amino acids. These starting compounds are acted upon by computational rules. The starting compound's structure is assessed, and the rule base is searched. If a rule applies, the starting structure is acted on to generate a product which can in turn be taken to another product. If this is successful after an iterative process, a metabolic pathway is generated.

BESS faces the same constraints as the META system or indeed any prediction system. First, the system is only accurate if the rules accurately reflect biodegradation reactions which occur in nature. While knowledge of biodegradation reactions is surely incomplete, any system should draw on the entire breadth of information which is currently known. Second, the problem of combinatorial explosion is addressed by using a knowledge of metabolic logic, as discussed in chapter 8. The logic of metabolism is based on a synthesis of chemical and evolutionary considerations. Chemically, a reaction must be plausible based on the types of reactions which enzymes can catalyze. Evolutionarily, there is most often a conservation of metabolism; it is beneficial to transform a compound using the smallest number of reactions and by entering intermediary metabolism in the fewest steps.

The basic principles needed for prediction are defining (i) the metabolic trunk pathways that initial biodegradation reactions funnel into and (ii) the enzymatic mechanisms by which different organic functional groups are metabolized. These key issues in predicting biodegradation will be discussed in greater detail below.

Defining the Trunk Pathways

There are two major goals in developing a system to predict microbial biodegradation pathways. The first goal is to define the core metabolism. This can be said to consist of intermediary pathways and biodegradation trunk pathways as defined in chapter 8. The second goal is to decide how organic functional groups, which are not common in intermediary metabolism, can be transformed into functional groups typically found in intermediary or trunk pathways. This necessitates a definition of the key organic functional groups and the enzymatic reactions that could transform them into more common functional groups.

The compounds typically denoted intermediary metabolites are now well covered in World Wide Web databases, for example, in the Kyoto Encyclopedia of Genes and Genomes (5). Moreover, there is an ongoing project using the databases Klotho and Atropos to better depict this metabolism (http://www.ibc.wustl.edu/klotho/). Intermediary and trunk metabolites become the targets for starting compounds, with the goal of reaching those metabolic intermediates in the smallest number of steps. Microbial genomes are very streamlined, and evolutionary pressure invariably favors the smallest gene complement to attain the most metabolism (8). A large body of experimental evidence indicates that there are many common trunk pathways with a limited number of enzymatic reactions added in individual bacteria that funnel into the existing pathways.

It is instructive to think of bacterial metabolism as an electrical circuit diagram with the compounds as nodes and the enzyme-catalyzed reactions as connecting lines (Fig. 8.1). This way of thinking about metabolism will become more common as we increasingly use genomic data to reconstruct metabolism and to model fluxes through pathways. It is also useful here to consider that the reactions of the glycolytic and tricarboxylic acid cycles are the "core" of metabolism. This has more than one meaning. First, these reactions constitute a core in that most prokaryotes carry out these reactions as part of their overall metabolism. Second, the reactions are the central pathways which provide precursor compounds for biosynthesis and the core intermediates into which compounds funnel after catabolism of more complex molecules.

In the latter context, one can think of biodegradation reactions as those which are peripheral to core intermediary metabolism, as indicated in Fig. 8.2. Only some prokaryotes contain these peripheral pathways. Other prokaryotes contain different pathways that would feed into the same or other core intermediary metabolites. One can think of this overlaid metabolism as modular and fluid. There can be many variations, but as long as the metabolism funnels into a core compound, it can be oxidized further to obtain energy and can be used for biosynthesis. This was illustrated in chapter 8 on metabolic logic. For example, it was illustrated that many single benzenoid ring compounds are metabolized aerobically via a catechol intermediate (see Fig. 8.6). In another example, C_2 organic pollutants often funnel into glyoxylate or acetate (see Fig. 8.4), common core intermediates in many bacteria.

One can think of a second layer of core biodegradation reactions which funnel into core intermediary metabolism. The former are illustrated on the University of Minnesota Biocatalysis/Biodegradation Database metapathway map (http://umbbd.ahc.umn.edu/meta/meta_map.html). It is readily apparent that many compounds in anaerobic benzenoid ring metabolism feed into the common intermediate benzoyl-coenzyme A (CoA). Benzoyl-CoA is further metabolized reductively, and the ring is cleaved to produce 3-hydroxypimelyl-CoA. This is a second-layer core metabolite which can be further metabolized to acetoacetyl-CoA and ultimately acetyl-CoA.

Another way to think of "layers" of catabolism is to think in terms of the prevalence of different clusters of reactions, or pathways, within the pool of known bacterial species. We have already discussed the most common metabolic reactions, those in the glycolytic and tricarboxylic acid cycles. Then there are metabolic reactions which are less common but still reasonably prevalent, for example, those involved in the metabolism of β-ketoadipate and cyanuric acid. This allows bacteria possessing those pathways to feed certain metabolic intermediates into the more prevalent core metabolism. This metabolism is thus more peripheral, as well as less prevalent. Then, there are metabolic reactions which are rarer, and the metabolism is more peripheral too. For example, the reactions which comprise pathways for *s*-triazine pesticides are likely present in less than 1% of soil bacteria. The prevalence of certain pathways can only be estimated at present, but with more genomic analysis of soil bacteria, this practice can become more rigorous over time.

Defining the Organic Functional Groups Relevant to Microbial Catabolism

> The fate of chemical pollutants cannot be reliably predicted without knowing what types of catabolic reactions are available.
>
> — *Stanley Dagley*

There are a limited number of organic functional groups that take part in intermediary metabolism. These appear repeatedly and include alkyl hydrocarbons, alcohols, aldehydes, ketones, carboxylic acids, amines, benzenoid aromatic rings, nitrogen heterocycles, thiols, thioethers, ethers, amides, and esters (Fig. 9.2). There are on the order of 30 functional groups commonly found in intermediary metabolism.

The set of organic functional groups which are transformed by all known microbial metabolic pathways that we are aware of doubles the set

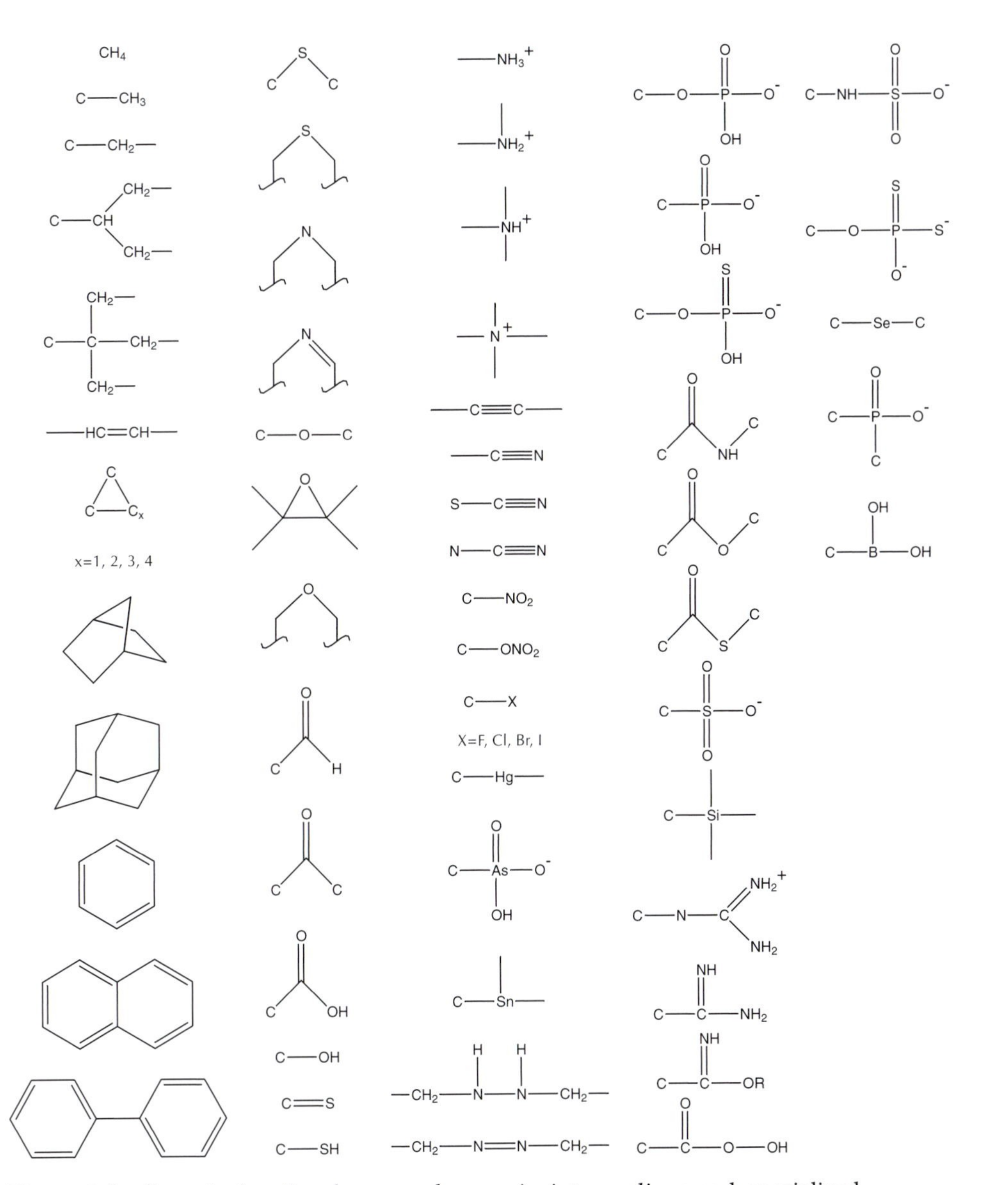

Figure 9.2 Organic functional groups known in intermediary and specialized metabolism.

of functional groups to consider, making a total of 58. As shown in Fig. 9.2, the novel functional groups include alkyne, cyano, thiocyano, nitro, nitrate ester, organohalide, organomercurial, organoarsenic, diazo, phosphonate, sulfonate, and cyanamide. Combined with those found in intermediary metabolism, this represents the entire range of organic functional groups for which catabolic reactions are known.

As the analysis in chapter 5 illustrated, the total set of biologically relevant functional groups is even greater. Table 5.2 alone presents over 90 organic functional groups which are found in nature, produced via biosynthesis or natural diagenic reactions. The gap between the total number of known naturally occurring functional groups (94) and those for which catabolic transformations are known (58) suggests that catabolic reactions capable of transforming at least 36 functional groups remain to be discovered. This is based on the assumption that all the naturally occurring functional groups are metabolizable by at least some small set of bacteria found globally.

How many organic functional groups must be considered in predicting biodegradation? First, let's look at the types of chemicals which are most heavily used by industry (Table 9.1). They are structurally simple, since billions of pounds must be synthesized or refined to meet industrial needs. Most of the compounds are composed of one or two of the functional groups depicted in Fig. 9.2. The most common groups are alkanes, alkenes, benzenoid aromatic rings, organohalides, and carboxylic acids.

There is a need for information on the biodegradation of the top 30 organic chemicals. In fact, the microbial metabolism of virtually all of them has been investigated experimentally. This makes them an excellent set for beginning to explore the principles of predicting biodegradation via functional-group analysis.

The Basis for Predicting Microbial Biocatalysis and Biodegradation

The general approach in functional-group analysis is to develop rules of transformation for each of the approximately 50 functional groups for which at least one example of metabolism is known. It is a manageable task, based on the published scientific literature and the current contents of the University of Minnesota Biocatalysis/Biodegradation Database, to indicate how each functional group could be metabolized. The chemicals in Table 9.1 composed of only one functional group include cyanide, ethylene, benzene, and natural gas (methane). For example, a cyano group might be hydrated with one water molecule to form an amide, hydrolyzed by two water molecules to form carboxylic acid and ammonia, or reduced by six electrons (for example, by nitrogenase) to yield a methyl group and ammonia. We estimate here that biological systems have evolved an average of two distinct enzymatic transformation mechanisms per functional group or a total of approximately 100 transformation reactions.

A larger but still quite manageable set of transformations deals with those compounds composed of two different functional groups juxtaposed. Compounds in Table 9.1 which include two functional groups include ethylene dichloride, vinyl chloride, ethylbenzene, styrene, xylene, terephthalic acid, phenol, acetic acid, propylene oxide, acrylonitrile, and acetone. The theoretical total combinations of two different functional groups juxtaposed, with 50 single groups, number 1,250. However, about half of those

Table 9.1 Top 30 organic chemicals in U.S. commerce (1995)[a]

Rank	Chemical	Wt (lb) in billions
1	Ethylene	47.0
2	Propylene	25.7
3	Methyl *tert*-butyl ether	17.6
4	Ethylene dichloride	17.3
5	Urea	15.6
6	Vinyl chloride	14.0
7	Ethylbenzene	13.7
8	Styrene	11.4
9	Methanol, synthetic	11.3
10	Formaldehyde	8.1
11	Terephthalic acid, dimethyl ester	8.0
12	Ethylene oxide	7.6
13	*p*-Xylene	6.3
14	Cumene	5.6
15	Ethylene glycol	5.2
16	Acetic acid, synthetic	4.7
17	Phenol, synthetic	4.2
18	1,3-Butadiene	3.7
19	Isobutylene	3.2
20	Acrylonitrile	3.2
21	Vinyl acetate	2.9
22	Acetone	2.8
23	Benzene	2.2
24	Cyclohexane	2.1
25	Bisphenol A	1.6
26	Caprolactam	1.6
27	1-Butanol	1.5
28	Isopropyl alcohol	1.4
29	Aniline	1.4
30	Methyl methacrylate	1.4

[a]Used with permission of the American Chemical Society (10).

can be discarded immediately as irrelevant to catabolism. For example, alkanes, alkenes, or arenes with a carbon-chlorine bond will be considered; however, compounds containing a chlorine atom bonded to a carbonyl carbon, known as acyl chlorides, will not be considered. In aqueous environments, acyl chlorides will rapidly and spontaneously undergo hydrolysis to yield a carboxylic acid and a chloride ion. The acyl chloride functional group is thus irrelevant to microbial metabolism. It is estimated here that approximately 600 combinations need to be considered.

Any one of the 600 bifunctional compounds may undergo reaction at one or both of the functional groups or undergo a reaction that cleaves the two functional groups apart. In the former case, the metabolism of each functional group has already been considered. We need only assign a priority to indicate which reaction is preferred. For example, with cyanobenzene (Fig. 9.3, left), we would predict that facile hydrolysis reactions are favored, leading to the formation of benzoic acid. Benzoic acid is a trunk metabolic compound or intermediary metabolite. In the case of chloromethane, we would predict that carbon-chlorine bond cleavage would oc-

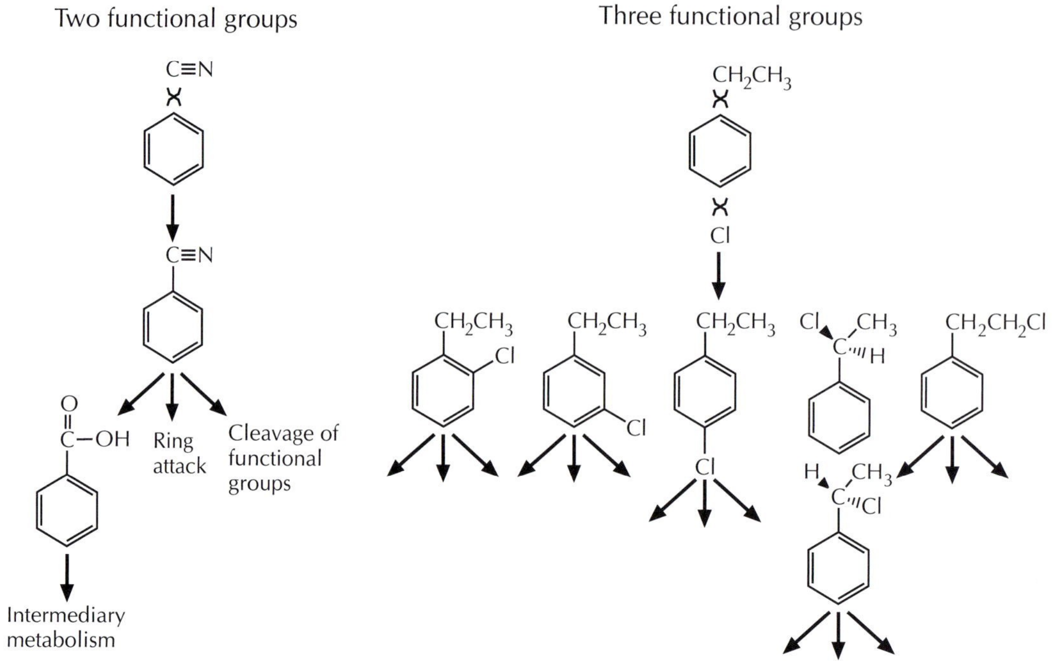

Two functional groups

Three functional groups

Intermediary metabolism

Ring attack

Cleavage of functional groups

Figure 9.3 Prediction of metabolism when two functional groups (left) or three functional groups (right) are fused, showing the increase in complexity.

cur preferentially. This could occur hydrolytically to yield methanol or oxygenolytically to yield formaldehyde, both trunk metabolites. The preference of reaction type is determined at each stage by a knowledge of the prevalence of a given reaction type and what yields the shortest pathway into a trunk metabolite. Assuming that there are an average of two reasonable pathways for each bifunctional compound, there will be 1,200 pathways to consider. With an average of two steps to a trunk pathway intermediate, the total number of reactions to consider is 2,400.

Compounds with three or more functional groups represent a more complex situation. The issue of combinatorial explosion becomes significant here. An example is shown in Fig. 9.3. Adding positional isomerism and stereoisomerism, for example with multiple functional groups on benzene rings and sp^3 carbons, a prediction system becomes informationally intensive. The information needs may become prohibitive for constructing a comprehensive prediction system by building up a prediction library based on a rigorous compilation of reactions with functional groups and all their reactions. To augment this approach, it will be necessary to do what humans do in complex situations: to use rules of thumb to take shortcuts though the maze of complexity.

Beyond Two Functional Groups: the Need for Heuristics

> . . . people follow rules of thumb, known technically as "heuristics." . . . By this means, the complex tasks of assessing probabilities and predicting outcomes are reduced to a few judgmental operations.
>
> — *Edward O. Wilson*

With three or more organic functional groups and with multiple functional-group occurrences and positional isomerism added, the number of possible compounds to consider becomes impossibly large. Is there a way to deal with this combinatorial problem?

We will use the analogy of chess. The number of permutations in a chess game with an average number of moves has been calculated to be 10^{50}. This number will go down as the game proceeds but will remain enormously large until a much simplified end game is reached. Examining every possible move is more than a chess expert, or even a supercomputer, can consider. Chess experts overcome this problem, and play credible games of chess, by using a set of guiding principles to allow them to focus on a small subset of possible moves. An example of such a rule is "develop your pawns, knights, and bishops to the middle of the board so that they control a maximum amount of space." Another rule of thumb is "do not give up a piece unless you capture a piece of equal or greater strength." However, as any chess enthusiast knows, the great chess players made their reputations by occasionally sacrificing a stronger piece for a weaker one only to gain a positional advantage that led them to victory. This is an example of a rule which is meant to be followed most of the time but which may be superseded by another rule based on the best judgment of the expert.

The experts in the example above are using heuristics, or rules of thumb, and this could also be applied by a biodegradation expert who was asked to consider the metabolic fate of a complex organic molecule. Perhaps unconsciously, the expert would apply his or her set of rules and make decisions, based on experience, about which rules are best to follow in a given situation. There are different levels of rules which might guide the biodegradation expert or the prediction software. First, there are general guiding principles which might be used to access the overall metabolism of a compound holistically. Some examples of these, called highly general rules, are listed in Table 9.2. They are largely rules which capture the metabolic logic of catabolism. For example, the idea that compounds are metabolized to intermediary metabolites in the smallest number of steps reflects the fact that biological systems are conservative. Fewer metabolic steps mean less energy expended in biosynthesizing enzymes and fewer

Table 9.2 Heuristics used with enzyme reaction rules for predicting biodegradation pathways: highly general rules

1. Large compounds are usually broken into smaller molecules at bonds that can be easily cleaved.
2. Compounds that contain repeating structures tend to be degraded first into these substructures.
3. Small compounds are preferentially converted to intermediary metabolites in the smallest number of catalytic steps.
4. Pathways are likely to be optimized for maximum yield of metabolic energy.
5. Novel pathways arise faster when fewer evolutionary changes to preexisting pathways are required.
6. Most (>60%) biodegradation reactions are oxidation-reduction reactions.
7. Under aerobic conditions, compounds containing only carbon and hydrogen are initially metabolized via oxygenases.
8. Chemically facile hydrolysis reactions generally have metabolic priority, e.g., ester, amide, and nitrile hydrolysis.
9. Polymers are generally poorly metabolized unless they contain readily accessible, chemically facile hydrolyzable groups.

genes to carry. A similar highly general rule pertains here: the idea that pathways arise faster when fewer evolutionary changes are needed.

After the highly general rules have been applied, it is time to look at the type of compound being dealt with and to apply some general rules (Table 9.3). For example, is it an ether, thioether, or secondary amine? If so, the rules for cleaving the respective carbon–heterocyclic-atom bond must be considered. This often occurs via oxygenation or perhaps dehydrogenation followed by hydration. Many such general rules can be applied to a broad class of compounds to guide the overall decision-making process.

Finally, a series of more specific rules would be used to deal with the intricacies of bond cleavage when the situation would be strongly influenced by the hybridization state of the carbon (sp^2 versus sp^3 versus aromatic) and the number of substituents attached to the carbon atoms. A good example of this is halogen substituents (Table 9.4). In that case, highly halogenated compounds are often metabolized by reductive dehalogenation. Some biological systems have even evolved the ability to use such compounds as their final electron acceptors. With less halogenated analogs, it becomes chemically more difficult for reduction reactions to occur and easier for hydrolysis reactions to occur. Moreover, the availability of sp^3 carbon-hydrogen bonds opens up the possibility of oxygenation of the compound.

Summary

An ideal biodegradation prediction system would use all available information to suggest all plausible metabolic routes for a given compound. As those same compounds are subjected to biodegradation experiments, some of the pathways may be verified to exist. The absence of discovery will not disprove alternative predicted pathways; nature is too vast to sample all microbes for the elucidation of all possible pathways. Nonetheless, predic-

Table 9.3 Heuristics used with enzyme reaction rules for predicting biodegradation pathways: general rules

1. Transferase reactions are rarer in biodegradation than in intermediary metabolism.
2. Many of the transferase reactions in biodegradation are catalyzed by glutathione transferases.
3. *s*-Triazine ring compounds are metabolized via cyanuric acid as an intermediate.
4. Ether carbon-oxygen bonds are cleaved via oxygenation of a carbon atom adjacent to the oxygen.
5. Thioether carbon-sulfur bonds are cleaved via oxygenation of a carbon atom adjacent to the sulfur.
6. Carbon-nitrogen bonds are cleaved by oxygenation of a carbon atom adjacent to the nitrogen or dehydrogenation to generate an imine followed by hydrolysis.
7. Aliphatic nitrate esters initially undergo N-O bond cleavage.
8. Anaerobically, benzenoid aromatic rings are carboxylated, transformed to Co-A esters, and then reduced.
9. Carbon-halogen bonds are cleaved by (i) two-electron reduction, (ii) hydrolysis, (iii) oxygenation of a carbon atom(s) followed by *gem* elimination, or (iv) β elimination.

Table 9.4 Heuristics used with enzyme reaction rules for predicting biodegradation pathways: specific rules for halogenated compounds

1. With halogenated alkanes containing three or more halogen substituents on a single carbon atom, the halogen is usually removed by reductive dehalogenation.
2. With halogenated alkanes containing two or fewer halogen substituents, the halogen substituent is typically removed by hydrolytic or oxygenative reactions.
3. With halogenated alkenes containing two or more halogen substituents on a single carbon atom, a halogen substituent may be removed by reductive dehalogenation or via oxygenation of the double bond and spontaneous elimination reactions.
4. With halogenated arenes containing four or more halogen substituents on a benzenoid aromatic ring, the halogens are most frequently removed by reductive dehalogenation.
5. Benzene rings containing halogen substituents and no vicinal carbon atoms containing hydrogen substituents do not typically undergo dioxygenation.

tive software, no matter how good, will never completely supplant the need for experiments. It should be clear from the discussion in chapters 5, 11, and 12 that there is much more microbial metabolism waiting to be discovered. Theory and experiment must continue to move forward hand in hand.

References

*1. **Alexander, M.** 1994. Effect of chemical structure on biodegradation, p. 159–176. *In Biodegradation and Bioremediation.* Academic Press, San Diego, Calif.

2. **Cowan, C. E., T. W. Federle, R. J. Larson, and T. C. Feijtel.** 1996. Impact of biodegradation test methods on the development and applicability of biodegradation QSARs. *SAR QSAR Environ. Res.* **5:**37–49.

*3. **Dagley, S.** 1979. Summary of the conference, p. 534–542. *In* A. W. Bourquin and P. H. Pritchard (ed.), *Proceedings of the Workshop on Microbial Degradation of Pollutants in Marine Environments.* U.S. Environmental Protection Agency, Washington, D.C.

4. **Howard, P. H., R. S. Boethling, W. Stiteler, W. Meylan, and J. Beauman.** 1991. Development of a predictive model for biodegradability based on BIODEG, the evaluated biodegradation data base. *Sci. Total Environ.* **109–110:**635–641.

5. **Kanehisa, M., and S. Goto.** 2000. KEGG: Kyoto encyclopedia of genes and genomes. *Nucleic Acids Res.* **28:**27–30.

6. **Klopman, G., and M. Tu.** 1997. Structure-biodegradability study and computer-automated prediction of aerobic biodegradation of chemicals. *Environ. Toxicol. Chem.* **16:**1829–1835.

7. **Klopman, G., Z. Zhang, D. M. Balthasar, and H. S. Rosencranz.** 1996. Computer-automated predictions of metabolic transformations of chemicals. *Environ. Toxicol. Chem.* **14:**395–403.

*8. **Parke, D., D. A. D'Argenio, and L. N. Ornston.** 2000. Bacteria are not what they eat: that is why they are so diverse. *J. Bacteriol.* **182:**257–263.

9. **Punch, B., A. Patton, K. Wight, B. Larson, P. Masscheleyn, and L. Forney.** 1996. A biodegradability and simulation system (BESS) based on knowledge of biodegradability pathways, p. 65–73. *In* W. J. G. M. Peijnenburg and J. Damborsky (ed.), *Biodegradability Prediction.* Kluwer Academic Publishers, Dordrecht, The Netherlands.

10. **Storck, W. J., P. L. Layman, M. S. Reisch, A. M. Thayer, E. M. Kirschner, G. Peaff, and J. F. Tremblay.** 1996. Facts and figures for the chemical industry. *Chem. Eng. News* **74**(26):42.

*11. **Wackett, L. P., L. B. M. Ellis, S. M. Speedie, C. D. Hershberger, H.-J. Knackmuss, A. M. Spormann, C. T. Walsh, L. J. Forney, W. F. Punch, T. Kazic, M. Kanehisa, and D. J. Berndt.** 1999. Predicting microbial biodegradation pathways. *ASM News* **65**:87–93.

10

Microbial Biotechnology: Chemical Production and Bioremediation

> The world we inhabit today teeters between becoming either the lovely garden or the barren desert that our contrary impulses strive to bring about. Our future is now closely tied to human creativity.
>
> — *Mihaly Csikszentmihalyi*

Humans have transformed the planet, and as Csikszentmihalyi points out, this can be both good and bad. We cannot go back to earlier times, however. The human population has reached 6 billion and continues to grow. As we push forward, it is important to do so with a spirit of creativity and inventiveness. Applied biodegradation and biocatalysis will likely play an increasing role in maintaining the desirable attributes of planet Earth as the global economy moves inexorably forward. The paths toward this goal are severalfold: (i) increasingly treating wastes at the source; (ii) using renewable resources, such as cornstarch, in preference to nonrenewable resources, such as petroleum; and (iii) increasingly recycling wastes back into production of saleable materials. The first two will be driven by governmental regulations and the costs of dwindling petroleum resources, respectively. The last point makes good business sense: less costly waste to deal with, less chance for negative publicity due to being labeled a "polluter," and higher product yield for the same raw-material cost. More than ever, biodegradation and microbial biocatalysis for manufacture will be inextricably linked. To understand how society has reached this point, we must consider how biocatalysis and biodegradation have been intertwined historically.

Historical and Conceptual Progress

> There is no such thing as Pure or Applied science, there is only Good or Bad science.
>
> — *Attributed to an unnamed British industrialist by John D. Bu'Lock*

As discussed in chapter 2, biodegradation had its origin in a search for the basis of spoilage. Its solid scientific underpinnings were derived from the study of an economically important biotransformation: Pasteur's research on spoilage during the alcohol fermentation of wine.

Most early applications of microbial biocatalysis were important in the production of foods or chemical additives used in food applications. In addition to alcoholic beverages, this included food fermentations producing lactic acid, for example, in yogurt and sauerkraut, and acetic acid, or vinegar. As discussed in chapter 2, fungi were harnessed to produce citric acid, which is a major food additive. More recently, the food industry has been on a long quest for new food and beverage sweeteners. This has led to the discovery of aspartame, or *N*-L-aspartyl-L-phenylalanine 1-methyl ester, which is produced by microbial catalysts, and high-fructose corn syrup (21). The latter is pervasive, because fructose tastes very similar to sucrose, with no undesirable aftertaste, but is 80 times sweeter per unit mass. It thus offers a low-calorie but high-quality substitute for sucrose, or common table sugar. High-fructose corn syrup, as the name implies, is not a pure compound but contains fructose as a major ingredient. It derives from cornstarch, which is now commonly hydrolyzed by heat-stable bacterial amylases. The resultant glucose is then transformed by glucose isomerase to yield fructose. In this application, enzymes are used rather than whole bacteria because the enzymes are relatively stable and cheap to mass produce. The transformation of starch and other plant polymers to glucose will be discussed in greater detail below.

The process for making acetone and butanol by using *Clostridium acetobutylicum* was discussed in chapter 2. This was an enormous success in Britain and Canada during World War I. However, this success was due to the unavailability of chemical feedstocks, which came primarily from Germany prior to the war. Thus, new resources for solvent production were required, and Chaim Weizmann turned to microbial fermentations using cornstarch as a starting material.

Certain large-scale fermentation processes persisted for some time, but after World War II, there was a tremendous rise in the size and influence of the commodity chemical industry. Petroleum became the dominant feedstock, and by the mid-1990s, over 95% of commodity organic chemicals were derived from petroleum. Of course, petroleum is not a renewable resource because millions of years are required to generate petroleum reserves naturally. In the mid-1970s, a large increase in the cost of petroleum and a brief disruption in availability increased the public perception that petroleum supplies will rapidly be consumed at current consumption rates. Industry has increasingly looked into alternatives, and slowly but surely, biotechnology is making inroads into commodity chemical production.

Perhaps one leading indicator of the swing to and from large-scale chemical processes is the production of ethanol in the United States in the 20th century. The numbers shown in Table 10.1 are striking. Ethanol was produced exclusively by fermentation in 1920 and by chemical synthesis in 1977; by 1991, it was produced almost exclusively by fermentation once again (11). The 20-year shift from chemical manufacture to fermentation was led by Brazil and driven by a combination of economic factors. In the mid-1970s, high oil prices, a glutted world market for sugarcane, and a desire to conserve foreign currency combined to prompt massive efforts to generate ethanol from cane sugar. The United States has followed this trend, as indicated in Table 10.1, driven by congressional subsidies and the policy of adding ethanol to gasoline.

Table 10.1 Percentages of ethanol produced biologically in the United States during the 20th century[a]

Yr	% Ethanol from microbes
1920	100
1935	90
1954	30
1963	9
1977	6.5
1982	55
1991	94

[a]From reference 21 with the permission of Cambridge University Press.

While ethanol is the largest-volume commodity product made almost exclusively by microbial biocatalysis in the United States, other small-molecule commodity chemicals are also being made at least partially by microbes. The compounds shown in Table 10.2 are oxygenated hydrocarbons and represent important targets for industrial biocatalysis. These examples highlight what many see as a trend away from nonrenewable resources to a sustainable industry based on biomass. This is discussed in greater detail below.

Recent Trends

An early triumph in the use of biotransformation reactions, also discussed in chapter 2, was Weizmann's fermentation of cornstarch to produce acetone for England during World War I. This presaged modern developments in biocatalysis in several ways. First, cornstarch is currently a key renewable resource which can be used directly or after conversion to glucose. Second, plant material was used to produce a commodity chemical, the solvent acetone, in huge fermentors. This required strong engineering expertise to run the fermentors and harvest the product. Third, there was a tension between the cost of chemical synthesis and the biotransformation process. This played out after World War I. When the chemical feedstocks became available again, the fermentation process to make acetone was supplanted by traditional chemical synthesis. Today, acetone is made largely by chemical synthesis. However, it is increasingly perceived that chemical feed-

Table 10.2 Commodity chemicals potentially derivable from glucose or other renewable feedstocks by microbial transformations

Chemical	Microorganism	Applications	U.S. market value[a] ($ million)
Ethanol	*Saccharomyces, Zymomonas*	Fuel; feedstock for esters and ethers	9,000
Acetic acid	*Acetobacter*	Solvent and intermediate	620
Isopropanol	*Clostridium*	Solvents: inks	500
Acetone	*Clostridium*	Solvent and intermediate	460
Acrylic acid	*Bacillus*	Polymers	360
Glycerol	*Saccharomyces*	Solvent; soaps; antifreeze	250
1,2-Propanediol	*Bacillus*	Antifreeze; mold inhibitor	220
1,3-Propanediol	*E. coli*	Carpets	200[b]

[a]From production and prices in 1979. Data from reference 11.
[b]Estimated market value.

stocks will become scarcer and hence more expensive. For an industry to be sustainable, it must inevitably shift to feedstocks from plant material in place of petroleum.

The chemical industry is anticipating this major transformation to a future in which biomass is used as a feedstock instead of petroleum. Under this scenario, natural products, principally glucose and other sugars, would serve as the starting materials for fermentation processes to produce important industrial compounds (Fig. 10.1). Because most bacteria are capable of catabolizing glucose, microbes can be isolated from nature for diverse biochemical properties and then used in glucose-fed fermentations to produce desired end products. *Escherichia coli* is often the standard organism for making high-value proteins by the use of recombinant bacteria because it is highly amenable to accepting and expressing foreign genes. For specialty chemical production, many different microorganisms are used, for example, *Streptomyces* sp. for antibiotics and *Corynebacteria* sp. for amino acids. With the heavy use of genomics techniques by industry, this trend will no doubt increase. A newly isolated microbe will be able to have its DNA sequenced and annotation accomplished quickly, thus shortening the time from discovery to application.

End Products of Fermentation: Pharmaceuticals

The shift to biocatalytic manufacture of pharmaceuticals is occurring slowly. In the early part of the 20th century, industrial biotransformation products were primarily high-value compounds which were difficult or impossible to synthesize chemically. The example given in chapter 2 was the manufacture of cortisone via combined biocatalysis and conventional chemical synthesis. Major examples of products made completely by biosynthesis are antibiotics.

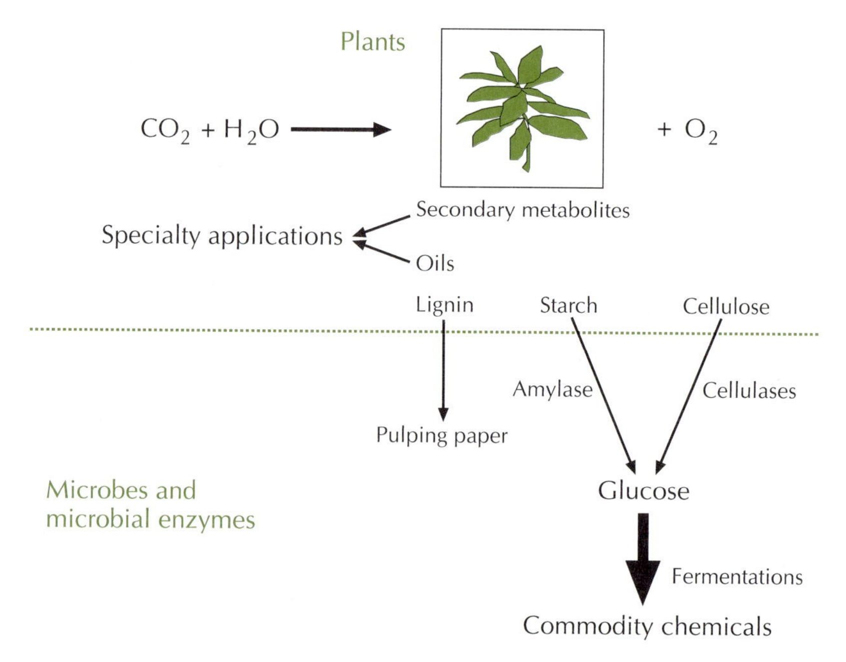

Figure 10.1 Plants are renewable sources of carbon, which can be transformed by microbes and their enzymes to produce many industrially useful compounds.

Converting Biomass to Glucose

There are two predominant biomass polymers: starch and lignocellulose. Corn, wheat, sorghum, and potato are starchy plants. Much other plant material is considered to be lignocellulosic. Starch and lignocellulosic material can be depolymerized via hydrolysis reactions (Fig. 10.1). This occurs in two enzymatic steps, which form the basis of large-scale industrial processes. First, heat-stable α-amylases, obtained from *Bacillus* sp. and other sources, are use to chop up the starch polymer into oligomeric glycodextrin fragments. Then, a second enzyme serves to hydrolyze the fragments to individual glucose molecules.

Most of the Earth's biomass is lignocellulosic, and much of that is composed of cellulose (2). At present, cellulose is a virtually unlimited resource. It is renewable and thus can form the bedrock of sustainable industrial processes. But there are some technical impediments to using cellulose. First, cellulose has β-1,4 chemical linkages joining the glucose monomers, which are more resistant to hydrolysis than the α-1,4 linkages in starch. Second, cellulose is often found in a crystalline form, making the linkages less accessible to enzymatic hydrolysis. Third, cellulose is tightly associated with hemicellulose and is bound to lignin, a tough aromatic polymer. Together, these factors make cellulose utilization difficult.

To make lignocellulose more utilizable, physical and chemical techniques can be used to disrupt its structure (29). In one example, the lignocellulosic material is subjected to high-pressure steam followed by a rapid drop in pressure to explode the polymer. Dilute acid may also be used to disrupt enough chemical linkages to open up the structure. These and other methodologies will prove critical in increasing the use of renewable biomass resources in chemical production industries, as they will influence the cost of the fermentable sugars.

A significant number of the compounds made by combined biocatalysis and chemical synthesis are in the classes of pharmaceutical products shown in Fig. 10.2. The list of examples includes amoxicillin, *S*-atenolol, naproxen, cispentacin, prostaglandin E_2, carbovir, oxamniquine, cortisone, D-ephedrine, nifenolo, amiloride, pyrizinamide, glipizide, and captopril. With medical products for which no synthetic alternatives exist, the market will bear almost any price, and the costs of research, production, and quality control can be covered. Of course, many medical compounds are also made by chemical synthesis, and almost every conceivable combination of synthesis and biocatalysis has been employed.

Microbial Catalysis To Produce Chiral Products

Perhaps, looking glass milk isn't good to drink.

— *Lewis Carroll*

A more recently emerging trend has been the reliance on biocatalysts for making chiral compounds in those cases where the stereochemistry of the product is important. This makes sense because chiral catalytic surfaces are intrinsic to enzymes, which are made exclusively of L-amino acids. The chiral surface of the active site has the potential to process a chiral or prochiral substrate in an enantiospecific manner. Methods in chiral chemical synthesis have progressed remarkably, but enzymes are still the catalysts of choice when a product is needed in 100% enantiospecificity. For example, the drug thalidomide has important medicinal value (8, 20). When it was introduced in the 1960s, it was sold as a racemic mixture, but only one enantiomer was medicinally active. At the time, it was thought that this merely diluted the medicine by one-half and that the nonactive enantiomer was inert. Tragically, this turned out not to be the case. Shortly after its introduction, infants born to pregnant thalidomide users were observed to have similar birth defects. The characteristic outcome was infants with fewer than 10 digits

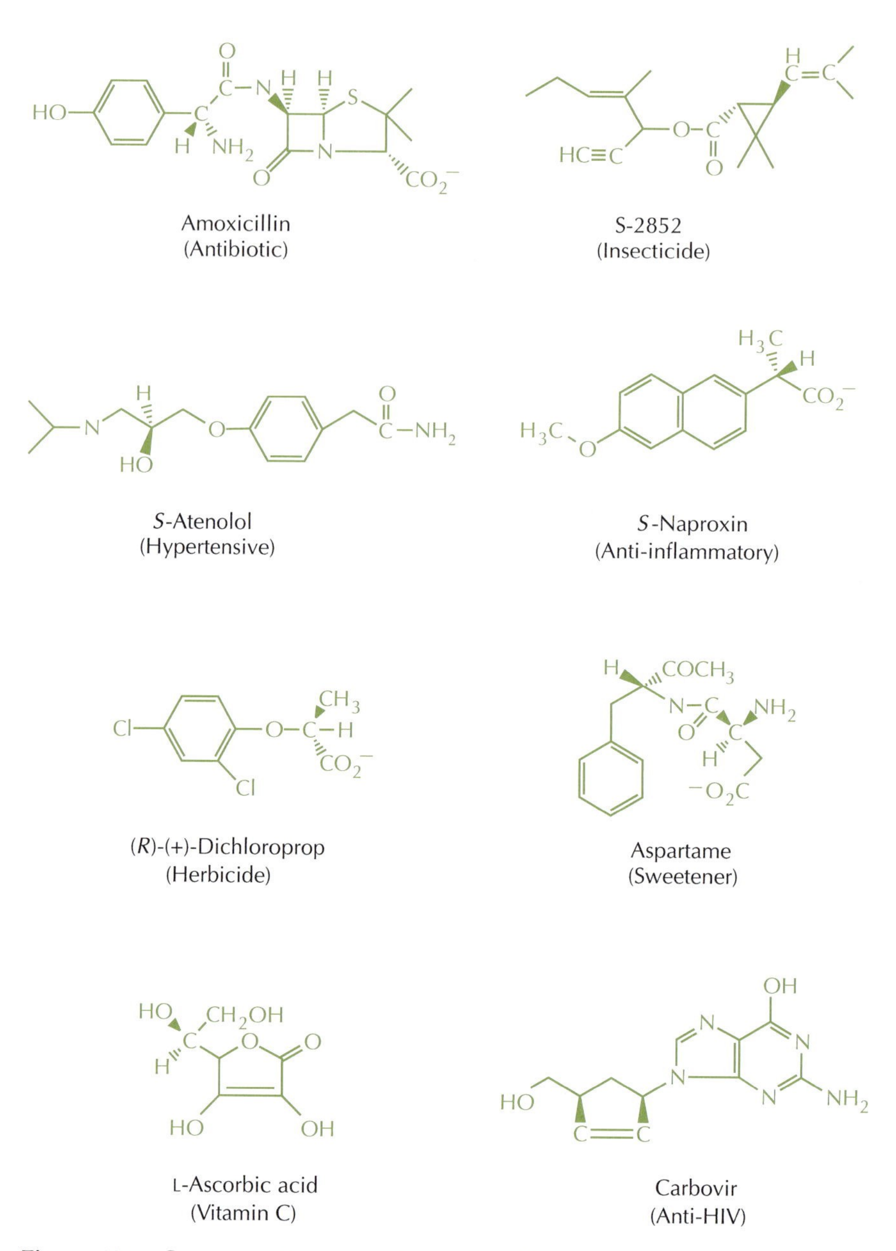

Figure 10.2 Commercial compounds produced by combined microbial-transformation and organic-synthesis steps. Chiral compounds are shown in green.

and reduction in limb size. It was learned that the birth defects were caused by the enantiomer opposite to the one with therapeutic effect. Thalidomide was quickly withdrawn from the market, but a profound lesson was learned: when a specific enantiomer is pharmacologically active, industry should strive to produce it enantioselectively. With this hurdle now overcome, thalidomide is being reintroduced as a pure enantiomeric drug. It has important medicinal value, but it will have to overcome the fears of a public aware of the problems associated with the racemic drug (12).

Of the 16 pharmaceutical products shown in Fig. 10.2, 13 have one or more chiral carbon atoms in the structure. In those cases, the decision to

Cispentacin
(Anti-*Candida*)

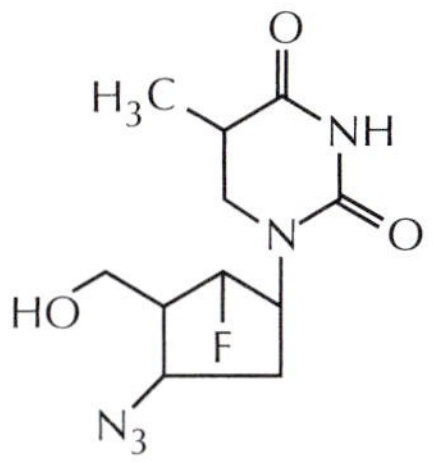

Experimental drug
(Anti-HIV)

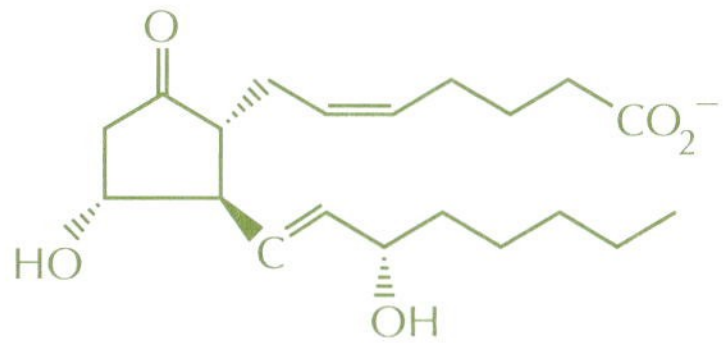

Prostaglandin E_2
(Anti-inflammatory)

(*R*)-Citronellol
(Fragrance)

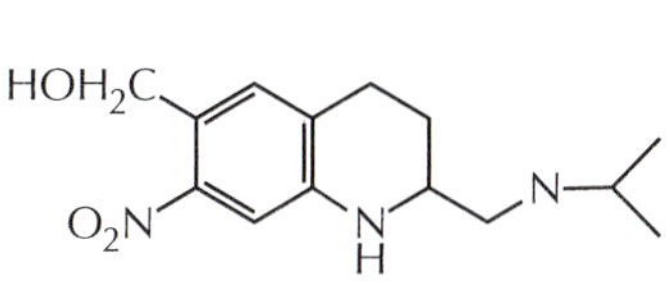

Oxamniquine
(Anthelmintic)

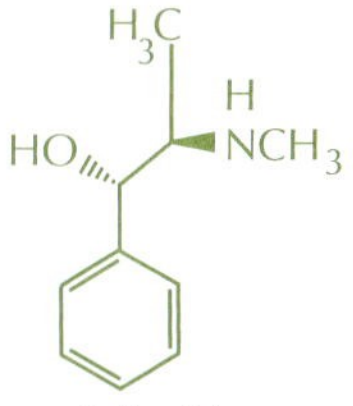

D-Ephedrine
(CNS stimulant)

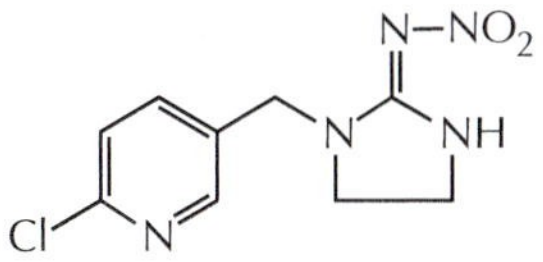

Imidacloprid
(Insecticide)

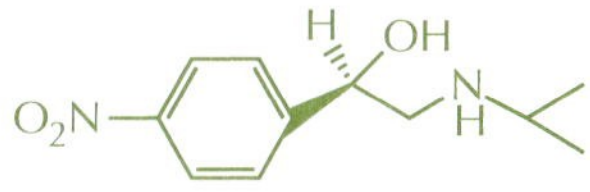

(*R*)-Nifenalol
(Beta-blocker)

Figure 10.2 *Continued.*

use a biocatalyst in synthesis was driven by the desire to obtain an enantiospecific product. The trend toward chiral drugs is a steep one. A strong indicator of this is the increase in sales of single-enantiomer chiral drugs. Their market value surged from $73 billion in 1996 to $88 billion in 1997, a 20% increase in 1 year (24). In 1997, single-enantiomer chiral drugs accounted for 28% of all pharmaceutical sales.

Enantiospecific drug synthesis will likely increase in scope with the increasing emergence of biocatalysis. Selling drugs as single enantiomers minimizes the potential for toxic side effects and allows efficient synthesis, since half the material is not wasted. Moreover, using less material mitigates environmental pollution. These points will be discussed in greater detail below.

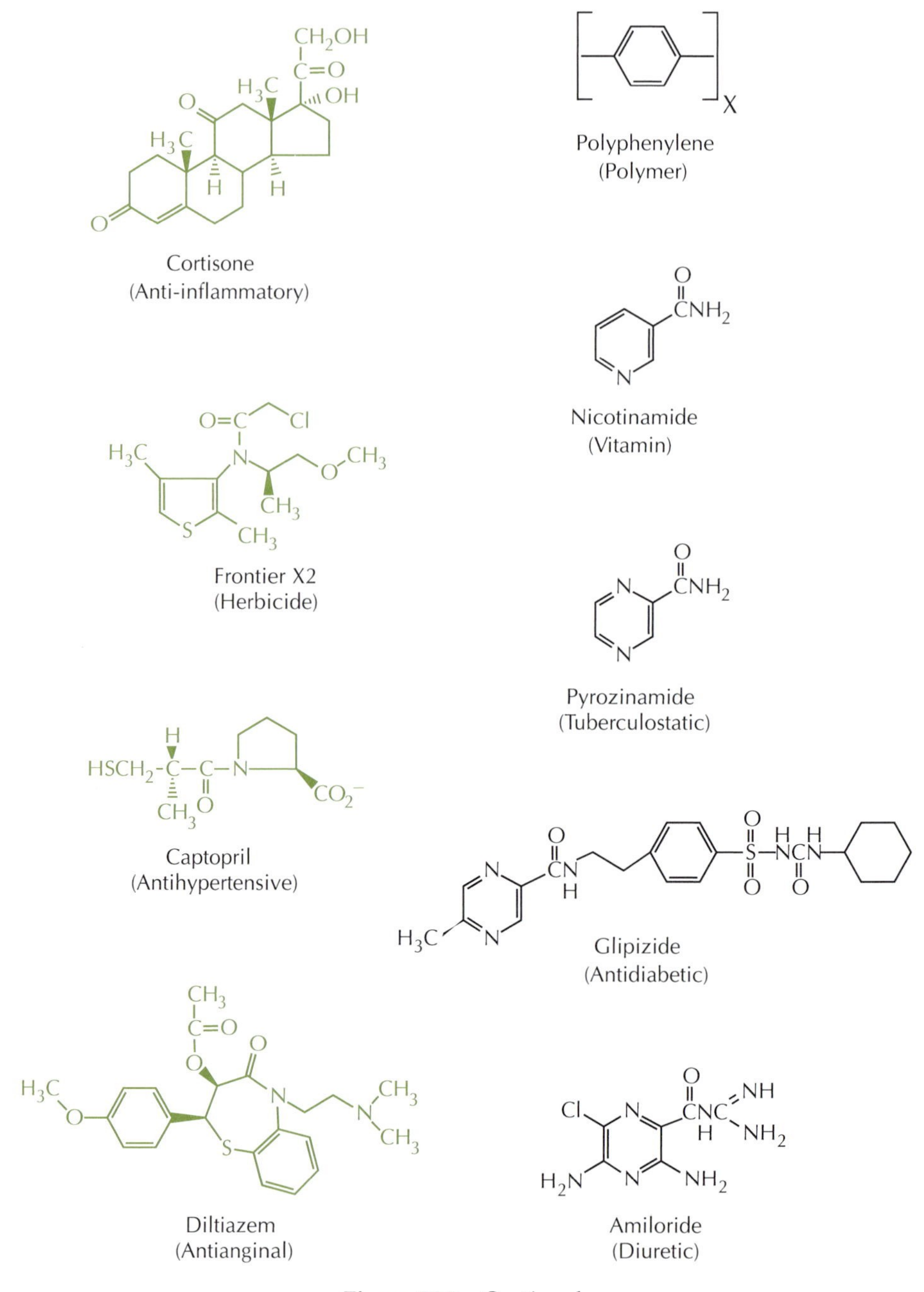

Figure 10.2 *Continued.*

Chiral Synthesis of Dichloroprop

Unlike pharmaceuticals, pesticides made by the chemical industry have rarely been manufactured as single-enantiomer products even if the molecule has one or more chiral centers. Enantiospecific manufacture is typically prohibited by the large volumes needed and the low selling price compared to pharmaceutical products. A farmer cannot pay any price to spray his field if he is only getting $2 per bushel of corn as a return on his investment.

One exception to this general rule is the recent introduction of a single-enantiomer herbicide to replace the corresponding racemic product. The herbicide dichloroprop is one member of an important class of phenoxypropanoic acid herbicides, which are structurally and mechanistically related to the herbicides 2,4-dichlorophenoxyacetic acid (2,4-D) and 2,4,5-trichlorophenoxyacetic acid (2,4,5-T). The latter was heavily used as a defoliant in the Vietnam War where it was known as Agent Orange. The structure of dichloroprop is shown in Fig. 10.2 and 10.3.

For many years the herbicide was marketed as the racemate, (*R*,*S*)-dichloroprop. The (*S*)-isomer is inactive as a herbicide. As a result, application of the racemate imposed an additional and unnecessary burden on the environment for the biodegradation of the herbicide. It was found that microbial metabolism of dichloroprop proceeds enantiospecifically: some bacteria seem capable of metabolizing only one of the enantiomers (32). A global soil study showed that different soils respond differently to the racemic mixture, with very marked but unpredictable preferences for one enantiomer or the other (13).

It was clearly desirable to apply only the active (*R*)-enantiomer to soils, but this had to await the development of a combined enzymatic and chemical-synthetic route to produce an economically viable process (Fig. 10.3). The key element was the discovery, by Slater and his coworkers in England and Soda and coworkers in Japan, of a class of hydrolytic dehalogenases which process chloropropionic acid enantiospecifically (16, 30). Chloropropionic acid is used in a reaction with 2,4-dichlorophenol to produce dichloroprop (Fig. 10.3). If the chloropropionic acid used is a pure enantiomer, the product will also be a pure enantiomer. The synthetic re-

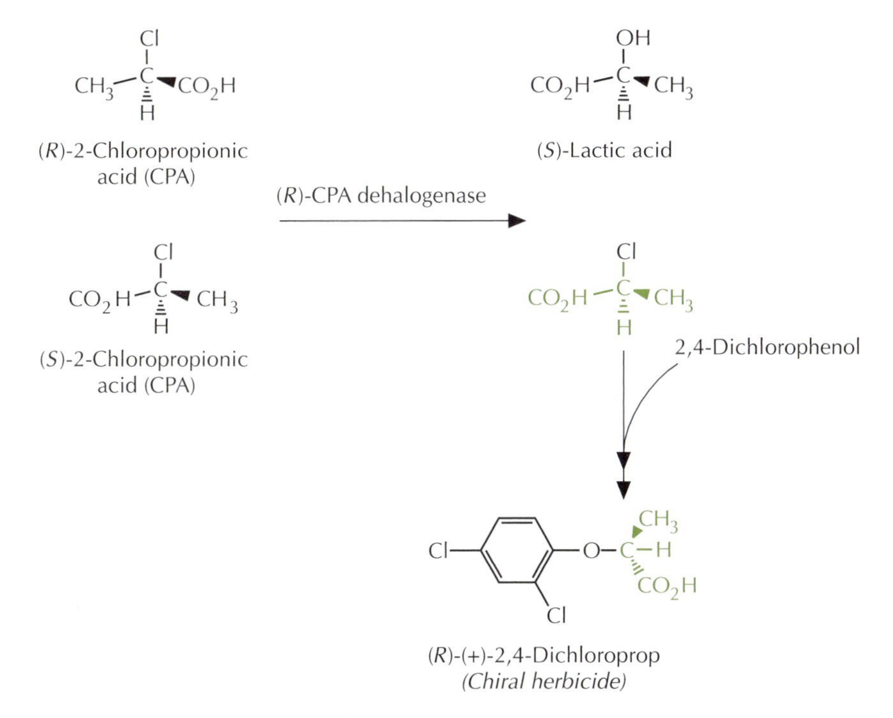

Figure 10.3 Enzymatic chiral dehalogenation of racemic 2-chloropropionic acid, leaving (*S*)-2-chloropropionic acid for synthesis of (*R*)-(+)-dichloroprop, a commercial herbicide now sold as a pure enantiomer.

action occurs via a nucleophilic displacement mechanism, which means (*S*)-chloropropionic acid is required to make (*R*)-dichloroprop.

This was recognized by a research group led by Stephen C. Taylor at ICI Biological Products in the United Kingdom. Their goal was to obtain a dehalogenase which would react with (*R,S*)-chloropropionic acid to produce (*S*)-lactic acid via conversion of all the (*R*)-chloropropionic acid but would leave (*S*)-chloropropionic acid untouched. This chlorinated acid and lactic acid are readily separable. Moreover, the pure isomer of lactic acid can be used in other applications. Thus, the end result is two valuable chiral products, one of which is used to make dichloroprop as an environmentally safer and more economical product.

The hurdle to be overcome was identifying a dehalogenase with the proper reactivity, which had not yet been described in the literature. As discussed in chapter 3 on enrichment culture, it is relatively easy to obtain suitable bacteria from the environment when a strong selection or screening method is applied. The ICI group used enrichment culture methods to obtain bacterial isolates which used chloropropionic acid as the sole source of carbon and energy. Of those, they found that some would use only the (*R*)- or (*S*)-enantiomer and that some would use both. One isolate, identified as *Pseudomonas putida,* was observed to produce two distinct dehalogenases, each with a specificity for one of the chloropropionic acid isomers (5). These enzymes were purified. One was shown to act on a racemic substrate to convert (*R*)-chloropropionic acid to (*S*)-lactic acid in near 100% yield based on the enantiomer acted on, and it left the (*S*)-chloropropionic acid in greater-than-98% enantiomeric excess. A mutant *P. putida* strain was obtained in which the (*S*)-specific dehalogenase was knocked out. ICI then had enzyme and whole-cell biocatalysts at their disposal, leading to the successful production and marketing of (*R*)-dichloroprop to replace the sale of the racemate.

More recently, a new herbicide, called Frontier (Fig. 10.2), has been synthesized enantiospecifically via a chiral amine generated by a lipase-catalyzed chiral resolution. The biocatalytic process, developed by BASF, uses a lipase from *Pseudomonas* sp. strain DSM 8246. As in the dichloroprop example, only one enantiomer has biological activity, and biocatalysis provides a route for making this specifically and economically.

Biocatalysis for Non-Medicinal, Non-Chiral Specialty Chemicals

Now that a strong niche for biocatalysis in producing pharmaceuticals and chiral chemicals has been established, other inroads are being made, largely in the production of other specialty chemicals. Specialty chemicals are those of moderate price and moderate volume. The cost of production must be low, so the feedstock must be cheap and the fermentation must be very efficient. The scale of production is usually on the order of 1,000 to 100,000 liters per batch.

One example of a product which has been explored for biocatalytic production to supplant chemical synthesis is the blue jean dye indigo. Indigo has a long and colorful history in human affairs (Table 10.3). Derived principally from plants of the genus *Indigofera,* it was first used in Asia about 4,000 years ago. The blue dye was unwittingly obtained via a microbial fermentation process. It was observed that the plants had to be buried in the soil and left for an extended period. Then, cakes of crude indigo

Table 10.3 History of human use of indigo

Event	Date
First recorded use of indigo	2000 B.C.
British tribes obtain indigo from woad; Romans learn process	0 A.D.
Europe-Asia indigo trade proliferates	1500
Baeyer elucidates structure of indigo	1883
Chemical synthesis of indigo commercialized	1890
Recombinant *E. coli* used to make indigo	1983

could be dug up and used for dyeing. Four thousand years later, we have learned that the plants make a glycoside of 3-hydroxyindole. When the plant is buried in the soil, microorganisms excreting nonspecific glycosidases serve to hydrolyze the glycosidic linkage. This liberates 3-hydroxyindole, or indoxyl, which is known to undergo spontaneous oxidation in air to yield the dimeric-ring product indigo.

While this molecular knowledge was unavailable, indigo nonetheless was a popular product in human commerce for thousands of years. The first report of it in Europe was in accounts by Roman invaders of the British Isles about frightening blue-colored natives. Though the Roman armies were initially afraid and fled, they quickly overcame their fears and conquered Britain. They discovered that the blue-colored flesh resulted from dyeing with indigo. The natives were extracting indigo principally from the woad plant; the Romans learned the practice and began taking indigo back to Rome, where it became greatly valued.

Throughout the next 2 millennia, indigo has been valued for its qualities of color and fastness of binding, particularly to cotton fabrics. It was prized during the period of European sea exploration and was one of the items sought during the quest for a shorter trade route between Asia and Europe. Another indigo dye also became greatly prized in Europe, but it was so scarce as to be reserved for the nobility. This was Tyrian purple, now known to be 5,5′-dibromoindigo. This dye was derived from a Mediterranean snail, but the yield per snail is so low that thousands of snails need to be ground up to obtain enough dye to color one robe. No wonder the dye for the king's garments became known as royal purple.

Thus, for almost 4,000 years, indigo was obtained via biocatalysis. This changed in 1883, when Adolf von Baeyer, one of the founders of the large German chemical industry, elucidated the chemical structure of indigo (Fig. 10.4). Over several years, he developed a feasible large-scale process for manufacturing indigo; this process, in its basic form, persists to the present time. Side products of the chemical synthesis are aromatic amines, which were not of concern at the time but are now known to be potent bladder carcinogens. Concern about this is the basis for the desire to return to a biological, but economical, production of indigo.

Interest in such a process was awakened in 1983, when it was reported that aromatic-hydrocarbon dioxygenases can dioxygenate indole at carbons 2 and 3, initiating a series of rapid spontaneous reactions which yield indigo (9) (Fig. 10.4). The enzyme of choice is naphthalene dioxygenase, the first enzyme in the biodegradation pathway of naphthalene and other polycyclic aromatic hydrocarbons (see Fig. 1.1 and Table 6.2). Recombinant *E. coli* strains expressing naphthalene dioxygenase were observed to produce in-

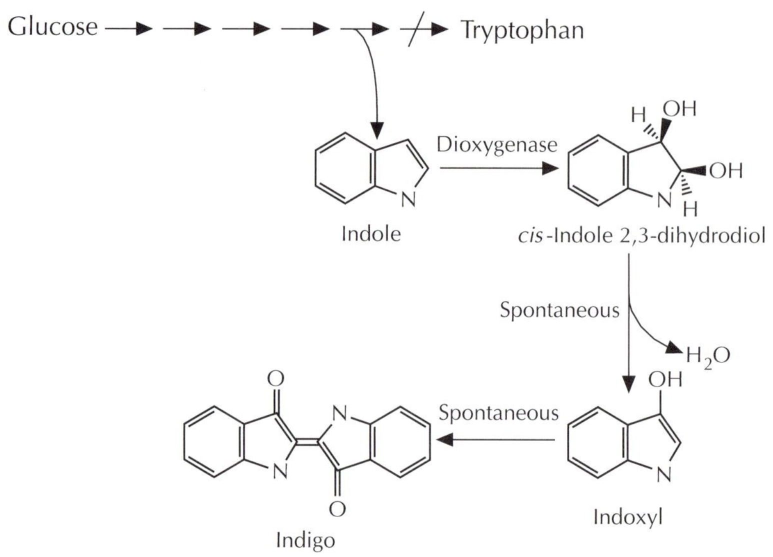

Figure 10.4 Biocatalytic route using glucose for production of indigo by recombinant *E. coli.*

digo spontaneously when grown on Luria broth or other media containing protein hydrolysates. It was quickly realized that, unlike *Pseudomonas* strains, *E. coli* strains produce the enzyme tryptophanase, which liberates indole from the amino acid tryptophan. However, indigo production from tryptophan or indole would not be commercially viable due to the cost of the starting material. A viable biocatalytic route needs to start with a cheap feedstock like glucose.

The bacterial dioxygenase-dependent indigo synthesis was protected for commercial development in a patent initially filed by the emerging biotechnology company Amgen. Researchers at Amgen took on the task of engineering *E. coli* strains which would transform a high flux of glucose into indole (17). This was accomplished by high-level expression of the tryptophan biosynthetic pathway enzymes and by making a clever modification in the last step of the pathway. The last enzyme, tryptophan synthase, removes the side chain of 3-indoylglycerol phosphate to release indole, which is condensed with serine to yield tryptophan. The intermediate indole is produced in one enzyme subunit and diffuses to the other subunit for the condensation reaction. The Amgen group successfully blocked the channel between subunits and deactivated the second subunit. The modified tryptophan synthase produces indole stoichiometrically.

Flush with success in the pharmaceutical area, Amgen decided to forgo specialty chemical developments. The patent rights were sold to Eastman Kodak and then transferred to Genencor. The indigo fermentation process has been further improved by the Genencor researchers. Recombinant *E. coli* strains with elevated indole synthesis and high-level expression of naphthalene dioxygenase are potentially useful in a commercial process for the environmentally benign production of indigo (Fig. 10.4). The recombinant strain can be fed glucose and grown to very high levels prior to switching on the naphthalene dioxygenase genes and converting the indole to indigo. In this way, indigo was produced in large-scale fermentations of up to 50,000 liters. It was observed that the dye color was tainted with a purplish tinge by a side reaction which produced a small amount of a structurally related red dye called indirubin. Indirubin was found to derive from oxindole (2-hydroxyindole) reacting with indoxyl (3-hydroxyindole). Oxindole likely comes from a minor amount of alternative spontaneous dehydration of *cis*-2,3-dihydroxydihydroindole. This problem was overcome by cloning an additional enzyme into the strain to transform oxindole back to indole. This in effect recycles the undesired side product, increases yield, and eliminates formation of the undesirable color.

Today, the process hovers at the borderline of commercial relevance, because indigo synthesis is cheap, and it is being produced chemically in countries with relatively relaxed environmental regulations. This process illustrates what biocatalysis can do and the ever-present tension between the development of clean processes using renewable resources and the needs of commerce for chemicals at the lowest possible price.

Biotechnological Waste Recycling

As industrial biocatalysis and biodegradation mature, there is an outstanding natural opportunity to combine the two for the betterment of society and industry. The ideal industry is one that makes an important product with minimum use of resources, at low cost, with little or no pollution. This ideal is often not met because side products of the chemical reactions drain raw materials, have the potential to become pollutants, and raise production costs, which are then passed on to the consumer. An obvious solution to this practical problem is to find ways to recycle the side products into production, either into the same process or to make a new product (Fig. 10.5). This is a niche problem that biocatalysis will increasingly be called on to solve. Solutions translate into higher profits and thus will be quickly implemented by companies. This approach will also stimulate more research and development in the pollution prevention arena.

An excellent example, currently under development, involves a large-scale process for making epichlorohydrin (27). Dow Chemical Company is the world's largest producer of epichlorohydrin, a compound used as a synthetic intermediate for pharmaceuticals and other classes of chemicals. During the synthesis, several percent of the product stream is 1,2,3-trichloropropane rather than the desired product, 2,3-dichloroethanol. There is no major industrial use for 1,2,3-trichloropropane, so it is discarded (Fig. 10.6).

Paul Swanson and his colleagues at the Dow Chemical Company Central Research Facility saw reports of hydrolytic dehalogenases, such as haloalkane dehalogenase, which act on 1,2-dichloroethane. They thought that it or similar enzymes might prove active with 1,2,3-trichloropropane. With 1,2-dichloroethane, the observed product is 2-chloroethanol. A similar reaction acting on 1,2,3-trichloropropane would yield 2,3- or 1,3-dichloroethanol. Either of these could be treated with a base to yield epichlorohydrin, thus recycling a waste product and boosting overall yield (Fig. 10.6).

Haloalkane dehalogenase is quite well understood. A high-resolution X-ray structure at 1.15 Å has been solved and key intermediates along the reaction pathway have been determined by B. Dijkstra, Dick Janssen, and colleagues (28). The enzyme from *Xanthobacter autotrophicus* GJ10 has a relatively low catalytic constant k_{cat} of 3.3 s^{-1} and a high K_m of 0.5 mM at 30°C and pH 8.2 with the substrate 1,2-chloroethane (22). With less pre-

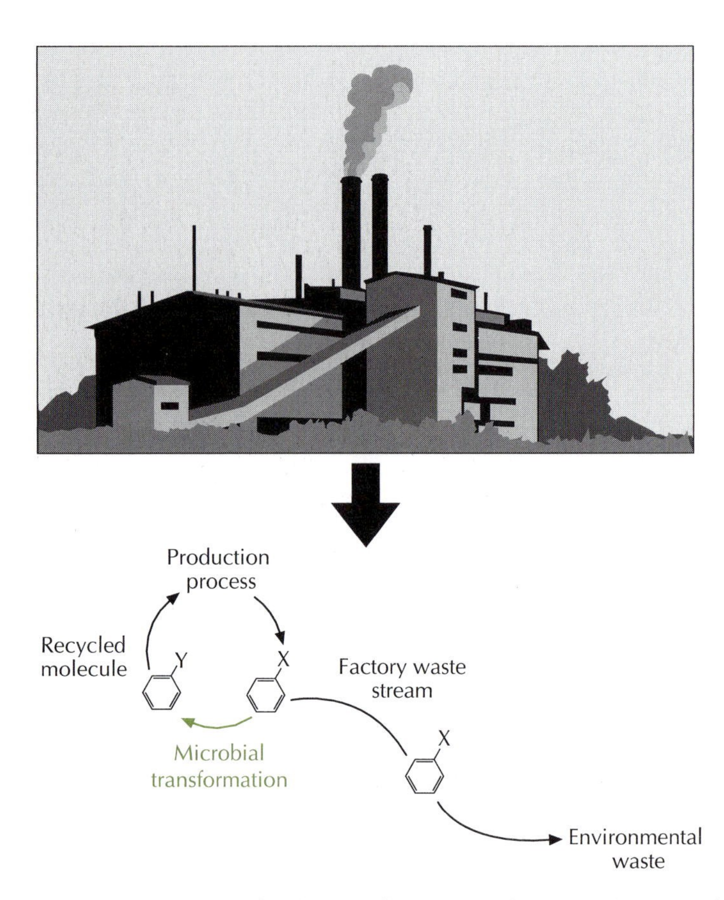

Figure 10.5 Microbial biotechnology will increasingly be used to recycle waste compounds within factories to simultaneously reduce environmental pollution and increase profits.

ferred substrates, like 1,2,3-trichloroethane, the catalytic constant is unacceptably low for developing a biocatalytic process for recycling the chemical on an industrial scale (6). In this context, several strategies are being pursued. One is to use enrichment culture techniques to isolate bacteria capable of growing on 1,2,3-trichloropropane or related compounds. It is anticipated that at least some of those organisms will have enzymes with more desirable kinetic properties. Another approach is to start with a haloalkane dehalogenase gene which could be evolved in the laboratory for improved activity. Directed-evolution techniques, such as DNA shuffling (23), can be useful here. A third useful method would be to screen soil gene libraries expressed in alternative hosts for new enzymes more active with 1,2,3-trichloropropane.

Research is ongoing to address this recycling issue, and it can point the way for new technologies aimed at making industry more environmentally friendly and competitive at the same time.

Case Study of Bioremediation: Atrazine in Soil

Even as industry moves to a more environmentally sound and sustainable future, chemicals will inevitably enter the environment. Some, such as agricultural chemicals, are made to be applied in the environment. Thus, there will always be a need for new methods of remediating undesirable chemical

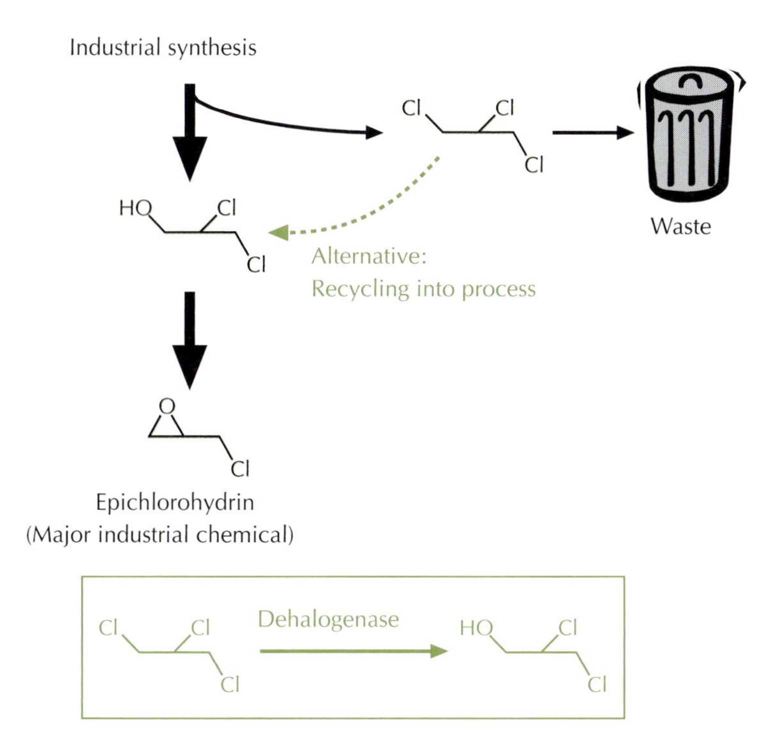

Figure 10.6 Recycling of a side product of chemical manufacturing into the production process. The product, epichlorohydrin, is widely used in the manufacture of pharmaceuticals and other important industrial compounds.

contamination of soil and water, and biological methods will be an important part of that repertoire of methods.

There are thousands of examples of industrial-scale soil and water remediation projects. Many of these are not reported in any detail; the goal is to do the job quickly, cheaply, and with as little fanfare as possible. Each remediation job is different and must be tailored to the site, the contaminants at the site, and the activity of the resident microbial flora if it plays an important part in the process. Often, the resident microbial flora alone is used to carry out the remediation, and the engineer's job is to stimulate the biodegradative activities in the proper manner.

On the other hand, there are thousands of reports in the literature of the genetic engineering of bacteria to biodegrade specific pollutants or pollutant classes. Rarely are the genetically engineered microbes used outside the laboratory, and if so, they are typically used in controlled tests away from the contaminated site.

One application for bioremediation is the treatment of pesticides, either trace residues in water or heavy concentrations in soil as the result of an accidental spill. One example is the remediation of the herbicide atrazine. Atrazine (2-chloro-4-ethylamino-6-isopropylamino-1,3,5-*s*-triazine) is a herbicide widely used in the United States for control of broadleaf weeds in corn, sorghum, and sugarcane. Atrazine may be transformed in soil by several different mechanisms. At typical soil pH, atrazine is only very slowly chemically hydrolyzed to produce hydroxyatrazine (3, 18). Microbial metabolism is likely to be the major mechanism for atrazine removal, and the major metabolic route is known to proceed through hydroxyatrazine as an intermediate (15, 19). The pathway, genes, and enzymes involved in bacterial atrazine metabolism have been discussed in chapter 7.

In soil, the most efficient method of transforming atrazine is by biostimulation or bioaugmentation (14). Biostimulation involves supplementing the contaminated soil to change the physical state of the contaminant, thereby converting it to a bioavailable form (4), or supplying a nutritional supplement or cosubstrate to increase the population of indigenous bacteria capable of catabolizing the contaminant (1). Bioaugmentation refers to the addition to the soil of nonindigenous microorganisms capable of catabolizing the contaminant (7).

The bioaugmentation of atrazine has been studied in the laboratory. In studies using nonsterile soil, the success of bioaugmentation was inversely related to the population levels of indigenous atrazine-degrading microorganisms (19, 26). In sterile soils devoid of indigenous atrazine-degrading bacteria, it has been reported that the atrazine concentration was reduced 70% (from 20 to 6 ppm) in 30 days (10) or eliminated (from 15 ppm) in 5 days (31).

In the spring of 1997, the authors had the opportunity to attempt the bioremediation of atrazine in soil as the result of an accidental spill of the herbicide. A 250-gallon tank containing a suspension of atrazine fell off a truck and broadcast the herbicide onto the roadside near an agricultural supply dealership in South Dakota. The contaminated soil was sequestered by placing it on a plastic liner and covering it. The containment was effective, as evidenced by the growth of atrazine-sensitive plant species within several feet of the soil pile. The 35 yd^3 of soil contained an average atrazine concentration of 11,500 ppm (milligrams per kilogram) and a maximum concentration of 29,000 ppm as measured some months after the spill. The options for the dealer were to have the soil removed by trucks to a distant landfill, to incinerate it, or to bioremediate it. The last was deemed to be least costly and would recover the soil in a usable form.

The goals of the bioremediation had to be further delineated with respect to the following points: (i) the treatments which would be allowed or not allowed, (ii) the acceptable endpoint atrazine concentration, and (iii) the final use of the soil after treatment. All reasonable options were initially considered, and several were presented to the South Dakota Department of Natural Resources and the U.S. Environmental Protection Agency. Part of the decision-making process required the acquisition of laboratory data, and these are described below.

As for other soil cleanup operations, both biostimulation and bioaugmentation were considered. To consider biostimulation, it was necessary to determine if atrazine-degradative microbial activity was indigenous to the soil and to test in the laboratory the conditions which would best enhance the native activity. A range of biostimulation conditions were tested: pH, carbon addition, and phosphate addition. A statistically designed central composite experiment revealed how best to stimulate biodegradation by resident microbial flora. Carbon addition decreased atrazine degradation ($P = 0.002$). While this would be surprising to some people who add carbon supplements without a second thought, this observation is readily explainable. Recently, bacteria which use atrazine as a sole carbon source have been identified in the soil, and the amended carbon probably suppressed their biodegradative activity. The effect of pH manipulation was statistically insignificant. However, phosphorus addition increased atrazine degradation ($P = 0.03$). Optimal atrazine degradation occurred with the addition of 300 ppm of phosphorus. This was further explained by soil chemical analysis, which revealed that the soil was very low in phosphorus and this was likely limiting the overall growth and metabolism of the resident microbes. Since atrazine contains carbon and nitrogen, the addition of phosphorus was essentially rounding out the diet of the soil microorganisms. Thus, it was decided that one treatment of the soil would be to add superphosphate fertilizer to attain the optimum concentration determined in the laboratory.

Bioaugmentation was also considered, to be used separately and in admixture with biostimulation, to learn the best range of treatments to determine an optimal protocol for dealing with atrazine spills of this type. The most active form of microbial activity for transforming atrazine that we have obtained is the genetically engineered strain *E. coli* (pMD4), which contains the *Pseudomonas atzA* gene to express high levels of atrazine chlorohydrolase in the recombinant host (see Fig. 7.4). *E. coli* (pMD4) transforms atrazine to hydroxyatrazine. Hydroxyatrazine is nonherbicidal and is considered completely safe from the standpoint of human health. Moreover, with our observation of indigenous atrazine degradation activity, it was felt that hydroxyatrazine would be further metabolized by indigenous microbial activity.

The use of recombinant organisms in field tests has been somewhat controversial, and we sought to avoid any problems in the following ways. First, the soil at the South Dakota site was screened for the presence of an *atzA* gene indigenous to the soil by using PCR methods. In fact, a homologous gene was amplified from the total soil DNA. The PCR-amplified DNA was sequenced, and some copies were found to be identical to the *atzA* gene sequence to be introduced. This demonstrated that a foreign gene would not be introduced by bioaugmentation, since the gene was already present. Second, we devised a method to kill the recombinant *E. coli* cells while preserving a significant amount of the atrazine chlorohydrolase activity. This was accomplished by treating concentrated cell suspensions with the chemical cross-linking agent glutaraldehyde. No viable cells could be detected after this treatment even when high cell densities were plated. Moreover, the method had the added benefit of greatly stabilizing the whole-cell atrazine chlorohydrolase activity from days to months. This could prove to be a very useful property in soil and other remediation treatment protocols.

In the South Dakota field test, it was decided to implement several treatments: biostimulation alone, bioaugmentation alone, combined biostimulation and bioaugmentation, and a control with water added to match the amount of water added to treated soils (25). Photographs and more details of the treatment can be viewed on the World Wide Web (http://biosci.umn.edu/cbri/lisa/web/).

The results of the test indicated the value of the combined approach. The atrazine distribution in the soil, even after extensive mixing by soil excavation equipment (Fig. 10.7), was not nearly so homogeneous as is typical in small soil experiments carried out in the laboratory. This caused large error bars in the data, which derived mainly from the large sampling error due to contaminant heterogeneity in the soil. The statistical software we used to deal with these large variations was MacAnova, a package developed at the University of Minnesota. From this statistical treatment, it

Figure 10.7 Site of a herbicide spill in the upper midwestern United States which was treated with recombinant *E. coli* expressing atrazine chlorohydrolase. (Top) The soil containing spilled atrazine was separated by earth-moving equipment into different soil mixtures for four different treatments. After separation, the same earth-moving equipment was used to mix phosphate fertilizer and recombinant bacteria homogeneously throughout the soil. (Bottom) The different soil treatments were contained in wooden boxes lined with water-impermeable plastic and covered with wood, with the top raised to allow air to enter but prevent potential rain washout of the herbicide. (By permission of L. Strong.)

was deduced that the non-enzyme-treated plots did not show significant atrazine degradation over the 14-week period. The enzyme-treated and enzyme-phosphate-treated soils showed 84 and 97% probability that the detected drop in atrazine concentration was significant. The enzyme-phosphate-treated soil showed a drop in atrazine concentration from 7,000 to 2,000 ppm.

The goal of the treatment was to attain a final atrazine level in the treated soils of 2,000 ppm or lower. At that point, the soil can be spread onto a cornfield by diluting the soil 1 part in 1,000, thus applying atrazine at 2 ppm, an approved field application concentration for the herbicide. This study demonstrated that a combination of laboratory pretesting, innovative technology, and regulatory agency-industry cooperation can result in the management of an accidental spill. This hopefully points the way toward further biological methods of treating soils so that they can be reused. This differs from landfill or incineration techniques, which sequester and destroy the soil, respectively.

Summary

In the past, applications of biodegradation and biocatalysis for chemical manufacture, even while using similar enzymes, have been carried out by different practitioners. This is changing as industry seeks cleaner practices and strives for greater competitiveness. Increasingly, wastes will be recycled, and it is likely that biological processes will play an important role. Generally, wastes are generated because the chemical and physical processes commonly used have reached maximum efficiency. Microbes and enzymes can have a role in increasing efficiency for the betterment of industry and society.

References

1. **Adriaens, P., and D. D. Focht.** 1990. Continuous coculture degradation of selected polychlorinated biphenyl congeners by *Acinetobacter* sp. in an aerobic reactor system. *Environ. Sci. Technol.* **24:**1042–1049.

2. **Arioli, T., L. Peng, A. S. Betzner, J. Burn, W. Wittke, W. Herth, C. Camiller, H. Hofte, J. Plazinski, R. Birch, A. Cork, J. Glover, J. Redmond, and R. E. Williamson.** 1998. Molecular analysis of cellulose biosynthesis in *Arabidopsis. Science* **279:**717–720.

3. **Armstrong, D. E., G. Chesters, and R. F. Harris.** 1967. Atrazine hydrolysis in soil. *Soil Sci. Soc. Am. Proc.* **31:**61–66.

4. **Atlas, R. M., and R. Bartha.** 1992. Hydrocarbon biodegradation and oil spill bioremediation. *Adv. Microb. Ecol.* **12:**287–338.

5. **Barth, P. T., L. Bolton, and J. C. Thomson.** 1992. Cloning and partial sequencing of an operon encoding two *Pseudomonas putida* haloalkanoate dehalogenases of opposite stereospecificity. *J. Bacteriol.* **174:**2612–2619.

6. **Bosma, T., E. Kruizinga, E. J. de Bruin, G. J. Poelarends, and D. B. Janssen.** 1999. Utilization of trihalogenated propanes by *Agrobacterium radiobacter* AD1 through heterologous expression of the haloalkane dehalogenase from *Rhodococcus* sp. strain M15-3. *Appl. Environ. Microbiol.* **65:**4575–4581.

7. **Brodkorb, T. S., and R. L. Legge.** 1992. Enhanced biodegradation of phenanthrene in oil tar-contaminated soils supplemented with *Phanerochaete chrysosporium. Appl. Environ. Microbiol.* **58:**3117–3121.

8. **Corral, L. G., and G. Kaplan.** 1999. Immunomodulation by thalidomide and thalidomide analogues. *Ann. Rheum. Dis.* **58**(Suppl. 1)**:**1107–1113.

9. **Ensley, B. D., B. J. Ratzkin, T. D. Osslund, W. J. Simon, L. P. Wackett, and D. T. Gibson.** 1983. Expression of naphthalene oxidation genes in *Escherichia coli* results in the biosynthesis of indigo. *Science* **222:**167–169.

10. **Fadullon, F. S., J. S. Karns, and A. Torrents.** 1998. Degradation of atrazine in soil by *Streptomyces. J. Environ. Sci. Health* **B33:**37–49.

*11. **Glazer, A. N., and H. Nakaido.** 1995. *Microbial Biotechnology: Fundamentals of Applied Microbiology.* W. H. Freeman and Company, New York, N.Y.

12. **Lary, J. M., K. L. Daniel, J. D. Erickson, H. E. Roberts, and C. A. Moore.** 1999. The return of thalidomide: can birth defects be prevented? *Drug Saf.* **21:**161–169.

13. **Lewis, D. L., A. W. Garrison, K. E. Wommack, A. Whittemore, P. Steudler, and J. Melillo.** 1999. Influence of environmental changes on degradation of chiral pollutants in soils. *Nature* **401:**898–901.

14. **Liu, S., and J. M. Suflita.** 1993. Ecology and evolution of microbial populations for bioremediation. *Trends Biotechnol.* **11:**344–352.

15. **Mandelbaum, R. T., L. P. Wackett, and D. L. Allan.** 1993. Soil bacteria rapidly hydrolyze atrazine to hydroxyatrazine. *Environ. Sci. Technol.* **27:**1943–1946.

16. **Motosugi, K., N. Esaki, and K. Soda.** 1982. Bacterial assimilation of D- and L-2-chloropropionates and occurrence of a new dehalogenase. *Arch. Microbiol.* **131:**179–183.

17. **Murdock, D., B. D. Ensley, C. Serdar, and M. Thalen.** 1993. Construction of metabolic operons catalyzing the de novo biosynthesis of indigo in *Escherichia coli. Bio/Technology* **11:**381–386.

18. **Plust, S. F., J. R. Loehe, F. J. Feher, J. H. Benedict, and H. F. Herbrandson.** 1981. Kinetics and mechanism of hydrolysis of chloro-1,3,5-triazines. *J. Org. Chem.* **46:**3661–3665.

19. **Radosevich, M., S. J. Traina, Y. Hao, and O. H. Tuovinen.** 1995. Degradation and mineralization of atrazine by a soil bacterial isolate. *Appl. Environ. Microbiol.* **61:**297–302.

20. **Ravot, E., J. Lisziewicz, and F. Lori.** 1999. New uses for old drugs in HIV infection: the role of hydroxyurea, cyclosporin and thalidomide. *Drugs* **58:**953–963.

*21. **Roberts, S. M., N. J. Turner, A. J. Willetts, and M. K. Turner.** 1995. *Introduction to Biocatalysis Using Enzymes and Microorganisms.* Cambridge University Press, Cambridge, United Kingdom.

22. **Schanstra, J. P., J. Kingma, and D. B. Janssen.** 1996. Specificity and kinetics of haloalkane dehalogenase. *J. Biol. Chem.* **271:**14747–14753.

23. **Stemmer, W. P.** 1994. Rapid evolution of a protein in vitro by DNA shuffling. *Nature* **370:**389–391.

*24. **Stinson, S. C.** 1998. Counting on chiral drugs. *Chem. Eng. News* **76**(38):83–104.

25. **Strong, L. C., H. McTavish, M. J. Sadowsky, and L. P. Wackett.** 2000. Field-scale remediation of atrazine-contaminated soil using recombinant *Escherichia coli* expressing atrazine chlorohydrolase. *Environ. Microbiol.* **2:**91–98.

26. **Struthers, J. K., K. Jayachandran, and T. B. Moorman.** 1998. Biodegradation of atrazine by *Agrobacterium radiobacter* J14a and use of this strain in bioremediation of contaminated soil. *Appl. Environ. Microbiol.* **64:**3368–3375.

*27. **Swanson, P. E.** 1999. Dehalogenases applied to industrial-scale biocatalysis. *Curr. Opin. Biotechnol.* **10:**365–369.

28. **Verschueren, K. H., F. Seljee, H. J. Rozeboom, K. H. Kalk, and B. W. Dijkstra.** 1993. Crystallographic analysis of the catalytic mechanism of haloalkane dehalogenase. *Nature* **363:**693–698.

*29. **Webster, L. C., P. T. Anastas, and T. C. Williamson.** 1996. Environmentally benign production of commodity chemicals through biotechnology, p. 198–211. *In* P. T. Anastas and T. C. Williamson (ed.), *Green Chemistry: Designing Chemistry for the Environment.* American Chemical Society, Washington, D.C.

30. **Weightman, A. J., A. L. Weightman, and J. H. Slater.** 1982. Stereospecificity of 2-monochloropropionate dehalogenation by the two dehalogenases of *Pseudomonas putida* PP3: evidence for two different dehalogenation mechanisms. *J. Gen. Microbiol.* **128:**1755–1762.

31. **Wenk, M., T. Baumgartner, J. Dobovsek, T. Fuchs, J. Kucsera, J. Zopfi, and G. Stucki.** 1998. Rapid atrazine mineralisation in soil slurry and moist soil by inoculation of an atrazine-degrading *Pseudomonas* sp. strain. *Appl. Microbiol. Biotechnol.* **49:**624–630.

32. **Zipper, C., K. Nickel, W. Angst, and H.-P. E. Kohler.** 1996. Complete microbial degradation of both enantiomers of the chiral herbicide mecoprop [(*RS*)-2-(4-chloro-2-methylphenoxy)propionic acid] in an enantioselective manner by *Sphingomonas herbicidovorans* sp. nov. *Appl. Environ. Microbiol.* **62:**4318–4322.

11

The Impact of Genomics on Microbial Catalysis

> . . . the era of molecular biology began . . . with the discovery of the DNA double helix. . . . most biologists interested in the mechanism of heredity quickly realized that the time had come to think about genetics in terms of large molecules that carry hereditary information.
>
> — *Gunther Stent*

Genomics, or the complete sequencing and analysis of an organism's DNA, is beginning to transform the way in which microbiologists approach their science. Genes, when their functions can be delineated, speak loudly about the roles of microbes in the world and their interactions with other microorganisms and larger organisms. Uncovering new genes with unknown functions provides new clues about the richness of microbial genetic diversity and gives us the pool from which novel biocatalysts will increasingly be drawn. This chapter will not cover all aspects of genomics. Rather, it is intended to provide a brief introduction and then focus on the projected major influence of genomics on the field of biocatalysis and biodegradation.

Genome Sizes and Organization

Some generalities regarding genome size can be made. Table 11.1 shows that prokaryotic genome sizes tend to cluster in the range of 1×10^6 to 10×10^6 bp. The smallest prokaryotic genomes (0.6×10^6 to 1.0×10^6 bp) are found in pathogenic organisms, and it has been suggested that an obligately parasitic lifestyle may require fewer genes, as there is substantial feeding off the host (7). The clustering of prokaryotic genomes within an order of magnitude contrasts with those of different multicellular eukaryotes, which vary by 4 orders of magnitude. While the lifestyles of prokaryotes differ widely, it is likely that they have an upper limit of genome size based on the need to replicate efficiently and thus have fairly compact genomes. Moreover, there is very little excess DNA above that of the regions encoding gene products: approximately 90% of the DNA is read.

History Leading Up to the Genomic Era

The stage for genomics was set in 1953, when the three-dimensional structure of DNA became known (1, 9, 10). The structure immediately suggested a general mechanism for DNA replication and thus solved an almost century-old mystery about how genetic traits were transferred.

The modern concepts of genetic traits, or phenotype, were formed in the classic experiments with pea plants completed by Gregor Mendel in 1864 (Fig. 11.1). Certain traits were observed in nearly precise fractional numbers of progeny, setting the foundations for the modern discipline of genetics. Genetics proceeded for decades without any concept of the nature of the genetic material. Meischer had isolated nucleic acids in 1869, but for most of the early 20th century, proteinaceous molecules were considered most likely to carry genetic traits.

The insightful studies of material from dead *Pneumococcus* cells conferring pathogenic traits on live cells strongly implicated DNA as the genetic material, and the elucidation of the structure of DNA in 1953 quickly led to a widespread acceptance of the role of DNA in heredity. At that time, the aggregate number of adenine, thymine, guanine, and cytosine bases could be determined in a DNA sample. However, the sequence, or order, of the bases on one or both of the helical strands could not be determined in 1953.

Coincidentally, the year 1953 also marked the first sequencing of a major nonrepeating biological polymer, the protein insulin (Fig. 11.1). Insulin is a relatively small protein of 51 amino acids, but determination of its complete sequence was a remarkable accomplishment at the time. Other proteins were sequenced in the ensuing decades. However, the methods were too arduous to be applied to many of the known proteins. Most of the proteins subjected to sequencing were also those for which X-ray structures were solved because the sequence information was essential to solve the structure from a set of diffraction patterns. More commonly, 5 to 30 of the N-terminal amino acid residues of a protein were sequenced as a means of comparing different proteins. This served to distinguish proteins and to generate a knowledge of some evolutionary relationships among them.

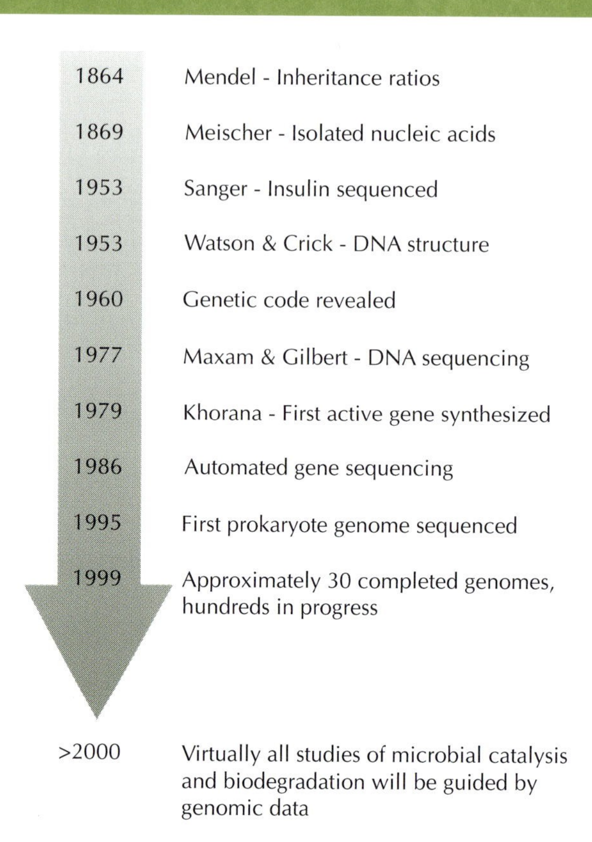

Figure 11.1 Time line showing several highlights of discovery in molecular genetics and genomics.

Watson and Crick considered that biological information was locked in the structures of macromolecules and flowed generally from DNA structure to RNA structure to protein structure. Figure 11.2 shows Watson's handwritten diagram from 1952 depicting this idea. The idea was confirmed in subsequent experiments by Francis Crick, Marshall Nirenberg, Severo Ochoa, and others which established the genetic code. From that point on, it was possible to deduce a protein's primary structure from its corresponding DNA sequence, or gene. In the early decades of molecular biology research (as defined from the date of elucidating the structure of DNA), proteins were sequenced in preference to genes. However, with the work of

November 1952

A Hypothetical Scheme of the Interrelationship between the Nucleic Acids and Proteins

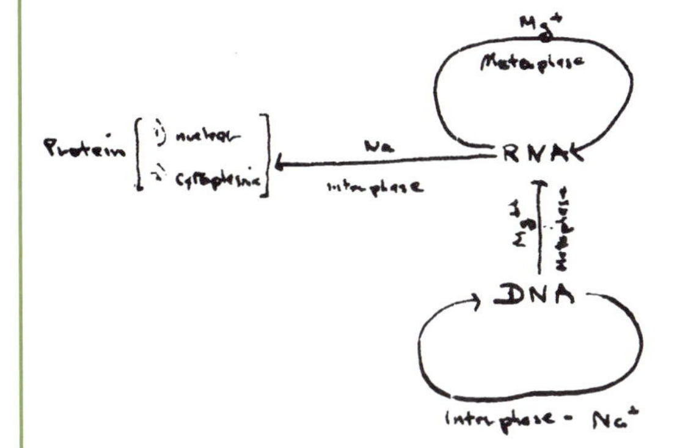

Consequences of scheme

1. RNA synthesis and DNA synthesis should not occur at the same time. Protein synthesis and DNA synthesis will occur simultaneously.

2. Nuclear RNA synthesis will occur only in dividing cells.

3. The total Mg^{++} concentration will increase toward metaphase and decrease during interphase

4. The content of nucleolar RNA may possibly remain constant during interphase. - Synthesis of nucleolar RNA occurs on the chromosomes during prophase-metaphase

Figure 11.2 Letter written by James Watson in 1952 indicating his thoughts on what has become known as the central dogma, or the flow of biological information from DNA to RNA to protein. (Reproduced from reference 8 with permission of James D. Watson.)

Walter Gilbert, Friedrich Sanger, and their colleagues, DNA sequencing became rapid enough that entire genes could be decoded, and the corresponding protein sequence could be obtained by theoretical translation using the genetic code.

With improvements in sequencing methodologies and the need to answer broader biological questions rather than merely deducing a protein's sequence, DNA sequences comprising two or more contiguous genes were determined. The molecular structures of operons were elucidated and added to our understanding of genetic regulation. The first complete genome of a biological entity, bacteriophage ϕX174, was sequenced in 1977 (6). It was 5,386 bp long.

Numerous bacterial plasmids have been sequenced because their importance in molecular biological research and susceptibility to restriction endonuclease digestion are sequence dependent. The first choices for sequencing were small plasmids from enteric bacteria which were important for research purposes or because they carried antibiotic resistance genes. A representative set of plasmids which have been sequenced is shown in Table 11.2. Most recently, catabolic plasmids have begun to be sequenced, one example of which is plasmid pNL1 from *Sphingomonas aromaticovorans* F199 (5). This 184,457-bp plasmid illustrates the clustering of catabolic functions for aromatic hydrocarbon and benzoic acid metabolism on a large plasmid. Based on sequence homology arguments, it is proposed that 12 oxygenases involved in aromatic-compound catabolism are encoded by this single plasmid. Consider this in light of the broad aromatic-substrate range for *Sphingomonas yanoikuyae* B1 described in chapter 4. The clustering of genes on a single replicon ensures the colocalization of genes in which pathways may overlap; that is, different dioxygenases can funnel many compounds into common catecholic and gentisic acid-like intermediates. The overall effect is to increase the total amount of metabolism accomplished by the combined pathways to more than the sum of the metabolisms of individual oxygenase-initiated pathways operating in isolation.

Table 11.1 Representative sizes of prokaryotic and eukaryotic genomes

Organism	Genome size (bp)
Prokaryotes	
Eubacteria	
Mycoplasma genitalium	0.6×10^6
Borrelia burgdorferi	1.0×10^6
Synechocystis sp. strain PCC6803	3.5×10^6
Escherichia coli	4.6×10^6
Pseudomonas aeruginosa PAO	5.9×10^6
Pseudomonas fluorescens	6.6×10^6
Streptomyces coelicolor	8.3×10^6
Myxococcus xanthus	9.4×10^6
Archaea	
Pyrococcus horikoshii	1.7×10^6
Methanococcus jannaschii	1.8×10^6
Archaeoglobus fulgidus	2.2×10^6
Eukaryotes	
Unicellular	
Saccharomyces cerevisiae	1.2×10^7
Schizosaccharomyces pombe	1.2×10^7
Giardia lamblia	1.4×10^7
Invertebrates	
Caenorhabditis elegans	8.8×10^7
Drosophila melanogaster	1.0×10^8
Some hemichordates	1.4×10^8
Plants	
Arabidopsis thalania	1.4×10^8
Oryza sativa (rice)	4.2×10^8
Tomato	1.0×10^9
Triticum aestivum (wheat)	1.6×10^{10}
Animals	
Human	3.0×10^9
Some amphibians	8.0×10^{11}

In contrast, the organization of prokaryotic genomes is not so homogeneous. At one time, it was thought that bacteria had a single chromosome and perhaps a few small plasmids. Now the genome organization is known to be more complex. For example, the *Burkholderia cepacia* AC1100 genome comprises five replicons of 4.0, 2.7, 0.53, 0.34, and 0.15 Mbp. There is evidence of gene rearrangement among the replicons in response to changing environmental conditions (3). *B. cepacia* AC1100 uses the herbicide 2,4,5-trichlorophenoxyacetic acid (2,4,5-T) as its sole source of carbon and energy. When strain AC1100 was grown in the absence of 2,4,5-T, mutants unable to grow on 2,4,5-T were obtained. In one mutant, genes involved in 2,4,5-T metabolism had translocated from one replicon to another, with the concomitant loss of some essential genes in the metabolic pathway.

Another unexpectedly diverse type of genome is found in *Borrelia burgdorferi* (2). As shown in Table 11.2, the organism contains a substantial amount of its DNA in replicons which are of the size generally considered to be plasmids. But in this case, essential genes are found on some of these

Table 11.2 Representative prokaryote plasmids which have been fully sequenced[a]

Organism	Plasmid	Size (bp)
Bacteria		
Acinetobacter sp. strain SUN	NC 000923[b]	6,076
Aquifex aeolicus	NC 001880	39,456
Arcanobacterium pyogenes	NC 001787	2,439
Bacillus pumilus	NC 001858	7,028
Bacillus stearothermophilus	NC 002062	1,883
Bacillus subtilis	NC 001764	7,837
	NC 001765	5,807
	NC 001766	8,737
	NC 002075	7,949
Borrelia burgdorferi	NC 001849	16,823
	NC 001850	24,177
	NC 001851	26,921
	NC 001852	29,766
	NC 001853	28,601
	NC 001854	27,323
	NC 001855	36,849
	NC 001856	38,829
	NC 001857	53,561
	NC 001903	26,498
	NC 001904	9,386
	NC 000948	30,750
	NC 000949	30,223
	NC 000950	30,299
	NC 000951	29,838
	NC 000952	30,800
	NC 000953	30,885
	NC 000954	30,651
	NC 000955	18,753
	NC 000956	52,971
	NC 000957	5,228
Buchnera aphidicola	NC 001910	7,967
	NC 001911	7,768
Burkholderia cepacia	pPPC1	1,641
Butyrivibrio fibrisolvens	NC 002059	2,804
Clostridium sp. strain MCF-1	NC 001772	2,450
Clostridium acetobutylicum	NC 001988	192,000
Corynebacterium glutamicum	NC 001456	3,054
	NC 001791	4,885
Deinococcus radiodurans	NC 000958	177,466
	NC 000959	45,704
Enterobacter aerogenes	NC 001735	53,339
Escherichia coli	pB171	68,817
	pCOIIb P-9	11,712
	pKL1	1,549
	pHly152	8,215
Fusobacterium nucleatum	NC 002002	6,281

(continued next page)

Table 11.2 *Continued*

Organism	Plasmid[b]	Size (bp)
Bacteria		
Helicobacter pylori	NC 001756	3,506
	NC 001843	1,222
Lactobacillus delbrueckii	NC 001670	7,921
Lactobacillus helveticus subsp. *jugurti*	NC 001379	3,292
Lactobacillus reuteri	NC 001757	5,113
Lactococcus lactis	NC 001949	60,232
	NC 000906	6,499
Microcystis aeruginosa	NC 001597	4,993
	NC 002060	2,287
Pantoea citrea	NC 001898	5,229
	NC 002070	3,661
Pasteurella multocida	NC 001774	5,360
Phormidium foveolarum	NC 002061	1,509
Prevotella ruminicola	NC 001760	3,130
Pseudomonas alcaligenes NCIB 9867	pRA2	32,743
Pseudomonas putida NCIB 9816-4	pDTG1	84,043[c]
Rhizobium sp. strain NGR234	NC 000914	536,165
Rhodothermus marinus	NC 001755	2,935
Ruminococcus flavefaciens	NC 001758	1,768
Salmonella enterica serovar Berta	NC 001848	4,656
Salmonella enterica serovar Typhimurium	NC 002056	9,263
Sphingomonas aromaticivorans	NC 002033	184,457
Staphylococcus aureus	NC 002013	2,910
	NC 001763	6,024
	NC 001767	4,439
	NC 001994	1,658
	NC 001995	1,551
Streptococcus agalactiae	NC 001797	6,437
Streptococcus thermophilus	NC 000937	9,531
	NC 000938	3,498
Streptomyces clavuligerus	NC 001738	11,696
Streptomyces nigrifaciens	NC 001425	10,992
Streptomyces phaeochromogenes	NC 001759	11,143
Synechococcus sp.	NC 001974	2,514
Yersinia pestis	NC 001881	9,610
	NC 001882	70,504
	NC 001883	100,984
	NC 001972	70,559
	NC 001976	100,990
Zymomonas mobilis	NC 001845	1,680
Archaea		
Halobacterium sp. strain NRC-1	NC 001869	191,346
Methanobacterium thermoautotrophicum	NC 001336	13,514
	NC 001337	11,014
	NC 000905	6,205
Methanococcus maripaludis	NC 001811	8,285
Pyrococcus abyssi	NC 001773	3,444
Sulfolobus islandicus	NC 001771	5,350

[a]Most of the data are from the National Center for Biotechnology Information website (http://www.ncbi.nlm.nih.gov/PMGifs/Genomes/eub_p.html). The accession number can be used to get information on individual plasmids.
[b]National Center for Biotechnology Information accession number.
[c]G. Zylstra, personal communication.

small replicons. Additionally, the various replicons are thought to undergo substantial interelement DNA rearrangement.

The Present Impact of Genomics

Genomics is just beginning to have an impact on biodegradation research. For the first few years, genome sequencing was almost exclusively focused on pathogenic bacteria. This was because the first sequencing projects were funded by the National Institutes of Health and were supposed to benefit research on human health. However, there is now a much broader base of support for whole-genome sequencing. Approximately 100 bacterial genome-sequencing projects are now completed or under way in the public domain (Table 11.3); it is likely that several hundred complete bacterial genomes will be available several years from now. This includes a significant number of phylogenetically diverse soil bacteria (Table 11.3).

Approximately 95% of the completed genomes are those of unicellular microbes, mostly prokaryotes. On the order of one-third of the genes sequenced cannot be assigned a biological function, and a fraction of those are distinct from other genes found in other genomes. While DNA sequence analysis is yielding important patterns, the greatest genomic richness results from assigning metabolic functions to individual genes and deriving a biological usefulness of the gene in the context of the organism and its environment.

For example, the authors of the *Bacillus subtilis* genome-sequencing paper expressed surprise that the bacterium had genes thought to encode the catabolism of certain plant natural products (4). Their surprise was due to a previous description of those genes in a taxonomically very different gram-negative soil organism. However, in one study using rRNA indicators, a gram of soil has been proposed to contain as many as 10,000 distinct species (types) of bacteria. Consider further that some plants make flavonoid compounds amounting to as much as 27% of the leaf mass, and this leaf matter provides the major new carbon input in soils under the plant. Thus, we would expect divergent taxa to contain genes encoding the catabolism of these compounds in temperate soils where the particular plant is found, and those genes might not be found in hot springs, arctic environments, or deserts. Therefore, we can begin to link gene clusters not only with single organisms but to biologically complex environments, thus making a start in deriving a global genomic composition. While this will never be close to complete, it will help to define a broader context for genomics, moving the information into the sphere of ecological research.

Functional Genomics in the Context of Microbial Biocatalysis

After gene sequences are obtained, similarities to known gene products (mostly proteins) are deduced, and the biological functions, or phenotype, are suggested (Fig. 11.3A). In this way, the genes are used to deduce what a bacterium can or cannot do. The identification of similarities between new sequences and known sequences, and the assignment of function based on this process, is often called gene annotation. The process requires a comparison of new DNA sequences with sequences in databases, and as such is a very computationally intense exercise. This is currently a major field of biologically focused computing. This focus stems from the fact that DNA sequencing is now relatively fast, and it is desirable to extract as much

Table 11.3 Microbial complete-genome-sequencing projects, both completed and ongoing, as of 1 January 2000[a]

Archaea	Eubacteria			Eukaryota
Aeropyrum pernix	*Actinobacillus actinomycetemcomitans*	*Legionella pneumophila*	*Salmonella enterica* serovar Typhi	*Candida albicans*
Archaeoglobus fulgidus	Actinomycetes	*Listeria innocua*	*Salmonella enterica* serovar Typhimurium SGSC1412	*Dictyostelium discoideum*
Halobacterium salinarum	*Aquifex aeolicus*	*Listeria monocytogenes*	*Salmonella enterica* serovar Typhimurium TR7095	*Emiliania huxleyi*
Halobacterium sp.	*Bacillus anthracis*	*Methylobacterium extorquens*	*Shewanella putrefaciens*	*Leishmania major*
Methanobacterium thermoautotrophicum	*Bacillus* sp. strain C-125	*Mycobacterium avium*	*Shigella flexneri* 2a	*Plasmodium falciparum*
Methanococcus jannaschii	*Bacillus subtilis*	*Mycobacterium bovis*	*Staphylococcus aureus*	*Pneumocystis carinii* sp. *carinii*
Methanococcus maripaludis	*Bartonella henselae*	*Mycobacterium leprae*	*Staphylococcus aureus* (MRSA)[b]	*Pneumocystis carinii* sp. *hominis*
Methanosarcina mazeii	*Bordetella bronchiseptica*	*Mycobacterium paratuberculosis*	*Staphylococcus aureus* (MSSA)[c]	*Saccharomyces cerevisiae*
Pyrobaculum aerophilum	*Bordetella parapertussis*	*Mycobacterium tuberculosis*	*Streptococcus agalactiae*	*Schizosaccharomyces pombe*
Pyrococcus abyssi	*Bordetella pertussis*	*Mycoplasma genitalium*	*Streptococcus mutans*	*Trypanosoma brucei*
Pyrococcus furiosus	*Borrelia burgdorferi*	*Mycoplasma mycoides*	*Streptococcus pneumoniae*	*Trypanosoma brucei* var. *rhodesiense*
Pyrococcus horikoshii	*Campylobacter jejuni*	*Mycoplasma pneumoniae*	*Streptococcus pyogenes*	*Ustilago maydis*
Sulfolobus solfataricus	*Caulobacter crescentus*	*Mycoplasma pulmonis*	*Streptomyces coelicolor*	
Thermoplasma acidophilum	*Chlamydia trachomatis*	*Neisseria meningitidis*	*Synechocystis* sp.	
Thermoplasma volcanium	*Chlorobium tepidum*	*Pasteurella haemolytica*	*Thermotoga maritima*	
	Clostridium acetobutylicum	*Pasteurella multocida*	*Thermus thermophilus*	
	Clostridium difficile	*Photorhabditis luminescens*	*Thiobacillus ferrooxidans*	
	Clostridium sp. strain BC1	*Porphyromonas gingivalis*	*Treponema denticola*	
	Corynebacterium diphtheriae	*Pseudomonas aeruginosa*	*Treponema pallidum*	
	Dehalococcoides ethenogenes	*Pseudomonas putida* KT2440	*Ureaplasma urealyticum*	
	Deinococcus radiodurans	*Pseudomonas putida* PRS	*Vibrio cholerae*	
	Desulfovibrio vulgaris	*Ralstonia eutropha*	*Xanthomonas citri*	
	Enterococcus faecalis	*Ralstonia solanacearum*	*Xylella fastidiosa*	
	Escherichia coli K-12	*Rhodobacter capsulatus*	*Yersinia pestis*	
	Escherichia coli O157:H7	*Rhodobacter sphaeroides*		
	Francisella tularensis	*Rickettsia conorii*		
	Haemophilus influenzae	*Rickettsia prowazekii*		
	Helicobacter pylori	*Salmonella enterica* serovar Paratyphi		
	Lactobacillus acidophilus			
	Lactococcus lactis			

[a]From the following databases and websites: http://genome.bnl.gov/; http://www.tigr.org/; http://www.sanger.ac.uk/Projects/Microbes/; and http://www.cbc.umn.edu/ResearchProjects/AGAC/Pm/. The taxonomic assignment was derived from the National Center for Biotechnology Information taxonomy database (http://www3.ncbi.nlm.nih.gov/Entrez/).

[b]MRSA, methicillin-resistant *S. aureus*.

[c]MSSA, methicillin-susceptible *S. aureus*.

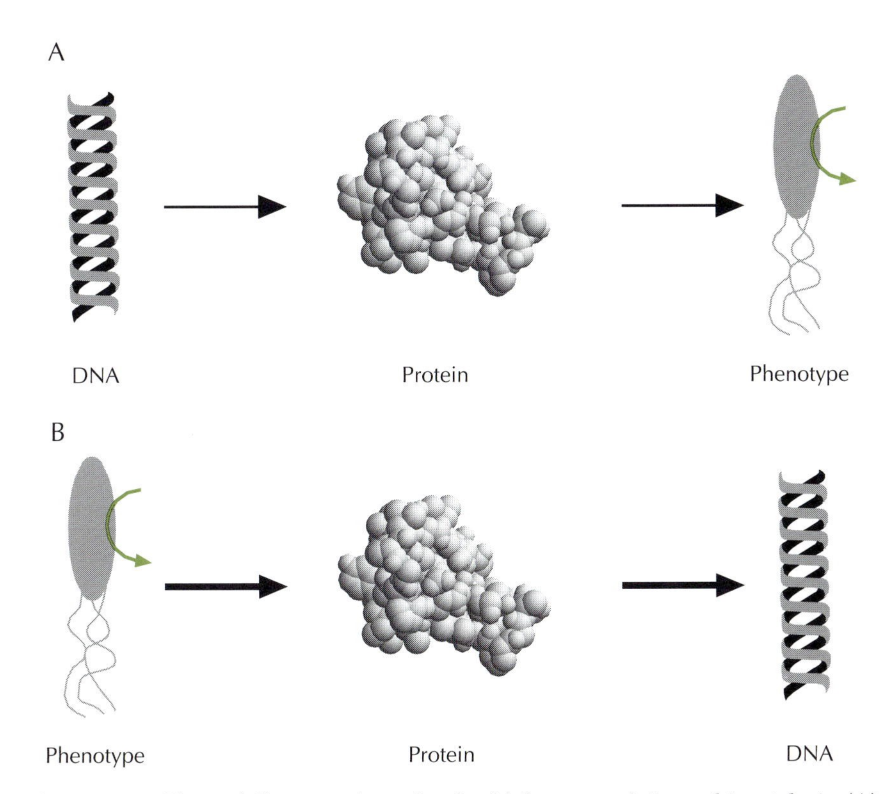

Figure 11.3 Flow of discovery in molecular biology pertaining to biocatalysis. (A) In current genomics studies, one often starts with DNA sequences and attempts to deduce the protein sequence and then the function of the protein, or the biological phenotype. (B) Reverse functional genomics, in which discovery flows from phenotype to protein to gene (DNA).

information as possible from a given sequencing project. The ultimate goal is to map a sequence back to one whose biological function has been well established by a variety of methods. If the annotation is correct, it is then unnecessary to repeat all of the biological and chemical experiments with all the new organisms being sequenced; the functions can be inferred from the sequence.

The order of events shown in Fig. 11.3A assumes that a complete genome sequence provides an unambiguous blueprint of a biological system, but this analogy is only apt if the gene sequence can be translated into a biological function with very high fidelity. Unfortunately, this is not the case. Most assignments of new gene sequences match known sequence types with high confidence about 60% of the time. At that point, the gene is put into the "known" box as indicated in Fig. 11.4. These genes are put into neat categories in the known box, based on the putative function they carry out. For example, a given gene(s) might encode (i) a transporter protein, (ii) a soluble ATPase, or (iii) a cytochrome P450 monooxygenase. But how neat are those functional classes, and what is really known about the biological function conferred on a bacterium containing that specific gene? For example, if the protein is a transporter, its biological role is to ferry one or more impermeable compounds across the membrane, but from the gene sequence alone, it is often not known with a high degree of certainty what the transporter is transporting. Hence, its biological role is still, in an im-

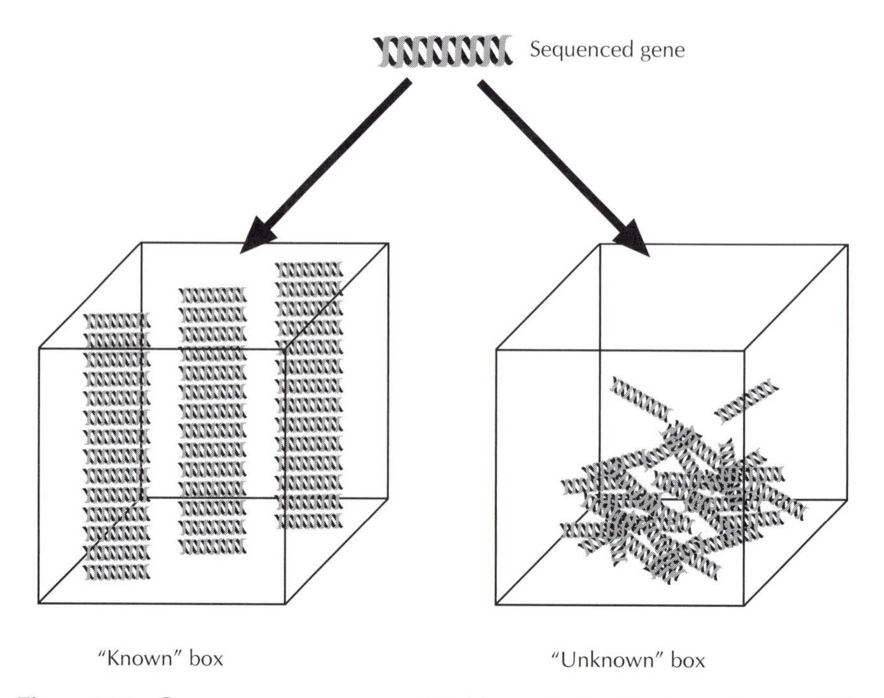

Figure 11.4 Gene sequences are annotated by assigning function when possible. A gene may be put into the known box, for example, by deducing that the gene encodes a cytochrome P450 monooxygenase. Genes that cannot be assigned a function go into the unknown box. Over time, the hope is to move as many proteins as possible from the unknown box to the known box.

portant way, unknown. Consider another example of a gene sequence identified as encoding a cytochrome P450 monooxygenase. This tells the annotator that it is a heme protein with a thiolate ligand to the heme iron. But cytochrome P450 monooxygenases collectively are known to oxidize thousands of different substrates. It is not yet possible to computationally deduce the active-site fold of a protein in great detail and predict what substrate(s) it will bind. In lieu of this, the actual physiological function of the cytochrome P450 must still be said to be unknown, although in annotation scorekeeping it is considered a known at such time as it can be said to be a cytochrome P450 protein. Considered in this light, with current annotation technology, it is likely that about 80% of newly sequenced genes are not being assigned a precise physiological function.

But with better bioinformatics tools, won't this problem go away? Gene sequences with both known and unknown functions are available on public databases, and the presumption is that, over time, all the genes in the unknown box will migrate to the known box (Fig. 11.4). Again, let us look more closely at this widespread perception of functional genomics. The assumption it is based on is that most gene functions are known. This perception is most widely held with metabolism. While it is recognized that metabolism is diverse with respect to overall structure, it is often believed that the fundamental reaction types have largely been discovered and studied.

To categorize reaction types, we use the definition here of novel reactions transforming distinct organic functional groups as developed in chapters 5 and 9 (Table 11.4). The 58 functional groups shown in Fig. 9.2 mainly cover the compounds in presently defined metabolism. While there are

Table 11.4 Organic functional groups represented in known microbial metabolism

Organic functional group	Representative substrate
Alkane, primary	*n*-Octane
Alkane, secondary	*p*-Cymene
Alkane, tertiary	Methyl-*tert*-butyl ether
Cycloaliphatic ring	Cyclohexanol
Bicycloaliphatic ring	(+)-Camphor
Alkene	Propylene
Alkyne	Acetylene
Monocyclic aromatic hydrocarbon	Toluene
Polycyclic aromatic hydrocarbon	Phenanthrene
Biphenyl-type benzenoid ring	4-Chlorobiphenyl
Oxygen ether	Tetrahydrofuran
Thioether	Dimethyl sulfide
S-Heterocyclic ring	Dibenzothiophene
N-Heterocyclic ring	Nicotine
O-Heterocyclic ring	Dibenzofuran
Ketone	Methylethylketone
Thioketone	Carbon disulfide
Alcohol	1,3-Dichloro-2-propanol
Thiol	Methanethiol
Amine, primary	2-Aminobenzoate
Amine, secondary	Glyphosate
Amine, tertiary	Nitrilotriacetate
Aldehyde	3-Hydroxybenzaldehyde
Carboxylic acid	3-Phenylpropionate
Carboxylic acid ester	Butyrolactone
Carboxylic thioester	Benzoyl-*S*-coenzyme A
Amide	Caprolactam
Nitrile	Acrylonitrile
Thiocyanate	Thiocyanate anion
Nitro	Nitrobenzene
Nitrate ester	Pentaerythritol tetranitrate
Diazo	4-Carboxy-4′-sulfoazobenzene
Organohalide	Tetrachlorethylene
Organomercurial	Methylmercury chloride
Organoarsenical	Arsenoacetate
Organosilicon	Octamethylcyclotetrasiloxane
Organotin	Tri-*n*-butyltin
Organophosphate ester	Paraoxon
Thiophosphate ester	Parathion
Phosphonic acid	Glyphosate
Sulfonic acid	Methanesulfonic acid
Sulfate ester	Dodecyl sulfate

thousands of individual compounds transformed by known metabolism, they largely contain mixtures of those functional groups shown in Fig. 9.2. The examples of substrates given in Table 11.4 for most of these functional groups are from the University of Minnesota Biocatalysis/Biodegradation

Database functional group page (http://umbbd.ahc.umn.edu/search/FuncGrps.html).

Inadvertent Deception in Modern Biochemistry Textbooks

> . . . biochemists have made a great deal of progress toward defining the basic reactions that are common to all cells and how those reactions are interrelated.
>
> — *Horton, Moran, Ochs, Rawn, and Scrimgeour, Principles of Biochemistry*

> . . . the chemical basis of many central [biological] processes are now understood.
>
> — *Stryer, Biochemistry, 3rd ed.*

> The number of reactions in metabolism is large, but the number of kinds of reactions is small.
>
> — *Stryer, Biochemistry, 3rd ed.*

> . . . organisms show marked similarity in their major pathways of metabolism. Given the almost unlimited possibilities within organic chemistry, this generally would appear most unlikely.
>
> — *Garrett and Grisham, Biochemistry*

> . . . by the time the human genome is completely sequenced, coding regions will be identifiable with almost perfect accuracy, and most new genes will carry in their sequence immediately recognizable clues about function.
>
> — *Cantor and Smith, Genomics*

The statements above are, by and large, true. Biochemists have made a great deal of progress, and intermediary metabolic reactions are relatively well understood. There are many similar biochemical reactions found in different living things. These ideas are presented in textbooks to convey the unity of biochemistry. However, as Alfred North Whitehead stated, "Seek simplicity and mistrust it." One should not be deceived into thinking that metabolism is essentially "worked out" and that we just need to better understand how it is all integrated to make a living cell work. As discussed below and in chapter 12, this has important implications for how genomics as a science will be pursued.

There is a need for continued new discovery in microbial biocatalysis. The situation in genomics may be analogous to the European perception of geography in the mid-15th century. The outlines of Europe, Asia, and Africa were reasonably well defined. Westward travel from Europe was anticipated to yield a fast sea route to the Pacific coast of Asia. The perception was found to be incorrect, but the significant land masses of North and South America made for an exciting period of discovery.

The Case for Reverse Functional Genomics and New Discovery in Biocatalysis

The pathway to discovery in functional genomics is usually taken to be that shown in Fig. 11.3A. This makes the assumption that genes will ultimately be assignable to boxes, as depicted in Fig. 11.4. Genes can be assigned to a previously elucidated function based on sequence homology arguments.

However, if a gene is discovered which encodes an enzyme catalyzing a fundamentally new reaction, that biological function will never be deduced from the sequence alone. The significance of this statement is dependent on the prevalence of new reactions in biochemistry. If that number is very small, better computational methods will lead to functional assignment of almost all new genes. If, however, many new reaction types are harbored by microbes around the globe, their genomes will be found to contain sequences which will never yield a correct annotation. This point will be discussed in greater depth in the next chapter when we discuss the limits of microbial catalysis.

Summary

Microbial genomics will continue to expand in depth and scope. It will have an impact on many areas of microbiology, particularly industrial applications of biocatalysis. The science will advance most quickly by better connecting gene sequence annotation with large-scale enzyme functional analysis. It is crucial to avoid the pitfall of thinking that life's complete set of biochemical reactions has largely been uncovered. The functional-group analysis of chapter 5 shows that much biochemistry remains to be discovered. Genome-sequencing data are consistent with this; on average, 25% of the genes in the first 25 prokaryotic genomes sequenced were found to be unique to each organism. It is clear that gene discovery and new enzyme discovery must go forward hand in hand.

References

1. **Franklin, R. E., and R. G. Gosling.** 1953. Molecular configurations in sodium thymonucleate. *Nature* **171:**740–741.
2. **Fraser, C. M., S. Casjens, W. M. Huang, G. G. Sutton, R. Clayton, R. Lathigra, O. White, K. A. Ketchum, R. Dodson, E. K. Hickey, M. Gwinn, B. Dougherty, J. F. Tomb, R. D. Fleischmann, D. Richardson, J. Peterson, A. R. Kerlavage, J. Quackenbush, S. Salzberg, M. Hanson, R. van Vugt, N. Palmer, M. D. Adams, J. Gocayne, J. Weidman, T. Utterback, L. Watthey, L. McDonald, P. Artiach, C. Bowman, S. Garland, C. Fujii, M. D. Cotton, K. Horst, K. Roberts, B. Hatch, H. O. Smith, and J. C. Venter.** 1997. Genomic sequence of a Lyme disease spirochaete, *Borrelia burgdorferi*. *Nature* **390:**580–586.
3. **Hubner, A., C. E. Danganan, L. Xun, A. M. Chakrabarty, and W. Hendrickson.** 1998. Genes for 2,4,5-trichlorophenoxyacetic acid metabolism in *Burkholderia cepacia* AC1100: characterization of the *tftC* and *tftD* genes and locations of the *tft* operons on multiple replicons. *Appl. Environ. Microbiol.* **64:** 2086–2093.
4. **Kunst, F., N. Ogasawara, I. Moszer, A. M. Albertini, G. Alloni, V. Azevedo, M. G. Bertero, P. Bessires, A. Bolotin, S. Borchert, R. Borriss, L. Boursier, A. Brans, M. Braun, S. C. Brignel, S. Bron, S. Brouillet, C. V. Bruschi, B. Caldwell, V. Capuano, N. M. Carter, S.-K. Choi, J.-J. Codani, I. F. Connerton, N. J. Cummings, A. Danchin, R. A. Daniel, F. Denizot, K. M. Devine, A. Dusterhoft, S. D. Ehrlich, P. T. Emmerson, K. D. Entian, J. Errington, C. Fabret, E. Ferrari, D. Foulger, C. Fritz, M. Fujita, Y. Fujita, S. Fuma, A. Galizzi, N. Galleron, S.-Y. Ghim, P. Glaser, A. Goffeau, E. J. Golightly, G. Grandi, G. Guiseppi, B. J. Guy, K. Haga, J. Halech, C. R. Harwood, A. Hnaut, H. Hilbert, S. Holsappel, S. Hosono, M.-F. Hullo, M. Itaya, L. Jones, B. Joris, D. Karamata, Y. Kasahara, M. Klaerr-Blanchard, C. Klein, Y. Kobayashi, P. Koetter, G. Koningstein, S. Krogh, M. Kumano, K. Kurita, A. Lapidus, S. Lardinois, J. Lauber, V. Lazarevic, S.-M. Lee, A. Levine, H. Liu, S. Masuda, C. Maul, C. Medigue, N. Medina, R. P. Mellado, M. Mizuno, D. Moestl, S. Nakai, M. Noback, D. Noone, M. O'Reilly, K. Ogawa, A. Ogiwara, B. Oudega, S.-H. Park, V. Parro, T. M. Pohl, D. Portetelle, S. Porwollik, A. M. Prescott, E. Presecan, P. Pujic, B. Purnelle, G. Rapoport, M. Rey, S. Reynolds, M. Rieger, C. Rivolta, E. Rocha, B. Roche, M. Rose, Y. Sadaie, T. Sato, E. Scanlan, S. Schleich, R. Schroeter, F. Scoffone, J. Sekiguchi, A., Sekowska, S. J. Seror, P. Serror, B. S. Shin, B. Soldo, A. Sorokin, E. Tacconi, T. Takagi, H. Takahashi, K. Takemaru, M.**

Takeuchi, A. Tamakoshi, T. Tanaka, P. Terpstra, A. Tognoni, V. Tosato, S. Uchiyama, M. Vandenbol, F. Vannier, A. Vassarotti, A. Viari, R. Wambutt, E. Wedler, H. Wedler, T. Weitzenegger, P. Winters, A. Wipat, H. Yamamoto, K. Yamane, K. Yasumoto, K. Yata, K. Yoshida, H. F. Yoshikawa, E. Zumstein, H. Yoshikawa, and A. Danchin. 1997. The complete genome sequence of the gram-positive bacterium *Bacillus subtilis. Nature* **390:**249–256.

*5. **Romine, M. F., L. C. Stillwell, K. K. Wong, S. J. Thurston, E. C. Sisk, C. Sensen, T. Gaasterland, J. K. Fredrickson, and J. D. Saffer.** 1999. Complete sequence of a 184-kilobase catabolic plasmid from *Sphingomonas aromaticivorans* F199. *J. Bacteriol.* **181:**1585–1602.

6. **Sanger, F., A. R. Coulson, T. Friedmann, G. M. Air, B. G. Barrell, N. L. Brown, J. C. Fiddes, C. A. I. Hutchison, P. M. Slocombe, and M. Smith.** 1978. The nucleotide sequence of bacteriophage ϕX174. *J. Mol. Biol.* **125:**225–246.

*7. **Shimkets, L. J.** 1998. *Structure and Sizes of the Genomes of Archaea and Bacteria.* Chapman and Hall, New York, N.Y.

8. **Stent, G. S. (ed.).** 1980. *The Double Helix: A Personal Account of the Discovery of the Structure of DNA,* by James D. Watson. Norton Critical edition. W. H. Norton and Company, New York, N.Y.

*9. **Watson, J. D., and F. H. C. Crick.** 1953. A structure of deoxyribosenucleic acid. *Nature* **171:**737–738.

10. **Wilkins, M. H. F., A. R. Stokes, and H. R. Wilson.** 1953. Molecular structure of deoxypentose nucleic acids. *Nature* **171:**738–740.

12

The Extent of Microbial Catalysis and Biodegradation: Are Microbes Infallible?

> We are all taught what is known, but we rarely learn about what is not known, and we almost never learn about the unknowable. That bias can lead to misconceptions about the world around us.
>
> — *Ralph E. Gomory*

> . . . it is part of the adventure of science to try to find a limitation in all directions and to stretch the human imagination as far as possible everywhere.
>
> — *Richard P. Feynman*

When it comes to human understanding of nature, how far might we ultimately explore, and what knowledge may remain beyond our grasp indefinitely? For microbial biocatalysis, as indeed for all fields, it is important to consider what the limits of knowledge may be. In some scientific disciplines, the practitioners may seek some defining goal, a Holy Grail of sorts. For example, in neuroscience, the defining question many ask is, "How does the brain work?" In particle physics, the question might be, "What are the number and characteristics of the elementary particles?" It is important to devise a similar question for the domain of microbial catalysis, such as, "What organic molecules are transformable by microbial catalysis?" It is often speculated that every nonpolymeric compound conceived of by humans and occurring in nature will be metabolized by some microorganism somewhere in the soil or water of the Earth. Is this true? Can it be studied systematically? In this chapter we pose these questions and begin to work on the answers. The beginnings of an answer are first addressed with existing knowledge. Then, ideas are presented as to how the far reaches of microbial metabolism can be further identified, based on the concepts of functional-group metabolism previously developed in this book. This effort is also linked to the widespread enterprise of microbial functional genomics. The total extent of genomic diversity is unknown, but comparative genomic analysis to date suggests that genetic diversity in prokaryotes is greater than previously anticipated, as discussed in chapter 11. This, along with other evidence presented in this chapter, suggests that there is much catalytic diversity yet to be discovered.

Microbial Enzyme Diversity

It was emphasized at the end of chapter 11 that new types of biochemical reactions will not be discovered by DNA sequence analysis alone. Thus, both reaction and gene screening need to occur in parallel.

What has been the general course of discovery for new biochemical reactions versus new gene sequences? In general, the curves of discoveries versus dates are as shown in Fig. 12.1. During the first half of the 20th century, when no DNA sequences had yet been determined, research on microbial metabolism was experiencing an intense period of discovery. As discussed in chapter 2, the outlines of intermediary metabolic reactions were largely elucidated with microorganisms, and some specialized metabolism was also revealed. However, this activity has slowed at the same time that enormous effort has gone into gene sequencing.

Still, new types of reactions and their corresponding genes are being described in the biodegradation literature. When new enzymatic reactions are discovered and the protein sequence is deduced later (the pathway of discovery shown in Fig. 11.3B), how related is the gene to known genes? While we cannot discuss future discovery, it is possible to examine an unbiased set of novel biocatalyses. This data set comes from the University of Minnesota Biocatalysis/Biodegradation Database (UM-BBD). Microbial reactions have been added to the UM-BBD based on the novelty of the biochemistry and without any prior knowledge of the gene sequence and its

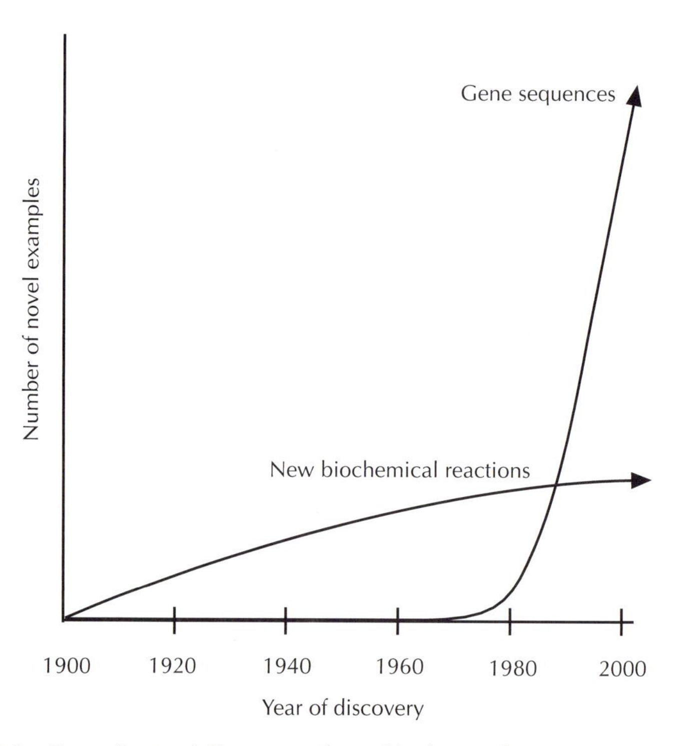

Figure 12.1 General rate of discovery of new biochemical reactions versus the rate of discovery of new genes over the last century.

homology to other genes in the data banks. Some of those novel enzymes are shown in Table 12.1. In this analysis, none of the sequenced genes shows very strong sequence relatedness to other enzymes (most less than 30%). When homology is observed, it is largely in regions that helped define the cofactors involved in the reaction. For example, benzylsuccinate synthase shows some detectable sequence identity around the region of a putative glycyl radical, and thus it is hypothetically linked to enzymes which use a glycyl radical in the reaction pathway of the enzyme (4). Since benzylsuccinate synthase catalyzes a carbon-carbon bond-forming reaction, it was chemically plausible to expect a radical intermediate in the reaction.

If the pathway of discovery had been inverted and the gene was discovered first, it would not have been possible to hazard even a reasonable guess that the benzylsuccinate synthase gene encoded an enzyme catalyzing functionalization of toluene at the benzylic carbon atom. It might only have been suspected that a glycyl radical was involved. The other enzymes in Table 12.1 buttress this point. The enzymes either have no significant homolog or the enzymes showing low homology (typically 15 to 30%) catalyze different reactions. It would be difficult or impossible to predict that a newly discovered gene—for example, that encoding atrazine chlorohy-

Table 12.1 Enzymes in the UM-BBD encoded by genes with no discernible homology to other known genes or with homology to genes encoding different functions[a]

Enzyme name	Enzyme type	Homologous enzyme	UM-BBD ID
Acetylene hydratase	Molybdenum-iron-sulfur	Unknown	r0591
Tetrachloroethene reductive dehalogenase	Cobalamin-iron-sulfur	No significant homology	r0349
Cyanamide hydrolase	Unknown	Closest sequences are hypothetical genes	r0668
Epoxide carboxylase	Hexameric zinc enzyme	Novel; multicomponent	r0050
Benzylsuccinate synthase	Flavin-glycyl radical	Homology in cofactor regions only	r0328
Benzoyl-CoA reductase	Iron-sulfur-flavin	2-Hydroxyglutaryl-CoA dehydratase	r0190
2-Ketocyclohexane	Dihydroxynaphthoate synthase	Carboxyl-CoA hydrolase	r0193
Pentaerythritol tetranitrate reductase	Flavoprotein	Old yellow enzyme	r0025
Phosphotriesterases	Binuclear metal center	Amidohydrolases	r0066
Organomercurial lyase	No cofactors	No significant homology	r0405
Mercuric reductase	Flavin-disulfide	Low homology; nonmetal reductases	r0406
Tetrachlorohydroquinone reductive dehalogenase	Glutathione transferase	Glutathione-dependent isomerase	r0314
4-Hydroxybenzoate decarboxylase	No cofactors	Homology to hypothetical proteins only	r0159
4-Hydroxybenzoyl-CoA reductase	Molybdenum-flavin-iron-sulfur	Dimethyl sulfoxide reductase	r0158
Styrene epoxide isomerase	No cofactors known	No significant homology	r0034
Atrazine chlorohydrolase	Iron enzyme	Low homology to amidohydrolases	r0113
1-Aminocyclopropane-1-carboxylate deaminase	Pyridoxal phosphate	Low homology to pyridosal phosphate enzymes	r0357
2-Aminobenzoyl-CoA reductase	Flavin-enzyme	Low homology to salicyclate hydroxylase	r0342

[a]Reactions catalyzed by these enzymes can be found by using the following URL: http://umbbd.ahc.umn.edu:8007/umbbd/servlet/pageservlet?ptype=r&reacID=rXXXX (where rXXXX is the UM-BBD identification number [ID] given in the table). CoA, coenzyme A.

Enormous Metabolic Diversity Has Yet To Be Discovered

There are other lines of evidence supporting the contention that many novel microbial enzymes await discovery. Some of this builds on points raised in previous chapters on microbial diversity (chapter 4), evolution (chapter 7), and the diversity of chemical functional groups made by biological systems (chapter 5). The main arguments are summarized in Table 12.2. There is clearly enormous microbial diversity, and many niches are not adequately described. Microbial diversity is reflected by genomic diversity; in the first 21 microbial genomes sequenced, 25% of the genes in each organism are different enough to be considered unique, and most of those are unknown genes. How many might encode enzymes catalyzing novel reactions is hard to predict. It is intriguing that reactions previously thought to be unique to the organic chemist's repertoire are continually being discovered in nature. For example, catalytic antibodies were prepared to catalyze a Diels-Alder reaction, and this was proposed to be a unique reaction for proteins (3). More recently, Diels-Alder adducts were discovered in plants (6). Other named organic reactions discovered to be catalyzed by biological systems include the Beckmann rearrangement (5), Raney nickel-catalyzed desulfurization (8), the Bamberger rearrangement (7), and the Kolbe-Schmidt rearrangement (2).

The most compelling argument for previously uncharacterized enzyme reactions comes from the wealth of organic functional groups reported to be produced by biological systems, as discussed in chapter 5 and depicted in Tables 5.2, 5.3, and 5.4. The biological production of such functional groups mandates a biological mechanism for their synthesis. Moreover, when biological systems biosynthesize molecules, a corresponding biodegradative ability can be expected to evolve. This is particularly likely since a reasonable number of the exotic functional groups are found in antibiotics and there would be strong selective pressure to evolve a metabolism to degrade the toxic agent.

Table 12.2 Evidence for undiscovered unique organic-functional-group metabolism

1. The vast majority of microorganisms have yet to be isolated and characterized.
2. Genome sequencing of 25 microbes has shown that approximately 25% of each organism's genes are unique.
3. Approximately 40% of genes in genomics projects are unidentifiable by current annotation techniques.
4. When reactions have been stated to be foreign to biology, e.g., a Diels-Alder condensation, someone has later found the reaction to be enzymatically catalyzed.
5. At least 94 organic functional groups are found in natural products, but only a little more than half have been studied with respect to their biosynthesis or biodegradation.
6. A brief survey of organic functional groups new to biochemistry has found microbes capable of metabolizing them.

drolase or phosphotriesterase, with low homology to proteins of the amidohydrolase superfamily—would catalyze completely different reactions: carbon-chlorine bond hydrolysis and phosphotriester hydrolysis, respectively. It would be necessary to show the function of the gene by direct biochemical experiments. This supports the contention that the pace of discovering new reactions must be increased to better match the pace of new gene discovery. The discovery of novel reactions will increasingly lead the researcher to genes of unknown function in the databases.

The limits of enzyme catalysis cannot be predicted at present, and that lofty goal may remain elusive. The catalytic power of enzymes derives first from 21 amino acid side chains, of which 10 (cysteine, selenocyteine, histidine, aspartate, glutamate, lysine, arginine, serine, threonine, and tyrosine) are principally important for catalysis. Add to that a broad range of catalytically relevant transition metals: iron, manganese, cobalt, zinc, nickel, molybdenum, vanadium, and the divalent metals calcium and manganese. Additional catalytic punch can be brought to bear on a substrate via organic cofactors: flavin, pterin, thiamine, and biotin ring structures. Additionally, there are organometallic factors which are important in biocatalysis: heme, cobalamin, molybdopterin, and nickel-coenzyme F430.

In this wealth of catalytic diversity possessed by microbial enzymes, two issues are relevant. First, it is unlikely that we have uncovered all of the catalytic potential of these known cofactors. Either variations of the cofactor or their concomitant juxtaposition to different amino acid groupings can potentially catalyze reactions not yet examined in biological systems. Moreover, it is unclear how many new cofactors, metals, and modified amino acids remain to be discovered in microbial systems. While we cannot predict what will be discovered, the analysis of recently discovered enzymes, shown in Table 12.1, argues that new biocatalyst discovery should proceed vigorously.

Experiments Suggest that Novel Biocatalytic Reactions Are Ubiquitous

To determine the potential for microbes to transform the exotic compounds for which metabolism is yet unknown, several compounds containing such functional groups were put into standard enrichment cultures in which the compounds were used as sole carbon or nitrogen sources to support growth. Most of these yielded actively growing cultures of microorganisms, and the target compound was transformed in the cultures, as demonstrated by high-pressure liquid chromatography (unpublished data). These preliminary data suggest that microbial activities might be much broader than generally anticipated. Moreover, the data suggest a pathway to new discovery via reverse functional genomics (Fig. 12.2).

Screening for new microbial transformation reactions prior to identifying the responsible gene(s) is not a rare activity. In fact, there are biotechnology companies, such as Diversa, Inc., which specialize in screening soil gene libraries for functional activity after cloning the genes into surrogate microorganisms, such as *Escherichia coli*. However, unlike the examples discussed above (Fig. 12.2), most industrial screening programs seek variants of well-studied enzymes and reactions. As discussed in chapters 4 and 10, a few hydrolytic enzymes dominate the microbial-enzyme market. These include proteases like subtilisin and starch-hydrolyzing enzymes like α-amylase. Many industrial screens are set up to obtain variants of these enzymes which function optimally under other conditions, such as high temperature or high alkalinity. Thus, screens are not typically set up to discover reactions which have not previously been described.

Why haven't such activities been found before? It is likely that they have never been looked for, or perhaps the substrates were too toxic or complex, or the right source of microorganisms was not screened. Like a mathematical proof, the demonstration of a metabolic activity existing in one microorganism somewhere around the globe demonstrates the biological feasibility of the reaction.

How many new reactions will be discovered? It is highly likely that metabolism of all the functional groups produced by plants and microbes will be found. From the analysis in chapter 5, we expect that the more than 90 functional groups found in nature are metabolizable. Other organic functional groups are used by organic chemists, and some make their way into the environment or might be metabolizable by existing biochemical reaction mechanisms.

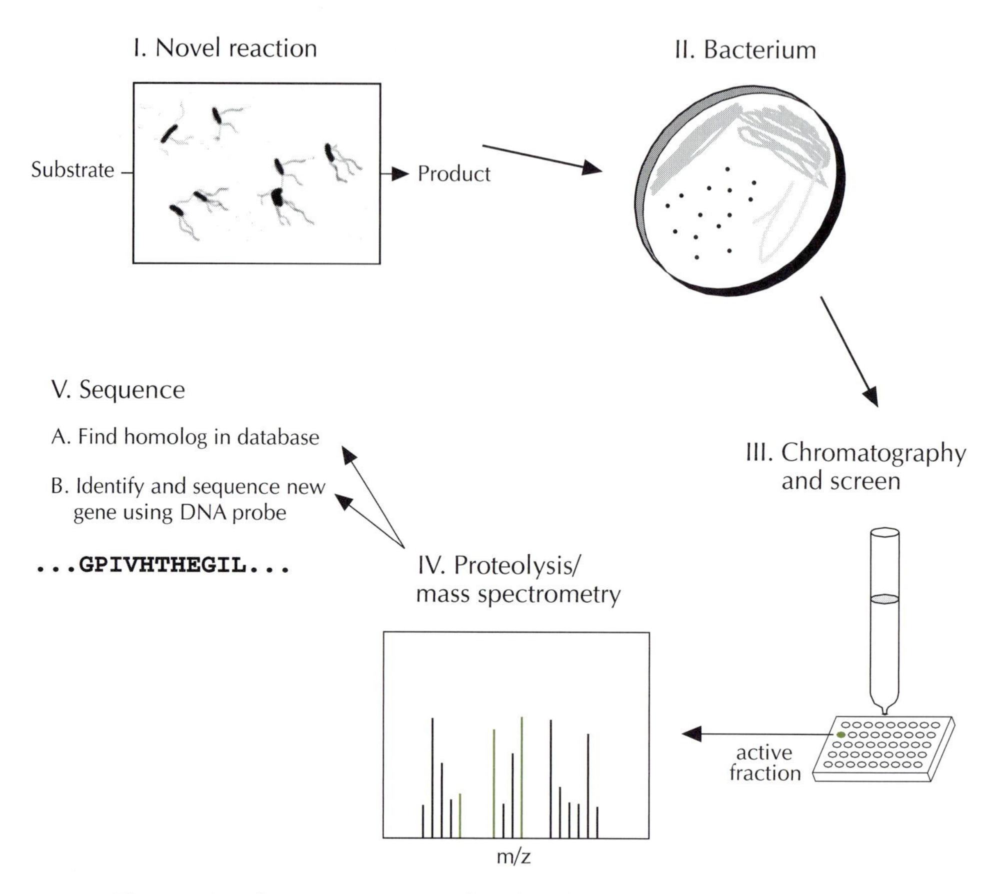

Figure 12.2 Screening program for identifying new biocatalytic reactions in microbes. The order is (I) enrichment culture, (II) bacterial isolation, (III) chromatographic separation of the enzyme(s), (IV) identification of protein sequence information by mass spectrometry, and (V) use of the sequence to identify possible homologs and to clone the gene(s) responsible for the activity.

How many unknown genes will prove to encode new reactions? There is insufficient information to predict a percentage. However, the evidence presented here suggests that functional genomics must be more than an exercise to put genes into categories with the assumption that all the categories are known. Clearly, there are a significant number of new reactions to be discovered, and microbes offer the best chance for their speedy discovery. Moreover, some of the enzyme reactions may be unique to microbial systems.

Enzyme Plasticity and New Biocatalysts

If we wish to consider enzyme mechanistic plasticity on a grander scale, we can look at enzymes clearly known to be evolutionarily related but which nonetheless catalyze what appear to be very different reactions. For example, as shown in Fig. 12.3, enzymes which catalyze glutathione transfer, dioxygenolytic ring cleavage, and a keto-enol isomeric shift were observed to bear low sequence identity to each other (1). The enzymes were, respectively, fosfomycin detoxifying enzyme, catechol 2,3-dioxygenase, and glyoxylase I. In detailed mechanistic studies, the enzymes were all observed

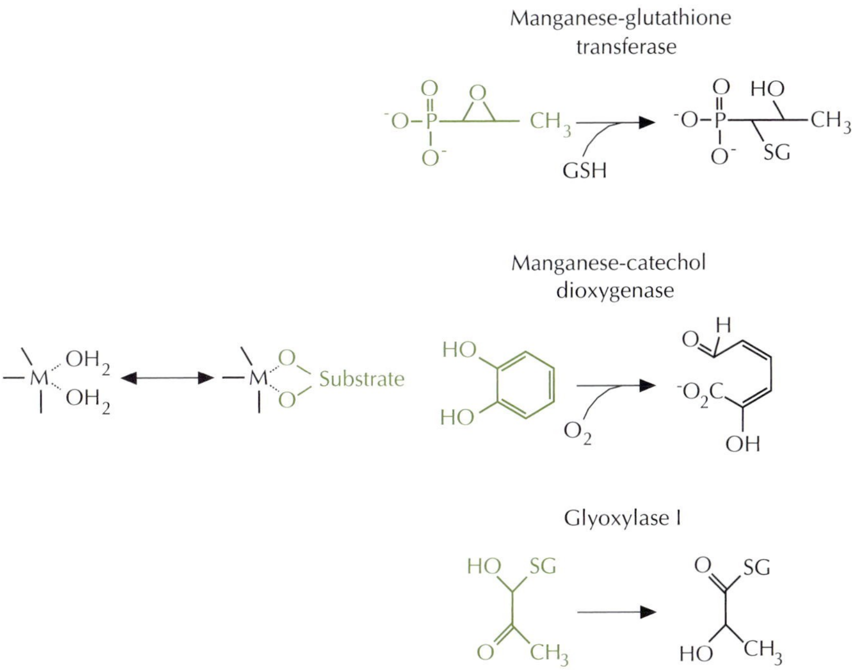

Figure 12.3 Evolutionarily and mechanistically related proteins catalyzing overall very different reactions. The enzymes share sequence identity, a common coordination environment for a metal atom, and a bidentate coordination of substrate oxygen atoms during the reaction (left). However, the enzymes are (right, from top to bottom) a transferase, a dioxygenase, and an isomerase.

to share similarities in the coordination environment of an active-site metal. It was discovered that the key to each mechanism revolves around tridentate protein amino acid coordination of the metal and the presence of loosely bound water molecules which can be shed in order to coordinate and activate the substrate. Thereafter, the mechanistic pathways of the three enzymes diverge. One catalyzes glutathione addition to an oxirane ring to detoxify an antibiotic. One opens another metal coordination site for diatomic oxygen to jointly activate oxygen and a catechol substrate for reaction. The third coordinates vicinal keto and alcohol oxygen atoms to effect an overall isomeric shift. Thus, a common structural feature, derived from a common history, provides a scaffolding for the evolution of overall different catalysis. It will be difficult to anticipate all the reactions which a given active-site scaffolding can achieve. Again, this augurs well for new functional genomics screening of the type described in Fig. 11.3B.

Summary

From the points raised in this chapter and throughout the book, several fundamental conclusions can be reached.

1. The limits of microbial biochemistry are much greater than is generally appreciated. More research to discover new microbial reactions is warranted.
2. A significant number of new microbial genes which are considered to have unknown function will prove to catalyze novel reactions.

3. Microbial metabolic diversity is likely to reflect microbial taxonomic diversity.
4. The discovery of new catabolic reactions will expand microbial biotechnology.

References

1. **Babbitt, P. C., and J. A. Gerlt.** 1997. Understanding enzyme superfamilies: chemistry as the fundamental determinant in the evolution of new catalytic activities. *J. Biol. Chem.* **272:**30591–30594.
2. **Genthner, B. R., G. T. Townsend, and P. J. Chapman.** 1989. Anaerobic transformation of phenol to benzoate via *para*-carboxylation: use of fluorinated analogues to elucidate the mechanism of transformation. *Biochem. Biophys. Res. Commun.* **162:**945–951.
3. **Gouverneur, V. E., K. N. Houk, B. de Pascual-Teresa, B. Beno, K. D. Janda, and R. A. Lerner.** 1993. Control of the exo and endo pathways of the Diels-Alder reaction by antibody catalysis. *Science* **262:**204–208.
4. **Leuthner, B., C. Leutwein, H. Schulz, P. Horth, W. Haehnel, E. Schiltz, H. Schagger, and J. Heider.** 1998. Biochemical and genetic characterization of benzylsuccinate synthase from *Thauera aromatica:* a new glycyl radical enzyme catalysing the first step in anaerobic toluene metabolism. *Mol. Microbiol.* **28:**615–628.
5. **Mangold, J. B., B. L. Mangold, and A. Spina.** 1986. Rat liver aryl sulfotransferase-catalyzed sulfation and rearrangement of 9-fluorenone oxime. *Biochim. Biophys. Acta* **87:**37–43.
6. **Shirota, O., K. Takizawa, S. Sekita, M. Satake, Y. Hirayama, Y. Hakamata, T. Hayashi, and T. Yanagawa.** 1997. Antiandrogenic natural Diels-Alder-type adducts from *Brosimum rubescens. J. Nat. Prod.* **10:**997–1002.
7. **Spiess, T., F. Desiere, P. Fischer, J. C. Spain, H. J. Knackmuss, and H. Lenke.** 1998. A new 4-nitrotoluene degradation pathway in a *Mycobacterium* strain. *Appl. Environ. Microbiol.* **64:**446–452.
8. **Wackett, L. P., J. F. Honek, T. P. Begley, S. L. Shames, E. C. Niederhoffer, R. P. Hausinger, W. H. Orme-Johnson, and C. T. Walsh.** 1988. Methyl-S-coenzyme M reductase: a nickel-dependent enzyme catalyzing the terminal redox step in methane biogenesis, p. 249–274. *In* J. Lancaster (ed.), *Bioinorganic Chemistry of Nickel.* VCH, New York, N.Y.

13

The Questions and Some Thoughts on Their Ultimate Answers

Summary

Big Questions and Future Prospects

> We live on an island of knowledge surrounded by a sea of ignorance. As our island of knowledge grows, so does the shore of our ignorance.
>
> — *John Archibald Wheeler*

At the end of any successful scientific investigation, we are inevitably faced with more questions than we had before we started. This is a healthy state of affairs for science and scientists, as long as we accept the fact that we are largely ignorant of the world around us. In this context, we choose to end the present book in the most scientifically optimistic manner we know: that of asking the next generation of questions. We call these the big questions in microbial biocatalysis and biodegradation. Others with different interests may frame the big questions differently. These were selected for their important impact on applications in biocatalysis and biodegradation and on our basic understanding of microbial physiology, genomics, and ecology.

The Questions and Some Thoughts on Their Ultimate Answers

1. *What is the extent of microbial metabolism?*

It is generally agreed that we have cultivated less than 1% of prokaryote species. The extent of genomic diversity is unknown, but comparative genomic analysis suggests that the genetic diversity in prokaryotes is greater than previously anticipated (see chapter 11). This suggests that there is much catalytic diversity yet to be discovered, as discussed in chapter 12. We do not have enough information to estimate what percentage of Earth's total biochemical repertoire we have discovered so far, with our current inability to translate a DNA sequence into a knowledge of the specific reaction(s) encoded by that gene.

2. *Are there clear linkages between the diversity and types of biocatalysis and phylogenetic lines? Stated another way, is biodegradative metabolism largely confined to a small percentage of the total prokaryotic diversity, or is this ca-*

pability distributed broadly, to be discovered as we either cultivate new organisms or extract their genes to be expressed in surrogate hosts?

In this book, we take the position that microbial catalytic diversity is more broadly distributed in different phylogenetic groups than is generally appreciated. The preponderance of *Pseudomonas* and related species in the biodegradation literature is due, at least in part, to the conditions used for their enrichment and cultivation. However, we acknowledge that, with >99% of the prokaryotic world largely uninvestigated from the standpoint of phylogeny, this view could change once more information is available. Hopefully, continued discovery of new metabolic phenotypes will help resolve this issue.

3. *Can a knowledge of microbial biocatalysis be used reliably to predict the environmental fate of existing chemicals and to design new materials which have the desired properties of biodegradability engineered into their structures?*

Industrial scientists already take biodegradability into account when designing new polymers. For example, some polylactate polymers are designed to have a specific lifetime for their particular market application and then biodegrade. Undoubtedly, this type of design with biodegradability in mind will become more and more common.

4. *What percentage of the metabolism of soil bacteria such as* Pseudomonas *and* Sphingomonas *is directed toward the catabolism of compounds found in the external environment?*

This is a question which we will begin to answer in the near future as we proceed to completely sequence the genomes of organisms found in soil and water and as we improve our ability to identify the functions of more divergent genes. The latter step is important; sequence information alone will not be enough. Comparative genomics is beginning to show, not unexpectedly, that the amount of catabolic metabolism encoded in the genome varies considerably, depending on whether the bacterium is an obligate parasitic pathogen or a soil organism known to have extensive biodegradative capabilities, such as *Pseudomonas* spp. and *Sphingomonas* spp.

5. *Can we use broad-based genomic and metabolic information to deduce the ecological niches of individual microbes?*

Genome projects suggest that virtually all of the DNA in most prokaryotes is transcribed into RNA and protein. The DNA thus yields clues to the lifestyles of prokaryotes, which differ widely. For example, *Deinococcus radiodurans* contains a disproportionately large number of genes encoding highly active enzymes involved in DNA repair. A knowledge of the DNA repair phenotype is consistent with discoveries of *Deinococcus* species in the feces of desert animals, an environment in which rapid desiccation would lead to extensive shearing of DNA, thus requiring extensive repair mechanisms for survival.

6. *How much will the chemical industry come to rely on biocatalysis?*

In the future, industry will need to use processes which consume fewer nonrenewable resources. Thus, chemical feedstocks will increasingly need to derive from renewable resources. It is hard to conceive of nonbiological chemical reactions being harnessed to transform carbon dioxide into complex molecules via the direct capture of energy from sunlight. This is done efficiently by plants, and the near future will likely see more reliance on plant materials as chemical feedstocks. Microbial reactions will increasingly be used for transforming glucose and other plant materials into more valuable chemicals.

7. *How rapidly can new metabolism evolve in natural environments?*

There is clear evidence for gene transfer in nature and much anecdotal evidence that new enzymes evolve in bacteria in natural environments. It is unclear how rapidly new metabolism can evolve, practically and theoretically. One approach that promises to shed light on this is the use of directed evolution in the laboratory. This can reveal the plasticity of microbial enzymes and pathways and point the way toward a better understanding of what happens in the soils of the world. Moreover, laboratory evolution will increasingly be used to develop new industrial biocatalysts.

8. *Do new microbial enzymes evolve largely from precursor enzymes with no activity toward new organic molecules, or are there preexisting enzymes able to metabolize virtually any organic pollutant?*

The latter view is analogous to the human immune system, in which antibodies are produced in sufficient diversity to combine with many antigens to which the person has never been exposed. Then, the B cell making a specific antibody proliferates in response to an infection by an agent containing the appropriate antigen. A parallel idea is that at least one bacterium capable of metabolizing almost any given compound exists somewhere on Earth and proliferates in response to the selective pressure imposed by a chemical in the environment. Another view is that although no metabolism may exist for a given compound, random mutation ultimately generates an enzyme or enzymes able to carry out the metabolism. An intermediate view is that enzymes able to handle most or all organic compounds are present but their fortuitous activities are very low. Still, they are able to confer some selective advantage on the organism, and the activity can then improve by mutation and selection until it reaches a respectable catalytic constant and K_m more in line with enzymes typically studied with their preferred substrates.

9. *Will it be possible to cultivate a much greater diversity of prokaryotes in the future?*

This remains a major problem in environmental microbiology. Methods exist for sampling soil to extract, clone, and sequence DNA from nonculturable organisms. However, there is still much that can only be learned from growing an organism and studying its physiology directly. Much broader cultivation of microbes might result from using a broader range of culture conditions. The vast majority of enrichment and pure-culture studies use liquid medium and agar plates, media which are not likely to mimic the soil matrix. The use of solid suspensions and other strata, as well as a greater knowledge of nutritional requirements and cell regulation, some of which might emerge from genomics, could broaden the range of organisms which are culturable in the laboratory.

10. *Will there be a limit on the chemical substances whose biodegradation we need to understand?*

The range of possible organic structures, with even a few noncarbon elements, such as hydrogen, nitrogen, sulfur, and oxygen, is enormous. With combinatorial chemistry in full swing, many more organic compounds will be made, and some will have useful properties leading to commercial development. It seems very likely that even more challenges will be posed for microbial metabolism, and we will need to continually discover new microbial metabolism to match the discovery of new organic compounds.

Summary

While we offer answers to the questions posed above, readers are encouraged to seek their own answers and to ask new questions. It is only in this way that the science of biocatalysis and biodegradation, or any science, will continue to evolve.

> Science is the belief in the ignorance of experts.
>
> — *Richard P. Feynman*

APPENDIX

Books and Journals Relevant to Biodegradation and Biocatalysis

Books

Biodegradation

Agteren, M. H., S. Keuning, and D. B. Janssen, 1998. *Handbook on Biodegradation and Biological Treatment of Hazardous Organic Compounds.* Kluwer Academic Publishers, Dordrecht, The Netherlands.

Alexander, M. 1999. *Biodegradation and Bioremediation,* 2nd ed. Academic Press, San Diego, Calif.

Gibson, D. T. (ed.). 1984. *Microbial Degradation of Organic Compounds.* M. Dekker, New York, N.Y.

Leisinger, A., M. Cook, R. Hutter, and J. Nuesch (ed.). 1981. *Microbial Degradation of Xenobiotics and Recalcitrant Compounds.* Academic Press, New York, N.Y.

Lidstrom, M. E. (ed.). 1990. *Hydrocarbons and Methylotrophy. Methods in Enzymology,* vol. 188. Academic Press, San Diego, Calif.

Sigel, H., and A. Sigel (ed.). 1992. *Degradation of Environmental Pollutants by Microorganisms and Their Metalloenzymes: Metal Ions in Biological Systems,* vol. 28. M. Dekker, New York, N.Y.

Spain, J. C. (ed.). 1995. *Biodegradation of Nitroaromatic Compounds.* Plenum Press, New York, N.Y.

Young, L. Y., and C. E. Cerniglia (ed.). 1995. *Microbial Transformation and Degradation of Toxic Organic Chemicals.* Wiley-Liss, New York, N.Y.

Biocatalysis

Abramowicz, D. A. (ed.). 1990. *Biocatalysis.* Van Nostrand Reinhold, New York, N.Y.

Roberts, S. M., N. J. Turner, A. J. Willetts, and M. K. Turner. 1995. *Introduction to Biocatalysis Using Enzymes and Micro-Organisms.* Cambridge University Press, New York, N.Y.

General Microbial Biotechnology

Glazer, A. N., and H. Nikaido. 1995. *Microbial Biotechnology: Fundamentals of Applied Microbiology.* W. Freeman, New York, N.Y.

Journals

Biodegradation

Applied and Environmental Microbiology. American Society for Microbiology, Washington, D.C.

Biodegradation. Kluwer Academic Publishers, Dordrecht, The Netherlands.

Biocatalysis

Applied and Environmental Microbiology. American Society for Microbiology, Washington, D.C.

Biocatalysis and Biotransformation. Harwood Academic Publishers, Cooper Station, N.Y.

General Microbial or Environmental Biotechnology

Applied and Environmental Microbiology. American Society for Microbiology, Washington, D.C.

Current Opinion in Biotechnology, June 2000 issue on environmental biotechnology. Elsevier Scientific Publishers, New York, N.Y.

Journal of Industrial Microbiology and Biotechnology. Stockton, New York, N.Y.

APPENDIX

B

Useful Internet Resources in Biodegradation and Biocatalysis

Biodegradation

Databases

University of Minnesota Biocatalysis/Biodegradation Database (UM-BBD)

http://umbbd.ahc.umn.edu

The UM-BBD provides information on microbial nonintermediary catabolism organized in a hierarchical fashion from metabolic pathways to enzymes, genes, and microorganisms.

Biodegradative Strain Database (BSD)

http://bsd.cme.msu.edu

Under development at the time of writing, the BSD will be an excellent resource for information about bacterial strains known to biodegrade a wide range of substrates. Special emphasis is given to phylogenetic analysis and 16S rRNA data for these strains.

Database for Environmental Fate of Chemicals

http://www.aist.go.jp/RIODB/dbefc/index_E.html

This is a collection of data from various studies on the fate and distribution of chemicals in the environment.

Education and General Knowledge

Toxic Waste Site in the Microbe Zoo, Michigan State University

http://commtechlab.msu.edu/sites/dlc-me/zoo/zdtmain.html

The Microbe Zoo offers a basic description of the microbial world with a toxic waste section that focuses on bacteria being studied for a better understanding of biodegradation. It is designed for high school or introductory university level students.

Biodegradation of Wood

http:/helios.bto.ed.ac.uk/bto/microbes/armill.htm

This page offers an overview of the biodegradation of wood by wood decay fungi. It is part of the Microbial World site, which contains information on microbial ecology and environmental microbiology.

Lignin Biodegradation

http://www.ftns.wau.nl/imb/research/wrf.html

Description of lignin biodegradation by white rot fungi.

Biocatalysis and Biodegradation: Online Course, University of Minnesota

http://www.cee.umn.edu/dis/courses/MICE5309_5000_01.www/

This site describes a complete course available over the Internet covering the concepts of microbial biocatalysis and biodegradation. The first lesson, on microbial environmental biochemistry, is freely available.

Bioremediation Discussion Group

http://biogroup.gzea.com/

An e-mail discussion group for people interested in bioremediation and biodegradation. It includes archives as well as forms to sign up for the group.

Bioremediation Techniques

U.S. Geological Survey (USGS): Biodegradation

http://h2o.usgs.gov/public/wid/html/bioremed.html

This website contains general information on bioremediation and some examples of contaminated sites where the USGS participated in the cleanup.

Bioremediation Research Project Abstracts

http://www.hsrc.org/hsrc/html/abstracts/biorem-abstracts.html

Abstracts from research projects in bioremediation from the Hazardous Substance Research Centers.

Bioremediation in situ

http://water.usgs.gov/wid/html/bioremed.html

Studies of in situ bioremediation, focusing on the Hanahan Bioremediation Project.

CLU-IN, Hazardous Waste Clean-up Information site

http://www.clu-in.org/

Describes programs, organizations, publications, and other tools and treatments useful in hazardous waste remediation. The site is managed by the U.S. Environmental Protection Agency's Technology Innovation Office.

EPA REACH IT

http://epareachit.org/

From the U.S. Environmental Protection Agency's Technology Innovation Office; at this site one can search for remediation techniques based on the contaminating chemicals and conditions at the contaminated site.

Biodegradation Prediction

PREDICT-BT: Predicting Biotransformations Database

http://umbbd.ahc.umn.edu/predictbt/

PREDICT-BT deals with the prediction of microbial catabolic pathways for compounds not studied experimentally or where data are otherwise unavailable.

Biodegradability Evaluation of Chemical Structures; Dr. Bill Punch homepage

http://garage.cse.msu.edu/~punch/projects/biodeg.html

BESS (Biodegradability Evaluation and Simulation System) is an expert system which uses knowledge about biodegradation to predict pathways for the microbial metabolism of naturally occurring and industrial organic compounds.

Microbial Catalysis

Databases

The University of Minnesota Biocatalysis/Biodegradation Database (UM-BBD)

http://umbbd.ahc.umn.edu

The UM-BBD provides information on microbial nonintermediary catabolism organized in a hierarchical fashion from metabolic pathways to enzymes, genes, and microorganisms.

Microbial Biotransformation Products and Metabolites

http://mik.gbf.de/menu/products_text.html

A list of about 700 products of microbial biotransformation. The compounds listed are patented but are available from GBF.

BRENDA: The Enzyme Database

http://www.brenda.uni-koeln.de:80/

BRENDA contains information on the substrates and optimum conditions for a large number of enzymes, many of which would be useful in biocatalytic applications. The database is readily searchable by enzyme name or Enzyme Commission (EC) number.

Thermodynamics of Enzyme-Catalyzed Reactions

http://wwwbmed.nist.gov:8080/enzyme/enzyme.html

A collection of thermodynamic data on enzyme-catalyzed reactions ordered by EC code. Originally published in the *Journal of Physical and Chemical Reference Data.*

Case Studies and Techniques in Biocatalysis

Biocatalysis and Biomaterials—Fraunhofer Institute

http://www.igb.fhg.de/Biokat/home.en.html

A German biotechnology firm specializing in contract work for the pilot-scale production of native and recombinant enzymes, process development and optimization, enzyme optimization, and screening. Project descriptions of interesting applications in biocatalysis are available.

Catalysis and Biocatalysis Technologies—National Institute of Standards and Technology, Advanced Technology Program

http://www.atp.nist.gov/atp/97wp-cat.htm

A white paper outlining trends and opportunities in and significance of biocatalysis research.

Directed Enzyme Evolution

http://www.cheme.caltech.edu/groups/fha/Enzyme/directed.html

A paper describing the theory behind directed enzyme evolution. It includes a summary table of directed enzyme evolution projects.

Index

Page references in *italics* refer to tables.